Insect, Mite and Vertebrate Pests and their Management in Horticultural Crops

BY THE SAME AUTHOR

- Diseases of Horticulture Crops: Nematode Problems and their Management ISBN: 978-81-7233-543-4
- Integrated Nematode Management in Horticultural Crops ISBN: 978-81-7233-523-6
- Organic Farming for Sustainable Horticulture: Principles and Practices ISBN: 978-81-7233-501-4
- Plant Protection in Horticulture, Vol. 1-3 (Set) ISBN: 978-81-7233-618-9
- Fungal Diseases and their Management in Horticultural Crops ISBN: 978-81-7233-629-5
- Bacterial and Viral Diseases and their Management in Horticultural Crops ISBN: 978-81-7233-630-1

Insect, Mite and Vertebrate Pests and their Management in Horticultural Crops

P. Parvatha Reddy

Former Director and Former Head
Division of Entomology & Nematology
Indian Institute of Horticultural Research
Bangalore-560 089, Karnataka (India)

SCIENTIFIC
PUBLISHERS

Published by
SCIENTIFIC PUBLISHERS (INDIA)

Jodhpur –
5 A, New Pali Road
P.O. Box 91
Jodhpur - 342 001 INDIA

Delhi –
4806/24, Ansari Road
Daryaganj
New Delhi - 110 002 INDIA

Print: 2018

ISBN: 978-81-7233-628-8

Reprint from Plant Protection in Horticulture
Volume 1, 384 pp. (2010)
ISBN: 978-81-7233-615-8

Visit the Scientific Publishers (India) website at
http://www.scientificpub.com
Printed in India

Affectionately dedicated to
My colleagues

K. Srinivasan
P.N. Krishna Murthy
A. Verghese
N.K. Krishna Kumar

PREFACE

Horticulture in India is fast emerging as a major commercial venture, because of higher remuneration per unit area and the realization that consumption of fruits and vegetables is essential for health and nutrition. In the last one decade, export potential of horticultural crops has significantly increased attracting even multinationals into floriculture, processing and value added products.

During 2005-06 coverage of area under various horticultural crops in the country was 19.237 million hectares (which is about 13.08% of the total cultivated area) with production of 185.207 million MT. The horticulture sector contributed around 28% of GDP in agriculture. India is the largest producer in the world of cashew nuts, coconuts, tea, ginger, turmeric and black pepper. India is the second largest producer of fruits and vegetables (after China) with a production of 58.74 and 109.05 million MT and contributes 10.9 and 11.9% share in global fruit and vegetable production, respectively. India is the second largest producer of flowers after China. About 126,000 hectare of area is under floriculture, producing 694,000 million tonnes of loose flowers annually (2005-2006). The production of cut flowers has increased to 2,762 million numbers. India is still a marginal player in the world floriculture trade indicating strong potential that can be exploited in this sector. India is the third largest producer of coconut, fourth largest producer and consumer of rubber and sixth largest producer of coffee in the world.

Productivity of horticultural crops in India is relatively low compared to other countries. Of the several factors responsible for lower productivity of horticultural crops, pests (insect, mite and vertebrate pests) are considered as important limiting factors. Some pest problems are of national importance which causes significant losses to horticultural crops. The annual losses due to pests to all the crops in India was estimated at Rs.60,000 million in 1983, which at today's prices could exceed Rs. 200,000 million.

The information on pests (insect, mite and vertebrate pests) in horticultural crops (fruits, vegetables, plantation, spice, tuber, ornamental, medicinal and aromatic crops) is very much scattered. There is no book at present which comprehensively and exclusively deals with the above aspects on horticultural crops. The present book deals with integrated pest management in horticultural crops in detail using regulatory, physical, cultural, chemical, biological, host plant resistance and integrated methods. The book is extensively illustrated with excellent quality photographs enhancing the quality of publication. The book is written in lucid style, easy to understand language along with adoptable recommendations involving eco-friendly components of IPM.

In view of greater emphasis being given by the Government of India to horticulture by establishing horticulture mission and exclusive State Horticultural Universities, there is an urgent need for good text books to teach courses on different disciplines of horticulture. In this context the present book is of immense help to all those concerned with horticultural crops.

This book is a practical guide to practicing farmers of horticultural crops. Further, it is a useful reference to policy makers, research and extension workers and students. The material can also be used for teaching undergraduate and post-graduate courses. Suggestions to improve the contents of the book are most welcome (E-mail: reddy_parvatha@yahoo.com).

I profusely thank Dr. K. Srinivasan, Chief Agricultural Economist, TAFE, Chennai; Dr. A. Verghese, Dr. N.K. Krishna Kumar and Shri. P.N. Krishna Murthy, Division of Entomology and Nematology, Indian Institute of Horticultural Research Centre, Bangalore; and Dr. V.J. Shivankar and Dr. C.N. Rao, National Research for Citrus, Nagpur; who have kindly made available many of the photographs included in this book. The publisher, Scientific Publishers (India), Jodhpur, deserves commendation for their professional contribution.

Bangalore
August 15, 2009

P. Parvatha Reddy

CONTENTS

Chapter 1

HORTICULTURAL CROPS

1.1. Importance of Horticultural Crops

Agriculture is the mainstay of the Indian economy, and horticulture is a crucial component thereof. Horticulture development has been accorded high priority in the XIth Five Year Plan. The impact of enhanced investment in horticulture has been highly encouraging in terms of vastly improved quality production and export potential - an increase from 96.1 million tonnes in 1990-91 to 185.2 million tonnes during 2005-06 (Bijay Kumar, 2007) (Table 1.1).

Table 1.1. Area, production and productivity of horticultural crops in India (2005-06) (Bijay Kumar, 2007)

Horticultural crops	Area (million ha)	Production (million MT)	Productivity (MT/ha)	% Share of hortil. crops
Fruits	5.510	58.740	10.7	31.74
Vegetables	7.164	109.050	15.2	58.93
Flowers-loose	0.126	0.694	---	0.38
Flowers-cut	---	2762.000*	---	---
Mushrooms	---	0.040	---	0.03
Spices	2.858	5.108	1.8	2.76
Plantation crops	3.283	11.263	3.4	6.09
Nuts	0.121	0.134	1.1	0.07
Aromatic & Medicinal	0.266	0.220	---	---
Total	19.237	185.207	9.6	---

*Million numbers

India is the second largest producer of fruits next only to China contributing 10.9% of the total world production. Total production of fruits during 2005-06 has been estimated at 58.74 million MT from 5.51 million ha. (Bijay Kumar, 2007). India is the largest producer of mango, banana, and papaya. India produces 41.7% of the world mangoes, and 25.7% of the bananas. In grapes, India has recorded the highest productivity per unit area in the world. The per capita consumption of fruits has increased from 40 to 85g/day within the last one decade.

India is the second largest producer of vegetables next only to China contributing 11.9% of the total world production. Vegetables occupy an area of 7.164 million ha with a production of 109.05 million tonnes (Bijay Kumar, 2007). India produces 13.6% of the world onions. In vegetables, India occupies the first position in the production of peas and okra, second in onion, brinjal, cabbage, cauliflower and third in potato. The per capita consumption of vegetables has increased from 96 to 175 g/day within the last one decade.

Floriculture is estimated to cover an area of 126,000 ha with production of 694,000 MT of loose flowers. The area under cut flowers has increased significantly in recent years and the production has reached 2762 million numbers.

India has emerged as the largest producer of spices, tea, cashew and areca nut. Coconuts sustain an area of 2.029 million ha with a production of 101.90 million tonnes. Areca nut is cultivated in 0.381 million ha with a production of 0.483 million tonnes. Cashew nuts are grown in an area of 0.843 million ha with a production of 0.579 million tonnes. The annual production of spices is 5.108 million tonnes from 2.858 million ha.

Horticulture has emerged as an important sector for diversification of agriculture. Presently, horticulture has established its credibility in improving income through increased productivity, generating employment and in enhancing exports besides providing protective food, household nutritional security, precious raw material for various processing and ancillary industries and finally in enhancing the aesthetic aspects of life. The focused attention on investment in horticulture during the last decade has been rewarding in terms of increased production and productivity of horticultural crops with manifold export potential. Horticulture has played a significant role in women empowerment, providing employment opportunities through mushroom cultivation, floriculture, processing, nursery raising, vegetable seed production etc.

1.2. Export of Horticultural Produce

Horticulture in India is fast emerging as a major commercial venture, because of higher remuneration per unit area and the realization that consumption of fruits and vegetables is essential for health and nutrition. In the last one decade, export potential of horticultural crops has significantly increased attracting even multinationals into floriculture, processing and value added products.

The present annual growth rate of horticulture is more than 6.5% which contributes 24.5% from the mere 8% of area to the GDP and 54.55% to the export earnings in the agriculture sector (Table 1.2).

Table 1.2. Export of horticultural produce from India (2005-06) (Bijay Kumar, 2007)

Horticultural product	Export quantity (MT)	Value (Rs. Crores)	% Share of value
Floriculture & seeds			
Floriculture	35186.00	299.41	7
Fruit & vegetable seeds	7472.9	92.83	2
Total	42658.90	392.24	
Fruits & vegetables			
Fresh onions	960507.32	708.16	18
Other fresh vegetables	217280.59	267.69	6
Walnuts	5256.56	114.47	3
Fresh mangoes	69606.60	128.11	3
Fresh grapes	54049.87	214.61	5
Other fresh fruits	158339.26	225.67	5
Total	1465040.20	1658.71	
Processed fruits & vegetables			
Dried & preserved vegetables	124495.71	364.11	9
Mango pulp	134613.20	364.24	9
Pickle & Chutneys	135381.85	260.98	6
Other processed fruits and vegetables	107335.24	370.21	27
Grand Total	943568.59	2454.60	

The major destinations for export of different horticultural products and potential international markets for fresh and processed products are presented in Table 1.3.

Table 1.3. Major destinations for export of different horticultural products (Chamber, 2006)

Horticultural crops	Major destinations for export
Fruits	UAE, Saudi Arabia, Sri Lanka, Bangla Desh, The Netherlands, USA, Malaysia
Fresh grapes	UK, UAE, The Netherlands, Bangla Desh, Germany, Sri Lanka
Pomegranate	UAE, Saudi Arabia, Sri Lanka, Kuwait
Fresh guava	Saudi Arabia, Netherlands, USA, UAE, Kuwait
Gooseberries	UAE, Qatar
Mango pulp	UK, Canada, Bangle Desh, Bahrain, Netherlands, Japan, Saudi Arabia
Mango slices	UK, Saudi Arabia, Japan, USA
Dried grapes	UK, Malaysia, Netherlands, UAE

Mango pickles	Saudi Arabia, UAE, UK, Canada
Mango chutney	Oman, USA
Guava jam & jelly	UK, Germany, Hong Kong, USA, Netherlands, Saudi Arabia
Dried tamarind	Egypt, UAE, Pakistan, Saudi Arabia, UK
Dried figs	Nepal, Bangla Desh
Vegetables	UAE, UK, Saudi Arabia, Sri Lanka, USA, Singapore
Fresh vegetables	UAE, Saudi Arabia, Sri Lanka, USA
Fresh onion	UAE, Singapore, Sri Lanka, Bangla Desh, Malaysia
Fresh tomato	Bangladesh, Nepal, Sri Lanka
Gherkin	France, USA, Spain, Belgium
Dried vegetables	UAE, Sri Lanka, USA, UK, Singapore
Dehydrated onion	Germany, Romania, Sri Lanka, USA, Malaysia, Netherlands, UK
Vegetable seeds	USA, The Netherlands, France, Italy, Japan, Pakistan
Floriculture	USA, Japan, The Netherlands, UK, Italy

Horticulture today is not merely a means of diversification but forms an integral part of food and nutritional security, as also an essential ingredient of economic security. Adoption of horticulture both by small and marginal farmers brought prosperity in many regions of the country of which Maharashtra, Karnataka and Andhra Pradesh are the prime examples. A Golden Revolution through significant production increases in horticultural crops is already on the horizon.

Since India is now a signatory member of WTO, it will require its produce and products to be competitive both in the domestic and export markets. This would call for the use of improved HYV/F1 hybrids, recommended agro-techniques and plant protection measures, hi-tech intervention and scientific post-harvest management.

After the success of 'Green Revolution' in agriculture, India is now moving towards the **'Golden Revolution'** comprising fruits, vegetables and flowers. Commercialization of horticulture alongwith post harvest management, processing and marketing assume great significance and add value not only in the context of farmers' income, but also its contribution to the agriculture GDP. Commercialization of horticulture is largely dependent upon use of appropriate technology and market oriented production.

Chapter 2

FRUIT CROPS

A large variety of fruits are grown in India. Of these, banana, citrus, pineapple, papaya, sapota, jack fruit and avocado (tropical fruits); mango, grape, guava, litchi and loquat (subtropical fruits); apple, pear, peach, plum, apricot, almond and walnut (temperate fruits) and amla, ber, pomegranate, annona, fig and phalsa (arid zone fruits) are important. The leading fruit growing states are Maharashtra, Andhra Pradesh, Tamil Nadu, Kerala, Gujarat, Karnataka, Bihar, Uttar Pradesh, West Bengal and Orissa.

India is the second largest producer of fruits next only to China contributing 10.9% of the total world production. The present area under fruit crops is estimated to be about 5.510 million hectares with an estimated production of 58.740 million tonnes (Tables 2.1 and 2.2). Fruit crops and nuts contribute 31.81% share to the total production of horticultural crops. The nut crops are grown in 0.1209 million hectares with an annual production of 0.1335 million tonnes (Table 2.3). India is the largest producer of mango, banana and papaya. India produces 40.1% of the world mangoes, 24.5% of the bananas and 28.8% of world papaya. In grapes, India has recorded the highest productivity per unit area (25.4 tonnes/ha) in the world. The per capita consumption of fruits has increased from 40 to 85 g/day within the last one decade.

Table 2.1. Major fruit producing countries of the world (Bijay Kumar, 2007)

Country	Area (in '000 ha)	Production (in '000 MT)	Productivity (in MT/ha)	% share of world production
China	10456	89355	8.5	16.7
India	5510	58740	10.7	10.9
Brazil	2386	36952	15.5	6.9
USA	1215	28144	23.2	5.2
Italy	1219	18134	14.9	3.4
Spain	1806	15537	8.6	2.9
Indonesia	1167	15406	13.2	2.9

Mexico	1128	14903	13.2	2.8
Iran	1354	13848	10.2	2.6
Philippines	1073	13110	12.2	2.4
Turkey	1074	12998	12.1	2.4
Others	24781	219465	8.9	40.9
Total	53170	536591		

Table 2.2. Area, production and productivity of fruit crops in India (2005-06) (Bijay Kumar, 2007)

Fruit crop	Area ('000 ha)	Production ('000 MT)	Productivity (MT/ha)	% share of India production
Apple	235	1756	7.5	3.0
Banana	565	18702	33.1	31.8
Citrus	749	6326	8.4	10.8
Grapes	64	1631	25.4	2.8
Guava	182	1823	10.0	3.1
Litchi	63	381	6.0	0.6
Mango	2021	12538	6.2	21.3
Papaya	73	2317	31.7	3.9
Pineapple	85	1353	15.8	2.3
Pomegranate	116	849	7.3	1.4
Sapota	140	1117	8.0	1.9
Others	1217	9947	8.2	16.9
Total	5510	58740	10.7	

Table 2.3. Area, production and productivity of nut crops in India (2005-06) (Bijay Kumar, 2007)

Nut crop	Area ('000 ha)	Production ('000 MT)	Productivity (MT /ha)
Almond	21.1	14.6	0.7
Walnut	99.8	118.9	1.2
Total	120.9	133.5	

India has been exporting fruits and fruit products since several decades. However, in the recent past export promotional activities received greater attention (Table 2.4). Mango and grapes are the main items in fresh fruit export and some other fruits like pomegranate, sapota, banana, citrus, litchi, apple and strawberry also entered fresh fruit export market in limited quantities.

Table 2.4. Export of fruit produce from India (2005-06) (Bijay Kumar, 2007)

Fruit products	Quantity (in MT)	Value (in Rs. Million)
Fruit & vegetable seeds	7472.90	928.3
Walnuts	5256.56	1144:7
Fresh mangoes	69606.60	1281.1

Fresh grapes	54049.87	2146.1
Other fresh fruits	158339.26	2256.7
Mango pulp	134613.20	3642.4
Pickle & Chutneys	135381.85	2609.8
Other processed fruits & vegetables	107335.24	3702.1

The major potential international markets for fresh and processed fruit products are presented in Table 2.5.

Table 2.5. Potential international markets for fresh and processed fruits

Fresh & processed fruits	Potential international markets
Fresh grapes	UK, Netherlands, Sri Lanka, UAE
Pomegranate	UAE, Saudi Arabia, Sri Lanka, Kuwait
Fresh guava	Saudi Arabia, Netherlands, USA, UAE, Kuwait
Gooseberries	UAE, Qatar
Mango pulp	UK, Canada, Bangla Desh, Bahrain, Netherlands, Japan, Saudi Arabia
Mango slices	UK, Saudi Arabia, Japan, USA
Dried grapes	UK, Malaysia, Netherlands, UAE
Mango pickles	Saudi Arabia, UAE, UK, Canada
Mango chutney	Oman, USA
Guava jam & jelly	UK, Germany, Hong Kong, USA, Netherlands, Saudi Arabia
Dried tamarind	Egypt, UAE, Pakistan, Saudi Arabia, UK
Dried figs	Nepal, Bangla Desh

Pest Scenario in Fruit Crops

Fruits are amongst the most cherished bounties of nature. They are the major sources of Vitamins 'A' and 'C' and minerals like calcium and iron. Fresh fruits have low calorific values and are low in fats. Of the several factors responsible for lower productivity of fruits, insect pests, diseases and nematode parasites constitute important limiting factors.

The pest scenario in fruit crops during the last one decade has changed because of increase in area and change in agronomic practices like close planting, drip irrigation, regular pruning, etc. As a consequence, several insects are attaining major pest status in several fruit crops. For example in Maharashtra, the mango hopper (*Amrasca splendens*) attained serious status on new flush (Dalvi *et al.*, 1992), while a new mite (*Cisaberopterus kenyae*) has been reported on mango from Uttar Pradesh (Rai *et al.*, 1993). In a survey carried out in Krishna and West Godhavari districts of Andhra Pradesh during 1997-2000, the fruit borer (*Deanolis albizonalis*) was found attacking all stages of mango fruits, which was earlier reported by Zaherudden and Sujatha (1993). Recently a complex of lepidopteron larvae was found to be serious, feeding on mango inflorescence in many mango growing tracts of Andhra Pradesh and Karnataka (Verghese and Jayanthi, 1999). In grapes, the shot hole borer (*Xyloborus crassiusculus*) is a serious pest in Karnataka (Verghese and Tandon, 1995). In surveys during 1997-98, one of the main problems faced in grape was *Helicoverpa armigera* and *Spodoptera litura*. Survey conducted by scientists of the Indian Institute of Horticultural Research, Bangalore, showed severe incidence of scolytid (*Xyloborus* sp.) on pomegranate causing 20-100% loss in orchards of several districts of Karnataka.

Due to increased area, pest distribution is also getting widespread. For example, in Bangalore, the banana pseudo stem weevil (*Odioporus longicollis*), hitherto not a serious pest, has now begun to cause major loss (Jayanthi and Verghese, 1999b). This shows that change in pest status is going to be a continuous phenomenon. This calls for vigil, scientific forecast and proper management strategies.

2.1. BANANA

Globally, banana (*Musa* spp.) is one of the maximum distributed fruit crop grown in more than 130 countries in 4.306 million hectares with an annual production of 76.436 million tonnes (Table 2.6). It is the fourth most important food crop after rice, wheat and maize. It is a dessert fruit for millions and is used in different regions of the world as staple food. As a diet, banana is filling, easy to digest, nearly fat free, rich source of carbohydrate with calorific value of 67 calories/100g and is free from sodium, making it salt-free diet and also has zero cholesterol. Hence, regular use of it is also considered to be good for heart. Banana also contains several minerals such as Calcium, Magnesium, Potassium and Phosphorus. Banana fruit is particularly rich in Vitamin-C and contains significant amount of other vitamins such as Vitamin-A.

Table 2.6. Major banana producing countries of the world (Bijay Kumar, 2007)

Country	Area (in '000 ha)	Production (in '000 MT)	Productivity (in MT/ha)	% share of world production
India	565	18702	33.1	24.5
Brazil	491	6703	13.6	8.8
China	274	6670	24.4	8.7
Philippines	418	6298	15.1	8.2
Ecuador	221	6118	27.7	8.0
Indonesia	315	5178	16.4	6.8
Costa Rica	41	2353	57.2	3.1
Mexico	77	2250	29.2	2.9
Thailand	141	1865	13.2	2.4
Colombia	65	1765	27.2	2.3
Others	1698	18534	10.9	24.2
Total	4306	76436		

Banana is an important fruit crop having great socio-economic significance in India as it supports livelihood of millions of people. India is the largest producer of banana in the world with an annual production of 18.702 million tonnes from an area of 0.565 million ha (Table 2.7). Although, banana occupies 10.3 per cent of the area under fruits it contributes 31.8 per cent of the total fruit production in the country. The national productivity of banana is very impressive (33.1 tonnes per ha). Its productivity in commercial zones especially in Maharashtra, Gujarat, Tamil Nadu, Karnataka, Andhra Pradesh and Bihar is highly impressive and comparable to the productivity in other countries.

Table 2.7. Area, production and productivity of leading banana growing states of India (2005-06) (Bijay Kumar, 2007)

State	Area (in '000 ha)	Production (in '000 MT)	Productivity (in MT/ha)
Tamil Nadu	94.6	4647.6	49.1
Maharashtra	73.2	4608.5	63.0
Gujarat	49.2	2498.8	50.8
Andhra Pradesh	61.1	1528.7	25.0
Karnataka	56.4	1399.1	24.8
Bihar	28.0	920.0	34.3
Kerala	56.2	443.1	7.9
Assam	47.0	699.4	14.9
Madhya Pradesh	15.4	615.6	40.0
West Bengal	27.8	544.9	19.6
Others	56.2	756.9	13.5
Total	565.1	18701.9	33.1

Research and development efforts made in the last two decades have resulted in commendable increase in productivity (144 per cent in the last decade) and overall production of banana in the country. The improved production technologies include higher plant densities (4500 plants per ha), improved crop geometries coupled with adoption of drip irrigation, more judicious fertilization management through balanced use of organic and inorganic fertilizers, appropriately timed split application of inorganics, use of proper sized rhizomes and well timed plantings, commercial production of *in vitro* plants and development of effective integrated pest management strategies. By adoption of these improved technologies, productivity of several individual farmers in different parts of the country has resulted in substantial increase (60 to 65 tonnes/ha).

2.1.1. Insect Pests

2.1.1.1. Rhizome Weevil, *Cosmopolites sordidus*

The rhizome weevil is one of the most destructive pests of banana. In India, it is widely distributed and has been reported as a major pest from Delhi, Assam, Kerala, Tamil Nadu, Karnataka and Maharashtra.

i) Damage: Infestation at early stage of plant results in cessation of growth and gradual death of the plant, while incidence at late stage affects the vigour and yield. The grubs feed on rhizome while the adult weevils feed on the pseudo-stem (Fig. 2.1). The eggs are laid singly on the collar region (above ground or on rhizomes under ground). After hatching, the grubs bore into the rhizome by making tunnels. Pupation occurs in these tunnels. The plants become weak, eventually rot and collapse particularly in monsoon.

Fig. 2.1. Rhizome weevil damage on banana pseudo stem.

ii) Management

- Use healthy, uninfested suckers or rhizomes for planting.
- Wash the suckers and dip in a solution of Chlorpyrifos 20 EC at 2.5 ml/l before planting.
- Follow clean cultural practices and maintain sanitation in the orchard.
- In case of post-planting infestation, spray the pseudo stem and drench around the base of the tree with Chlorpyrifos 20 EC at 2.5 ml/l.
- After a week, spray and drench with Malathion 50 EC at 2 ml/l.
- Banana cvs. Poovan, Kadali, Kunnan, Chetti, Basrai, Poomkalli, Sawaii and Pisang Seribu are less susceptible.

2.1.1.2. Pseudo Stem Borer, *Odoiporus longicollis*

This is another serious curculionid weevil pest of banana. It has been recorded from Assam, West Bengal, Bihar, Delhi and Uttar Pradesh.

i) Damage: The weevil breeds throughout the year but its incidence is at peak during summer and monsoon seasons. Both adults and grubs feed on the plant. Eggs are laid in air chambers cut through leaf sheaths. On hatching, the grubs start feeding on tissues around the air chambers and then bore inside the pseudo stems. These pseudo stems become weak and start rotting. Ultimately, as in the case of rhizome weevil these trees also break down with strong winds.

ii) Management

- Clean cultivation prevents the occurrence of this pest.
- In case of severe infestation, uproot and burn the plants.
- Spray Chlorpyrifos 20 EC at 2.5 ml/l for two or three times at three weekly intervals.
- If infestation is low, inject with 0.25% dichlorovos into the bored hole.
- Stem injection of azadirachtin at 10 ml/litre above the last visible hole in the pseudo stem is effective in reducing the weevil damage.

2.1.1.3. Aphid, *Pentalonia nigronervosa*

The banana aphid has been reported from Kerala, Tamil Nadu, Karnataka, Gujarat and Maharashtra. These aphids are very important as they are persistent vectors of bunchy top disease in banana.

i) Damage: The aphids congregate below the leaf base (Fig. 2.2) and their direct damage is fairly less. In case of severe infestation, there is progressive leaf dwarfing and leaf curling; fruit bunches become small and sometimes even the fruits are distorted.

Fig. 2.2. Aphid infestation on banana leaf.

ii) Management

- Rogue out the virus-affected plants before spraying.
- Spray dimethoate 30 EC at 2 ml/l or oxy-demeton methyl 25 EC at 2ml/l at fortnightly intervals.

2.1.1.4. Leaf Eating Caterpillars

Hairy caterpillars, *Pericallia ricini, Asura ila, Creatonotus gangis, Diacrisia obliqua, Euproctis fraterna*, and *Eupterote geminata* are found causing considerable injury to the leaves. While leaf rollers, bag worms, slug caterpillars and tobacco caterpillar (*Spodoptera litura*) are some of the minor lepidopterous pests feeding on banana leaves.

i) Management

- Hand picking and mechanical destruction of caterpillars in early stage of attack.
- In case of severe infestation, spray endosulfan 35 EC at 2ml/l or carbaryl 50 WP at 4 g/l
- In case of leaf rollers, remove and destroy the rolled leaves with larvae and pupae.

2.1.2. Vertebrate Pests

i) Damage: The banana fruits were depredated by monkey, barbet and bat. The barbet damaged ripening fruits. The barbets damaged one or two fruits/bunch and continued to forage in search of next ripening bunch. Damage on fruits began when the fruits showed yellow colour. Interestingly, birds attended on cv. Nendran when the fruits were still green. Isolated banana orchards were prone to heavier vertebrate pest damage.

ii) Management : It was found that by retaining damaged fruits on the plant, bird damage to banana was lessened. This was because birds revisited the damaged fruits to feed on the sweetened fruit portions. The fruits of banana cv. Mysore were least preferred because it secreted a mucilage ooze which rendered the fruit distasteful.

2.2. CITRUS

Citrus fruits (*Citrus* spp.) rank third in area and production after banana and mango with an estimated production of 6.323 million tonnes from an area of 0.749 million hectares (Table 2.8). There are many commercially grown *Citrus* spp. such as mandarins (*Citrus reticulata*), sweet orange (*C. sinensis*), acid lime (*C. aurantifolia*), lemon (*C. limon*), grape fruit (*C. paradisi*), pumello (*C. grandis*), etc.

Table 2.8. Area, production and productivity of leading citrus growing states of India (2005-06) (Bijay Kumar, 2007)

States	Area (in '000 ha)	Production (in '000 MT)	Productivity (in MT/ha)
Andhra Pradesh	166.2	2234.9	13.4
Maharashtra	248.8	1469.0	5.9
Punjab	26.0	365.4	14.0
Gujarat	30.5	303.3	9.9
Madhya Pradesh	16.7	266.9	16.0
Rajasthan	12.7	237.7	18.7

Karnataka	11.0	231.5	21.0
Orissa	26.5	205.3	7.7
Manipur	27.0	167.7	6.2
Assam	18.7	159.5	8.5
Uttarakhand	26.1	122.9	4.7
Bihar	16.8	112.3	6.7
Others	121.6	449.7	3.7
Total	748.8	6323.2	8.4

2.2.1. Insect Pests

Citrus is a highly sensitive crop and is attacked by number of insect pests. These play major role towards the low productivity of citrus in India with national average of 8.4 tonnes per hectare, as compared to 25-30 tonnes per hectare in other citrus growing countries.

2.2.1.1. Leaf Miner, *Phyllocnists citrella*

i) Damage: It is a serious pest of nursery and young plants but even attacks the grown up citrus trees all over India. The larvae feed on the epidermis of tender leaves making serpentine mines, which are silvery in colour (Fig. 2.3). The affected leaves become distorted and crumpled. The larvae may also mine the epidermis of tender shoots. Severe infestation may cause defoliation. Since new flush is attacked, the growth is severely hampered. In case of twig attack in young plants, 'die-back' also occurs. Ventral surface is preferred by the pest, but due to high population pressure, dorsal infestation is also seen. Citrus leaf miner helps in spreading mealy bug infestation and also acts as foci of citrus canker.

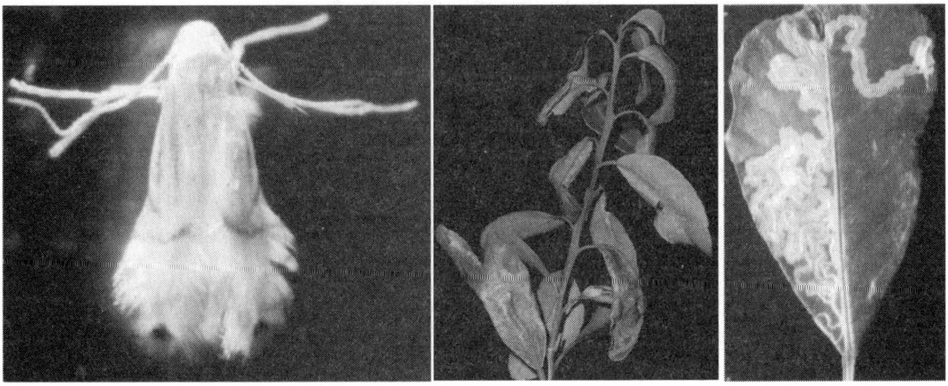

Fig. 2.3. Leaf miner adult, infestation on leaves of citrus (Courtesy: V.J. Shivankar & C.N. Rao, NRCC, Nagpur).

Of the total damage caused by the pest complex in citrus, 30% is claimed by the leaf miner alone. Moderate infestation of 1-2 larvae of leaf miners per leaf on seven year old trees was sufficient to reduce leaves and lower yields by 30-40% in the following year. A reduction in yield up to 50% and fruit weight from 120 to 70 g was observed.

ii) Biology: The pest completes its life cycle within 20 to 60 days depending upon climatic conditions with 9 to 13 overlapping generations in a year. The tiny silvery moth comes out during the dusk and lays eggs in leaf buds. The pest remains active throughout the year but causes heavy infestations during July-October and again in February-March.

iii) Management

a) Cultural Methods: Pruning of leaf miner infested branches is recommended, but not during active growth periods as it may induce further new flush and thereby allow the pest to have more number of generations. Pruning of only infested shoots from the inner canopy may be undertaken during winter if necessary.

Frequent irrigations and application of nitrogenous fertilizers in split dosages should be avoided which promote several new flushes and thereby provide continuous food to the pest.

b) Chemical Methods: Foliar spray of synthetic pyrethroids such as fenpropathrin and bifenthrin each at 0.01% gave 100% reduction of the pest infestation up to seven days after application. The conventional insecticides like monocrotophos, quinalphos and acephate all at 0.05%, however, reduced pest infestation up to 80% only (Shivankar and Singh, 1998). It was found that both the synthetic pyrethroids (fenpropathrin and bifenthrin) which were found most effective in controlling the pest were the cheapest treatments costing Rs. 2.40 and Rs. 2.10/tree/application as compared to acephate (Rs. 2.59) and monocrotophos (Rs. 3.20), respectively.

Insecticidal spray should be initiated as soon as the new flush is damaged up to 15% and should be continued at weekly intervals. Application of synthetic pyrithroid, fenvalerate at 0.01% may be under taken once in 45 days to bring the population below threshold limits. Spray should be directed at the new flush only.

c) Biological Methods: Among the bioagents, *Citrostichus phyllocnistoides* (Eulopidae: Hymenoptera) and *Cirrospilus quadristriatus* (Eulopidae: Hymenoptera) as major parasitoids and *Mallada boninensis* and *Menochiles sexmaculata* as major predators were found associated with the citrus leaf miner (Shivankar *et al.*, 2008).

The parasitoids, *C. phyllocnistoides* and *Ageniapsis* sp. are important causing up to 80% parasitism in Punjab. Release of *Mallada boninensis* at 20-50/plant depending upon the crop canopy, age and the pest density is recommended for reducing the leaf miner infestation (Singh, 1995).

Parasitoids like *C. phyllocnistoides*, *C. quadristriatus*, *Elasmus clarripennis*, *Bracon phyllocnistoides* and *Simpieses purpurea* have been recorded causing 30-45% parasitism. Black chalcid larvae, *Ageniapsis* sp. alone were found causing 80% parasitism of the moth (Anon, 1990 b).

Parasitism due to *C. phyllocnistoides* (8.5-10.2%) and *C. quadristriatus* (7.9-9.9) was significantly higher during October-November. Total per cent parasitism was recorded up to 22.4% on acid lime in the month of October. Among the parasitoids, *C. phyllocnistoides* was predominant followed by *C. quadristriatus* (Shivankar *et al.*, 2008).

The predatory population (*M. boninensis* and *M. sexmaculata*) on citrus was significantly higher in January and February months. The predator, *M. boninensis* was predominant followed by *M. sexmaculata. M. boninensis* larva consumed more number of leaf miner larvae reared on acid lime (149.3-151.4) than on Nagpur mandarin (140.3-148.8). *M. sexmaculata* adult consumed greater number of leaf miner pupae collected from acid lime (198.1-201.8) than on Nagpur mandarin (184.7-187.5) (Shivankar *et al.*, 2008).

d) Integrated Methods

- Under low to moderate infestation, spray 4% neem seed kernel extract during new flush emergence.
- Under heavy infestation, give sandwich sprays involving 4% neem seed kernel extract - cypermethrin 25 EC at 0.5ml/l - neem seed kernel extract 4% at fortnightly intervals.
- Combined release of *M. boninensis* (30 larvae/tree) and *Tamarixia radiata* (40 adults/tree) resulted in 23-26% reduction in leaf miner population.
- Spraying with fish oil resin soap and nicotine sulphate (1:1) + 50 parts of water in the early stages of infestation help in reducing the pest population.
- The parasitoid, *Tetrastichus phyllocnistoides* and *Ageniaspis* sp. cause up to 80% parasitism in Punjab.

2.2.1.2. Black fly, *Aleurocanthus woglumi*

This is a regional endemic pest in parts of Maharashtra and Karnataka. The epidemic proportions of the pest during late 70's and 80 have brought havoc on the citrus industry in Central India. The monitory loss was estimated to the tune of Rs. 25 to 50 million annually.

i) Damage: Both nymphs and adults suck cell sap and secrete voluminous honeydew on which sooty mold grows wildly that leads to fungal manifestation (*Capnodium* sp.) covering entire plant due to which photo-synthesis is affected. The adults lay eggs in spiral fashion on new leaves. The nymphs, which are black in colour, suck sap from leaves and devitalize the plants (Fig. 2.4). In severe cases, fruit bearing capacity of the tree is also affected. Fruits are rendered insipid in taste and blackened due to sooty mold. Such fruits fetch low price in the market.

For successful fruit set, a minimum of 2.2% organic nitrogen in leaf is must. Five to ten black flies/cm^2 area or 50 to 100 nymphs/leaf are sufficient to reduce leaf nitrogen level below 2.2%.

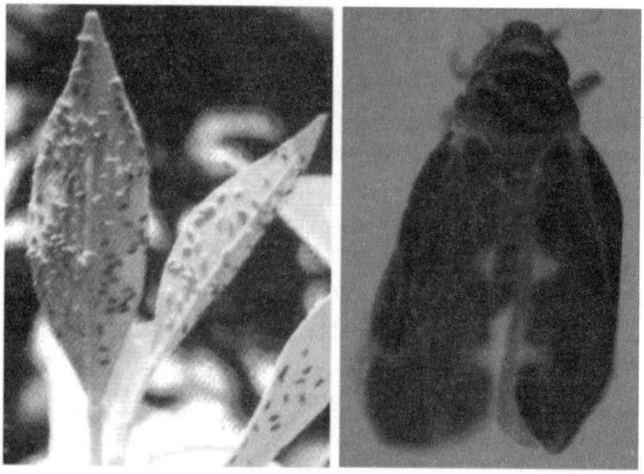

Fig. 2.4. Black fly on citrus (Courtesy: V.J. Shivankar & C.N. Rao, NRCC, Nagpur).

ii) Biology: The pest has three overlapping generations in a year corresponding to three flush seasons viz., *Ambia, Mrig* and *Hasta* and the life cycle takes 105, 120 and 135 days in respective seasons. Adults of these broods emerge during February-March, June-July and October-November and remain active for 1½-2 months. A female lays eggs on the underside of the young leaf in three round whorls, each ring comprising of 7 to 27 eggs. About 3,000 eggs are present on a single leaf in 150 whorls. Egg hatching takes two to four weeks during March-April and June-July while it continues for more than a month during November-December due to low temperature. Nymphal development takes 6 to 10 weeks whereas pupation takes about 8 weeks period up to first fortnight of May, August and January during *Ambia, Mrig* and *Hasta* broods, respectively.

iii) Pre-disposing Factors: Grown up orchards on heavy clay soils had evergreen canopies intermingling with each other thus creating a microniche underneath. Further, poor drainage in such soils adds to the dampness which together helps in pest buildup. The tall evergreen border shrubs aggravate the pest problem further by sheltering pest population and in a way provide alternative to main host citrus plants.

iv) Management

a) Cultural Methods: Guava, sapota and pomegranate are collateral hosts of black fly which should not be grown in the vicinity of the citrus orchard as for as possible. Close planting and water logging should be avoided. The affected shoots should be clipped-off and destroyed. Neem seed kernel extract is found better in reducing the blackfly population without affecting its natural enemy complex.

b) Chemical Methods: Spraying of monocrotophos at 1 ml or acephate at 0.8 ml or phosalone at 1.5 ml or dimethoate at 2 ml or phosphomidon at 0.7 ml/litre of water on the lower sides of leaves covering the entire tree canopy is effective. Second spray should be given after 15 days with any of the above insecticides or neem oil at 10 ml/litre of water. The stage of 50% egg hatching is the most ideal timing for spray application as it will ensure maximum coverage of the younger stages (first and second nymphal instars). This would coincide with first fortnight of April for *Ambia* season, second fortnight of July for *Mrig* bahar, and first fortnight of December for *Hasta* bahar seasons.

c) Biological Methods: The parasitoids, *Encarsia bennetti* (Aphelinidae: Hymenoptera) and *Eremoceros gunturiensis* and predators, *Serangium paracesetosum* and *Mallada boninensis* were predominant. Predators were observed feeding on eggs and nymphal stages and parasitoids parasitizing only the nymphal stages. Incidence of the parasitoid, *E. gunturiensis* was found high (10-20/leaf) during December. Field release of the predator, *M. boninensis* is recommended at 25-30 larvae/tree (19-22% predation). Field release of the parasitoids, *E. gunturiensis* and *E. bennetti* is recommended at 2000 adults/tree. Releases of *E. opulenta* have considerably brought down the blackfly population from 68% to 35% at Nagpur (Anon, 1987 b).

In Assam, *Encarsia divergence* and *E. merceti* are dominant accounting for 90% parasitism. *Amitus hesperidum* and *E. dypealis* cause 80-90% parasitism in Karnataka and Maharashtra.

Entomopathogenic fungi, *Aschersonia aleyrodis* and *Aegerita webberi* play an important role by attacking all stages of the black fly.

d) Integrated Methods

- Avoid close planting and water logging.
- The affected shoots should be clipped-off and destroyed.
- Avoid excessive nitrogen and irrigation.
- Spray dimethoate 30 EC at 2 ml/l during new flush and repeat at 10-12 days interval.
- Spray Neem seed kernel extract at 4% which does not affect natural enemies.
- The pest is regulated by parasitoids like *Amitus hesperidium* and *Encarsia clypealis* up to 80-90% in Karnataka and Maharashtra. At Gawahati, *Prospaltella divergence* and *Encarsia merceti* are dominant parasitoids accounting for 90% of the total parasitoids.
- *Chrysoperla carnea* or *Mallada boninensis* can also be released at 10-15 eggs/1st instar larvae/plant. Combined release of *M. boninensis* (30 larvae/tree) and *Tamarixia radiata* (40 adults/tree) resulted in 28-30, 42-45 and 23-26% reduction in black fly, psylla and leaf miner population, respectively.

2.2.1.3. Lemon Butterflies

Among eight species of papilionids namely *Papilio demoleus, P. helenus, P. helenus daksha, P. polytes, P. polyctor, P. polymnester, P. machaon asiatic, P. memnon* and *P. protenor* reported feeding on citrus leaves in India, of which *P. demoleus* is the major one.

i) Damage: The pests (Fig. 2.5) are active throughout the year except during winter and the infestation is more pronounced in nurseries and on young trees. Peak activity coincides with emergence of new flush. It feeds voraciously on the vegetative growth throughout the year. The more recent outbreak of the pest during July-August 1996 was reported from Vidarbha region of Maharashtra State on Nagpur mandarin (Shivankar and Singh, 1998b). The pest attacks mandarin and acid lime plantations almost throughout the year but is serious during July-August. Larvae being voracious feeders, cause severe defoliation of plants.

Fig. 2.5. Egg and larval instars and adult of lemon butterfly (Courtesy: V.J. Shivankar & C.N. Rao, NRCC, Nagpur).

ii) Biology: Eggs are laid singly on the tip of young leaves. Young brownish-black larva turns into full grown green caterpillar in 13 to 16 days and has a bifid 'Y' shape structure known as "osmaterium" which gives strong odour when disturbed. Larva pupates on the plant itself and pupa remains attached to the plant. Life cycle is completed in 18 to 40 days and may take up to 145 days in winter. There are 4 to 5 overlapping generations in a year.

iii) Management

a) Cultural Methods: Hand picking of various stages of the pest and their destruction is recommended as it is very easy to do in both nurseries and young orchards. Use of forceps for this purpose is an efficient method and in less than 4 hours one person can collect larvae from 0.4 ha nursery.

b) Chemical Methods: The pest can be controlled by foliar spray of any of the contact or systemic insecticides viz., dimethoate at 1.5 ml, fenitrothion at 1 ml, carbaryl at 2 g or endosulfan at 1.5 ml/litre of water at an early larval instar stage.

c) Biological Methods: *Trichogramma chilonis* and *Telenomus* sp. are the predominant egg parasitoids; a braconid *Apanteles flavipes* and ichneumonid *Melalophscharops* sp. are the major larval parasitoids; while a chalcid *Pteromalus puparium* and an ichneumonid *Holcojoppa coelopyga* are the dominant pupal parasitoids which could be utilized for effective management of lemon butterfly.

Release of *T. chilonis* at 500 adults/tree is recommended for the suppression of the pest. The parasitoids namely, *T. chilonis, Telenomus incommodus* and *Distatrix papilionis* account for a cumulative parasitism of

88% (Krishnamoorthy and Singh, 1988). Conservation of the indigenous braconid larval parasitoid, *Distatrix papilionis* through application of less toxic chemicals and botanicals in citrus IPM (Jayanthi and Verghese, 1999a) will ensure natural suppression of citrus butterfly.

Bacillus thuringiensis (Dipel) spray at 0.05% against early larval stages and nematode *Steinernema carpocapsae* (DD-136) gives good control of the pest. Entomopathogens like bacterium *Serratia marcescens* and fungus *Fusarium* sp. also reduce the pest population substantially.

d) Integrated Methods

- Collect and destroy larvae which look like bird droppings in the early stage.
- Spray endosulfan 35 EC at 2ml/l or carbaryl 50 WP at 3 g/l on new flush.
- Release *Trichogramma chilonis* at 500 adults per tree.
- Conservation of indigenous braconid larval parasitoid, *Distatrix papilionis* keeps the pest under check. The parasitoids, *T. chilonis* and *D. papilionis* account for a cumulative parasitism of 88%.
- Application of *Bacillus thurungiensis* var. *kurstaki* at 1 ml/l + spre-ader is very effective for the control of citrus butterflies. The susceptibility of citrus caterpillar to 'Bactospeine' has been reported.

2.2.1.4. Aphids, *Toxoptera* spp.

Two species of the pest namely, black citrus aphid *Toxoptera aurantii* and brown citrus aphid *T. citricida* were observed infesting orange orchards of Central India causing leaf curl and stunted growth of the plant.

i) Damage: The aphid incidence was observed throughout the year on young plants except in hot summer months of May and June. Most of the period the pest existed in nymphal stage except during September and October wherein the adults were seen. Presence of both the species was observed only during August. *T. aurantii* appeared only during August, while *T. citricida* was found throughout the year. *Hasta* and *Ambia* young flushes were heavily infested. Warmer days of February-March gave rise to the winged populations which migrated once the temperature got sufficiently warmer. Adults and nymphs suck the sap from tender leaves and shoots devitalizing the plants (Fig. 2.6). The pest secretes copious sugary solution on which sooty mold grows.

The aphids generally do not cause serious damage but are important as vectors of tristeza virus. The brown citrus aphid, *T. citricida* is responsible for vectoring citrus tristeza virus. The affected leaves get curled badly while fruits fall prematurely. Normally these aphids attack at the time of flowering but occasionally severe outbreak occurs when the rainy season is followed by dry weather.

Fig. 2.6. Aphids feeding on citrus leaves (Courtesy: K. Srinivasan, TAFE, Chennai).

ii) Biology: Citrus aphid produces about 5 young ones daily for a period of 1 to 3 weeks parthenogenetically which attain maturity in about 6 days. Day time warm temperature during February-March favours the pest population buildup. Winged colonizing adults called 'stem mothers' directly give birth to nymphs with no egg stage and no mating is required. Nymphs of black aphid give out red haemolymph and brown aphid, the yellow on squashing.

iii) Management

a) Chemical Methods: Spraying dimethoate or malathion or quinalphos all at 0.75 ml/litre of water or methyl demeton or monocrotophos at 0.5 ml/litre of water at fortnightly intervals checks the pest effectively.

b) Biological Methods: Parasitoids like *Lipolexis* sp., *Aphelinus* sp. and *Lysephlebus* sp. and the predators such as chrysopids, coccinellids, syrphids and spiders feed on this pest and reduce their incidence. *Cheilomenes sexmaculata* is observed in considerable numbers in clearing the aphid colonies (Singh, 1993). Release of *C. sexmaculata* at 50/plant is suggested.

c) Integrated Methods

- Spray dimethoate 30 EC at 2ml/l at new flush.
- In case of severe outbreak, spray imidacloprid 200 SL at 0.25 ml/l.
- The residual aphid population can be managed by the natural enemies.
- Release *Chielomenes sexmaculata* at 50 per plant, in the absence of natural enemies.

2.2.1.5. Scales, *Coccus hesperidum, C. viridis, Aonidiella aurantii, Icerya purchasi*

i) Damage: Nymphs and adults suck the sap from foliage, twigs and stems resulting in devitalization of plants. Fruits are also attacked. Excretion of honeydew favours development of black sooty mold affecting photosynthesis. Such trees exhibit yellowing, early shedding of leaves, bear less fruits and fruit quality deteriorates. Red scale is destructive in nature. Pest is active throughout the year but the spring, summer and winter flushes witness heavy incidence, populations attaining peaks particularly in autumn.

ii) Management

a) Chemical Methods: Pre-bloom spray of monocrotophos at 0.75 ml or quinalphos at 1.0 ml or cypermethrin at 0.5 ml/litre of water are most effective against scales. Spraying of methyl parathion at 1.0 ml or dichlorvos at 0.5 ml/litre of water during early hours of the day gives satisfactory control of scales. Chlorpyrifos at 4 ml, neem oil at 20 ml/litre of water gave protection for many days which is safe to natural enemies. Banding of tree trunk with monocrotophos and dimethoate at 10 ml/tree gives excellent control for 6 to 8 months.

b) Biological Methods: Periodic release of the lady bird beetle, *Rodolia cardinalis* can effectively control *Icerya purchasi.* The fascinating success of this scale control has established biological control as a valid method of pest control. The parasitoid, *Aphelinus* sp. parasitizes a high percentage of scales. Field release of aphelinid, *Aphytis melinus* and *Chilochorus nigritas* at 15 adults/tree show a good activity on different scales including *Aonidiella aurantii.* Release of the predator, *Chilicorus nigrita* at 10-15 beetles/tree controls red scale. Releasing of the parasitoid, *Aphytis melinus* at 2000/tree is recommended for the suppression of the red scale. Neem/pongamia oil 2% can be sprayed thrice at 15 days interval. The red scale, *Aonidiella aurantii* is found chiefly parasitized by *Aphytis* spp. and predated by *Chilocorous nigrita.* Release of *C. nigrita* is recommended at 10-15/tree for the control of citrus red scale. Releasing of *Aphytis melinius* at 2,000/tree is also suggested for the suppression of the pest (Krishnamoorthy, 1993).

The parasitoids, *Aneristus ceroplastae* and *Encyrtus lecanorium* cause up to 43% parasitism of green scale, *Coccus viridis* in the field (Tandon, 1985). Two coccinellid predators, *Chilicorus nigrita* and *C. circumdatus* play a major role in bringing down the population of green scale on acid lime. Release of *Cryptolaemus montrouzieri* at 10 beetles/tree was found to be highly effective in suppressing green scale.

In mixed planted orchards (citrus + coffee) with more shade and less light interception (900-1400 lux), spray of *Verticillium lecanii* at 10x10^6 spores/ml + 0.05% teepol was highly effective against green scale both in citrus and coffee. In pure citrus orchards, the combination was only effective during the rainy season. *V. lecanii* at 3x10^8 spores/ml is found effective in reducing the population of green scale in humid areas (Singh, 1993). Entomopathogenic fungus, *Nectria* sp. keeps purple scale under check in south India.

2.2.1.6. Mealy Bugs, *Planococcus citri, P. lilacinus, Nipaecoccus viridis, Maconellicoccus hirsutus, Ferrisia virgata*

Citrus mealy bugs are destructive in Peninsular India and NEH region. The pest causes heavy damage to nursery and grown up plants. Damage is severe in spring and autumn.

i) Damage: The leaves and tender shoots get deformed and twisted into knots and loops. The leaves become curled up. It feeds on stem ends of fruits and often causes heavy fruit drop. Clusters of white formations of the bug are found at the joints of twigs (Fig. 2.7). It extracts plant sap and reduces tree vigour. It also secretes honeydew which invites sooty mold. The severity of the problem and importance of the mealy bugs are frequently associated with an unsightliness of the plant and fruit caused by sooty mold. As a result, the growth of the tree is arrested, and the blossom and fruits fall off prematurely. They occur throughout the year but more in summer months and fruiting stage. The seedlings of *Murraya koenigi* maintained for psylla pure culture were found invariably and heavily attacked by the mealy bugs.

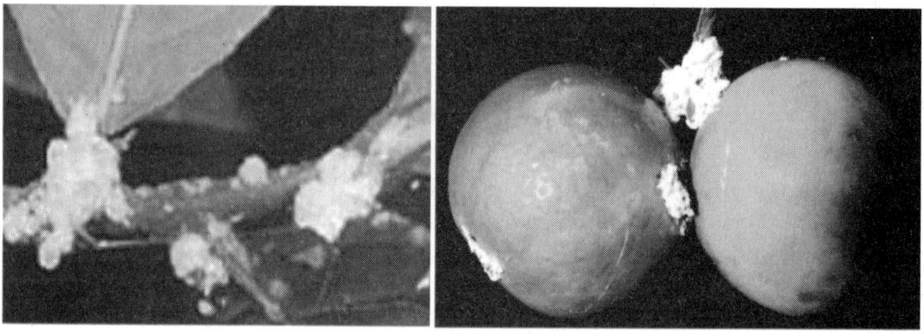

Fig. 2.7. Mealy bugs on citrus (Courtesy: K. Srinivasan, TAFE, Chennai; V.J. Shivankar & C.N. Rao, NRCC, Nagpur).

ii) Biology: Citrus mealy bug breeds continuously. Yellow eggs are produced in a loose colony of waxy filaments. About 600 to 800 eggs covered in an ovisac are deposited in a fortnight which hatches within 7 to 15 days. There are 3 molts of female nymphs and 4 molts of winged male nymphs completing 3 overlapping generations in a year.

iii) Management

a) Chemical Methods: Spraying of 150 ml dimethoate + 250 ml kerosene in 100 litres of water or 10 g carbaryl + 10 ml kerosene or 20 ml malathion in 10 litres of water checks mealy bugs effectively. Chlorpyriphos at 2 ml/litre of water also gives good control of the pest.

b) Biological Methods: The release of an exotic parasitoid, *Leptomastix dactylopii* at 5000/ha is recommended for suppression of *P. citri* population (Krishnamoorthy and Singh, 1987). The local parasitoid, *Tetracnemoidea indica* checks *P. lilacinus* population (Mani, 1993a). Release of *T. indica* or *Cryptolaemus montrouzieri* at 10 beetles/tree is recommended for the suppression of *P. lilacinus*. *C. montrouzieri* is also recommended for the control of *N. viridis*. *Coccidoxenoides peregrina* was found to parasitize both males and females of *P. citri* (nymphal stages including crawlers). The parasitoid was most effective at Hessaraghatta (Karnataka) in July-August. The mealy bug was completely suppressed by end of April mainly due to the activity of *C. peregrina* which appeared in February. From acid lime fruits infested with *P. citri*, only *C. peregrina* emerged in large numbers.

2.2.1.7. Psylla, *Diaphorna citri*

Citrus psylla is one of the most serious pests in North India especially in Punjab, Haryana, Himachal Pradesh and Maharashtra. It came into epidemic proportions during *Ambia* flush of 1998 in Central India. At times the pest alone causes the loss ranging from 83 to 95%.

i) Damage: The nymphs and adults suck the cell sap causing curling of leaves (Fig. 2.8), defoliation, heavy flower drop affecting the fruit set severely and even causing death of branches from tip to downwards i.e. dieback. It is suspected to secrete some toxins along with its saliva which kills even those branches not attacked. The nymphs also secrete whitish crystalline honeydew that covers the shoots and leaves giving the appearance of having been dusted on which the sooty mold grows and which in turn affects the market value of fruits. The fruits suffer from both quality and quantity, being undersized, poor in juice and insipid in taste. Psylla is active during spring and monsoon and less active during winter. Its fifth instar nymphs act as vector of the greening bacterium which accelerates the decline. Citrus plantations in Punjab and Assam have suffered the severe blow of this disease. Moreover, alternate hosts like *Murraya koenigi* which is the most favoured host for psylla, helps in quick build up of the pest within less time (George and Kandasamy, 1986).

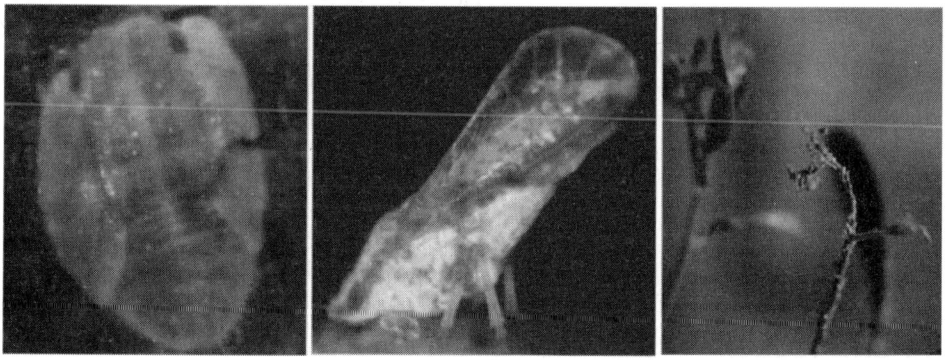

Fig. 2.8. Psylla nymph, adult and infestation on citrus (Courtesy: V.J. Shivankar & C.N. Rao, NRCC, Nagpur).

ii) Biology: *Ambia* and *Mrig* season flushes are congenial for this pest. Eggs are laid in the folds of half opened leaves or between the flower buds. Development of egg to adult takes 15 days in summer and 45 days in winter completing 9 to 10 overlapping generations in a year. Hibernating adults during cold and hot seasons may live for more than six months.

iii) Management

a) Cultural Methods: Pruning of affected and dried shoots and modifying the canopy structure in such a way that light interception below the canopy is maximum helps in reducing the incidence of psylla. The vegetative growth should be regulated by judicious application of irrigation and fertilizers. Collateral host like curry leaf (*Murraya koenigi*) should not be grown in the vicinity of citrus orchards as it may act as a breeding ground for psylla.

b) Chemical Methods: Foliar spray at bud burst stage or as and when infestation is noticed with phosphamidon, quinalphos and thiometon all at 0.025% concentration were found most effective in causing 100% mortality of the pest within three days after application. If required, second spray should be given after 15 days.

c) Biological Methods: The eulopid nymphal parasitoid, *Tamarixia radiata* is the most common parasitoid of citrus psylla causing up to 95% parasitism in Punjab and 30-90% in Maharashtra. Nymphal parasitization by *T. radiata* was maximum during March (91-93%) and December (90%). Among predators the chrysopids *Mallada boninensis, Apertochrysa crassinervis, Brumus suturalis* and the coccinellids *Coccinella rependox* prey upon nymphal psylla and can be used for the biological control of the pest.

2.2.1.8. White Fly, *Dialeurodes citri*

i) Damage: The nature of damage is identical to citrus black fly. Both nymphs and adults reduce plant vigour by sucking large quantities of cell sap (Fig. 2.9). As a result, foliage turns pale green to brown, curled and lastly shed down. Black sooty mold is attracted to the honey dew excreted by nymphs on leaves that affects the process of photosynthesis adversely. Fruiting capacity of the tree and quality of fruits is also greatly affected. Blackened fruits are insipid in taste and do not get good price in the market. Incidence is generally observed in the months of March-April, July-August and October-November coinciding with new flushes. It is serious on kinnow and sweet orange in Punjab.

ii) Management

a) Cultural Methods: Do not plant densely in the orchard. Provide good drainage.

b) Chemical Methods: Spraying alternately with monocrotophos at 1 ml, acephate at 1.25 g and phosalone at 1.5 ml/litre of water at 15 days interval in each flushing season gives effective control of the pest.

c) Biological Methods: Aphelinid parasitoids, *Encarsia lahorensis* and *E. bennetti* give satisfactory control of white flies. The immediate diffusion of chalcids, *Encarsia meritoria, Eretmocerus serius* and *E. gunturiensis* reduces the initial outbreaks of white fly. The parasitic fungi *Cladosporium* sp. and *Aschersonia aleyrodis* keeps the white fly population under check.

Fig. 2.9. White fly on citrus (Courtesy: K. Srinivasan, TAFE, Chennai).

2.2.1.9. Fruit Sucking Moth, *Ophideres materna. O. fullonica, O. ancilla, Achaea janata*

i) Damage: Fruit sucking moth is the regular and serious pest of ripening fruits causing heavy fruit drop in both mandarins and sweet orange, particularly during September-October. The robust beautiful moths are quite mobile and hence difficult to control. The adults puncture the ripening fruits (Fig. 2.10). Such fruits drop prematurely as a result of rotting due to fungal and bacterial infections introduced through punctures causing considerable fruit loss (40%). Adult moths are active during August-December with their peak activity during September-October.

Fig. 2.10. Fruit sucking moth on citrus (Courtesy: K. Srinivasan, TAFE, Chennai; V.J. Shivankar & C.N. Rao, NRCC, Nagpur).

ii) Biology: The pest does not breed in citrus orchard. The female moth lays about 200 to 300 eggs on wild plants like Gulvel, *Tinospora cardifolia, T. smilacina*; Vasanvel, *Cocculus villosus*; Chandvel, *Cirampelos pareira, Convolvulus arvensis, Trichisia pattens* and *Pericampylus glaucus* where larval stage survive. Therefore, younger stages of the pest are not at all harmful. Larva emerges after 3 to 4 days of incubation period, feeding through 5 instars within 13 to 17 days before full grown and pupates in soil for about 12 to 18 days. A generation is completed in 30 to 40 days.

iii) Management

a) Physical Methods: Generation of smoke in the orchard for two hours in the evening repels adult moths.

b) Cultural Methods: Clean cultivation and removal of alternate hosts are of primary importance. The fruit damage can be minimized by bagging the individual fruits with polmyrah baskets. Use light traps for attracting the adults and kill them. The fallen fruits should be destroyed by burying them deep in the ground.

c) Chemical Methods: Poison baiting with 20 g malathion WP or 50 ml diazinon + 200 g jaggery with some vinegar or fruit juice in 2 litres of water has been found effective. Two litre poison bait in two bottles is recommended for 25-30 trees. Spraying malathion or endosulfon on the weeds around the orchard where the young stages of the pest survive reduces the incidence of the pest.

2.2.1.10. Bark Eating Caterpillar, *Indarbela quadrinotata, I. tetraonis*

i) Damage: Infestation is seen in unhygienic and neglected orchards. It is spreading alarmingly even on the young plantations of Nagpur mandarin. As many as 17 tunnels were recorded on a single 8 year old tree. Such orchards give the declining look. Grubs feed on bark portion during night by making tunnels into the forked trunk and the branches near the joints resulting in cutting off the sap flow with low productivity. They hide in the tunnel during day time. Characteristic webbings consisting of wooden frass and faecal pellets of grubs hanging outside the tunnel is the indication of the pest infestation (Fig. 2.11). The infestation is common in old and neglected orchards. Bark eating caterpillar attack shortens the life span of tree and devitalizes the plant extensively.

ii) Biology: Female lays eggs in groups of 15 to 25 during May-June which remain attached to bark and hatch in 8-11 days. On hatching larva keeps on feeding under the bark until September and then bores tunnels in the wood. The larval stage lasts up to 21 to 41 days. Pupation takes place inside the tunnel which lasts for 21-41 days. Adult moths emerge during May-June. Only one generation is completed in a year.

Fig. 2.11. Bark eating caterpillar infestation on stem and larva (Courtesy: V.J. Shivankar & C.N. Rao, NRCC, Nagpur).

iii) Management

The pest is controlled by inserting a cotton wad soaked in petrol or kerosene both at 10 ml/tunnel or 0.1% dichlorvos or 0.02% monocrotophos into tunnels and plugging them by mud. Administering 4 -5 ml of insecticidal solution (4 ml dichlorvos or 2 ml monocrotophos dissolved in one litre of water) with the help of plastic injection syringe gives sure kill of the hiding larvae within 10 minutes. Spraying affected portion with the above mentioned insecticides after clearing the frass and faecal pellets is envisaged to keep the pest under check.

2.2.1.11. Trunk Borer/Stem Borer, *Monohammus versteegi, Chelidonium cinctum*

It is a serious pest of Khasi and Darjeeling mandarins in North Eastern States. In Assam, 15-60% of trees are attacked by this pest.

i) Damage: On hatching, the grub at first bores horizontally in the sapwood at ground level and then vertically tunnels upwards, taking circuitous course. Entire grub and pupal stages pass inside the hole. Adult comes out of the exit hole next summer which is 1.5 to 2.0 feet above the first hole made before undergoing pupation (Fig. 2.12). Appearance of resinous exudation and sawdust-like powder on tree trunks near the ground indicates the presence of citrus trunk borer. Borer feeds on leaf lamina along the midribs leaving the margins intact. Affected branches gradually dry up leading to citrus decline. Eggs are laid on the tree trunks and main branches, just above the ground level up to 2.5 m height.

ii) Management

a) Cultural Methods: Chipping off the affected new shoots, cutting and burning the affected branches help in reducing the incidence of borers.

Beetles should be removed by jerking the trees, collected and destroyed. A wad of cotton is soaked in kerosene or petrol, so as to absorb about 10 ml and then pushed into the mouth and plugged with mud.

b) Chemical Methods: Swabbing of stem tunnel up to 2 m height with 5 ml of carbaryl (2 ml/litre of water) or DDVP (1 ml/litre) or monocrotophos (1 ml/litre) gives effective control of trunk borer.

c) Biological Methods: The white muscardine fungus, *Beauveria bassiana* is highly effective against trunk borer.

Fig. 2.12. Entry holes on trunk and adult of trunk borer on citrus (Courtesy: K. Srinivasan, TAFE, Chennai; V.J. Shivankar & C.N. Rao, NRCC, Nagpur).

2.2.1.12. Thrips

i) Damage: Thrips are considered to be the serious pests on all citrus cultivars grown commercially (Fig. 2.13).

Fig. 2.13. Thrips infestation on leaves and fruits of citrus (Courtesy: V.J. Shivankar and C.N. Rao, NRCC, Nagpur).

ii) Management: Use of soft insecticides like soaps, botanicals such as *Sabadiella* sugar and even simply washing with clean water should be done in order to check the thrips menace in nurseries. Use of ryania and *Sabadiella* in combination with predaceous mites plays a greater role in the thrips control package.

2.2.2. Mite Pests

2.2.2.1. Mite, *Eutetranychus orientalis*

i) Damage: The citrus mite not only damages the leaves but also the fruits causing unpleasant brownish blemishes on the peel, deteriorating the quality that fetches very low price. The pest feeds on the upper surface of old leaves causing mottling appearance and dusty leaves (Fig. 2.14). The citrus mite extracts cell sap from leaves and fruits. In severe infestation, the stripping enlarges to dry necrotic areas. Generally leaves drop and die-back of twigs starts. New sprouting on grown up trees creates the congenial conditions for rapid and continuous multiplication of the pest leading to its outbreak.

Fig. 2.14. Mite infestation on citrus leaf, fruits and its adult (Courtesy: K. Srinivasan, TAFE, Chennai; V.J. Shivankar & C.N. Rao, NRCC, Nagpur).

ii) Biology: The pest is active throughout the year but has prominence during November- May having peak of 15 mites/leaf during February- June. Female lays up to 8 eggs per day mainly along the midrib of the leaf. It takes 10 to 15 days and 3 months for younger stages to develop fully and 17 to 20 days to complete the entire life cycle. The pest has a number of overlapping generations in a year.

iii) Management: Water stress aggravates the mite problem. Hence the trees should be well irrigated particularly during the stress in late summer.

Tetradifon at 0.025% was found most effective in controlling mites followed by phosphamidon at 0.025% with mortality of 96.49 and 87.57%, respectively. Foliar spray of dicofol at 1.5 ml, monocrotophos at 1.0 ml or oxydemeton methyl at 1.5 ml or wettable sulphur at 3 g/litre of water was also effective when mite incidence is noticed. Second spray, if required, may be given after a week.

The most important of citrus mite is a predaceous mite, *Euseius hibisci* and the predators are *Agistemus* sp. and *Amblyseius hibisci*.

2.2.3. Vertebrate Pests

i) Damage: In hill regions of Karnataka, mandarin oranges (*Citrus reticulata*) grown as shade trees in coffee estates served as nesting and roosting sites for more than 90 bird species. A fruit loss due to various species of vertebrates in Karnataka on oranges is 53% (Jungle crows-30%, Barbets-10%, Parakeets-5%, Thrushs-3%, Chloropsis-3%, Others-2%). Jungle crow is the dominant species feeding on oranges (Chakravarthy, 1993).

ii) Management : By watching, 42% fruit losses could be prevented and there is further scope for saving fruits by effectively watching orchards and scaring away the birds.

Table 2.9. Calendar of IPM activities in citrus

Month/s	Target pest	IPM activities
January	Bark eating caterpillar	Remove the wooden frass present in between the joints of the tree trunk and inject 5 ml of 0.1% dichlorvos 76 EC in the larval tunnel with the help of disposable syringe and push the cotton swab soaked in dichlorvos solution in the larval tunnel.
February	Leaf miner, Psylla, Blackfly	At the time of *Ambia* bahar, spray 1.5 ml monocrotophos or 2 ml quinalphos or 2 ml dimethoate in 1 litre of water. Against leaf miner, spray fenvalerate. Repeat application after 10 days in case of the infestation of leaf miner and citrus psylla persists. One spray may be given when blackfly attains peak of adult population.
	Bark eating caterpillar	Remove the wooden frass present in between the joints of the tree trunk and inject 5 ml of 0.1% dichlorvos 76 EC in the larval tunnel with the help of disposable syringe and push the cotton swab soaked in dichlorvos solution in the larval tunnel.
	Aphid	Spray quinalphos or dimethoate or malathion each at 0.05% and repeat spray after every 7 days.
	Leaf eating caterpillar	Spray 2 ml endosulfan or 1.5 ml dimethoate or 1 ml fenitrothion or 2 g carbaryl in 1 litre of water.
March	Psylla, Leaf miner	Spray 0.7 ml monocrotophos 36 SL or 1.5 ml dimethoate or 1 ml quinalphos in 1 litre of water. Repeat the spray after 15 days.
	Leaf eating caterpillar	Spray 2 ml endosulfan 35 EC in 1 litre of water.
April	Bark eating caterpillar	Remove the wooden frass present in between the joints of the tree trunk and inject 5 ml of 0.1% dichlorvos 76 EC in the larval tunnel with the help of disposable syringe and push the cotton swab soaked in dichlorvos solution in the larval tunnel.
	Blackfly	Spray 1 ml monocrotophos or 0.125 g acephate or 2 ml dimethoate or 1.5 ml quinalphos in 1 litre of water. Repeat spray after 15 days.
	Mite	Spray 1.5 ml dicofol 1.8 EC or 3 g wettable sulphur in 1 litre of water. Repeat the spray after 15 days.

May	Bark eating caterpillar	Remove the wooden frass present in between the joints of the tree trunk and inject 5 ml of 0.1% dichlorvos 76 EC in the larval tunnel with the help of disposable syringe and push the cotton swab soaked in dichlorvos solution in the larval tunnel.
	Mealy bug	The basin around tree trunk should be earthed up. Sticky band smeared with mobil oil should be tied around the tree trunk and destruction of ant nests should be done. Spraying of 2 ml chlorpyriphos 20 EC or 2.5 ml dichlorvos in 1 litre of water should be done on the tree hives and trunk.
	Snail	Use metaldehyde pellets. Spread poison baits at the base of tree trunk in the evening throughout the orchard (In order to prepare poison bait, dissolve 2 kg jaggery, 25 g yeast in 10 litres of water. Add wheat or paddy husk and mix well and keep for fermentation for 10-12 hrs. Add 100 g methomyl or 50 g thiomethazim). Snails will be attracted and will die after feeding on such poison bait.
July	Psylla, Leaf miner	Spray 0.7 ml monocrotophos 36 SL or 1.5 ml dimethoate or 1 ml quinalphos in 1 litre of water. Repeat the spray after 15 days.
August	Blackfly	Spray 1 ml monocrotophos or 0.125 g acephate or 2 ml dimethoate or 1.5 ml quinalphos in 1 litre of water. Repeat spray after 15 days.
	Fruit sucking moth	Destroy all fallen fruits by burying in deep pit. Poison baiting with 20 g malathion or 50 ml diazinon mixed with 200 g gur and some orange juice in 2 litres of water should be placed in a tray. Place a bulb of 60 Watts above this tray to kill the moths.
September	Fruit sucking moth	Destroy all fallen fruits by burying in deep pit. Poison baiting with 20 g malathion or 50 ml diazinon mixed with 200 g gur and some orange juice in 2 litres of water should be placed in a tray. Place a bulb of 60 Watts above this tray to kill the moths.
	Fruit fly	Use male attracting trap containing 0.1% methyl eugen- ol and 0.05% malathion. Use 25 traps/ha from 60 days before fruit harvest and change the bait after every 7 days.
October	Fruit sucking moth	Destroy all fallen fruits by burying in deep pit. Poison baiting with 20 g malathion or 50 ml diazinon mixed with 200 g gur and some orange juice in 2 litres of water should be placed in a tray. Place a bulb of 60 Watts above this tray to kill the moths.
	Fruit fly	Use male attracting trap containing 0.1% methyl eugenol and 0.05% malathion. Use 25 traps/ha from 60 days before fruit harvest and change the bait after every 7 days.

	Mite	Spray 1.5 ml dicofol 1.8 EC or 3 g wettable sulphur in 1 litre of water. Repeat the spray after 15 days.
November	Psylla	Spray 0.7 ml monocrotophos 36 SL or 1.5 ml dimethoate or 1 ml quinalphos in 1 litre of water. Repeat the spray after 15 days.
	Mite	Spray 1.5 ml dicofol 1.8 EC or 3 g wettable sulphur in 1 litre of water. Repeat the spray after 15 days.
December	Mite	Spray 1.5 ml dicofol 1.8 EC or 3 g wettable sulphur in 1 litre of water. Repeat the spray after 15 days.
	Blackfly	Spray 1 ml monocrotophos or 0.125 g acephate or 2 ml dimethoate or 1.5 ml quinalphos in 1 litre of water. Repeat spray after 15 days.
	Leaf miner	Spray 0.7 ml monocrotophos 36 SL or 1.5 ml dimethoate or 1 ml quinalphos in 1 litre of water. Repeat the spray after 15 days.

2.3. SAPOTA

It is cultivated in 0.140 million ha with annual production of 1.117 million tonnes (Bijay Kumar, 2007) (Table 2.10).

Table 2.10. Area, production and productivity of leading sapota growing states of India (2005-06) (Bijay Kumar, 2007)

States	Area (in '000 ha)	Production (in '000 MT)	Productivity (in MT/ha)
Gujarat	62.5	255.8	4.1
Maharashtra	22.8	249.8	11.0
Karnataka	24.9	235.7	9.5
Tamil Nadu	6.0	150.3	25.0
Andhra Pradesh	15.4	153.6	10.0
West Bengal	4.2	49.0	11.8
Orissa	3.3	14.3	4.4
Haryana	0.5	3.7	7.4
Tripura	0.1	1.4	11.3
Pondicherry	0.1	1.0	10.0
Others	0.3	2.5	8.3
Total	140.1	1117.1	8.0

2.3.1. Insect Pests

2.3.1.1. Bud and Fruit Borer, *Nephopteryx eugraphella*

This is the most serious pest of sapota, causing damage to leaves, buds, flowers and fruits.

i) Damage: The caterpillars (Fig. 2.15) generally feed on leaves but are often found on buds and flowers and sometimes on tender fruits as well. The caterpillars web together a bunch of leaves and feed within on chlorophyll, leaving behind a fine network of veins. They also bore inside the flower-buds and tender fruits which wither away and the caterpillars move on to the next bud or fruit. The infestation of this pest can be easily detected by the presence of clusters of dried leaves and flower buds hanging on webbed shoots. The pest is active throughout the year and their activity increase with the appearance of new shoots (April-May) and flower buds (May-June).

Fig. 2.15. Bud and fruit borer on sapota (Courtesy: K. Srinivasan, TAFE, Chennai).

ii) Biology: The oval shaped, yellowish white eggs are laid on ventral surfaces of leaves and flower buds. The larvae hatch in 2-5 days in summer and 4-11 days in winter. The larval period is completed in 13-26 days in summer and 31-60 days in winter. Pupation takes place in folded leaves and the grey coloured moths emerge after 8-13 days in summer and 15-29 days in winter. Total life cycle is completed in 26-38 and 49-92 days during summer and winter, respectively. There are 7 to 9 overlapping generations in a year.

iii) Management

- Remove and destroy all the infested clumps.
- Spray with Endosulfan 35 EC at 2ml/l or Chlorpyrifos 20 EC at 2.5 ml/l.

2.3.1.2. Green Scale, *Coccus viridis*

i) Damage: These are small greenish-yellow, sessile, oval shaped scales feeding on the leaves or tender twigs and shoots. As the pest excretes honey-dew, secondary fungal infection causes black sooty mold on the leaves and twigs. Generally the infestation is more in summer. Nymphs and adult females congregate on tender leaves, shoots and inflorescences and suck their sap. Severe infestation devitalizes the trees and affects their fruiting potential.

ii) Management

- Prune affected parts and burn at the early stage of infestation, during non-fruiting stage.
- In case of serious infestation, prune affected parts and spray with diazinon 20 EC at 4ml/l or Quinalphos 25 EC at 2ml/l during non-fruiting or early fruiting season.
- Apply methyl parathion 2% dust at 25 g around the base of the tree even when light infestation is noticed to prevent ascent of ants.
- Prune all adjoining branches of infested tree to prevent the easy spread of pests to neighbouring trees.
- Spray 2% starch to remove the sooty mold.
- Release of *C. nigrita* and *Cryptolaemus montrouzieri* in January/ February at 10 grubs/ tree suppresses the population of green scale (Mani and Krishnamoorthy, 1997b).

2.3.1.3. Green Shield Scale, *Chloropulvinaria psidii*

i) Damage: The Nymphs and adult females of green shield scale congregate and suck the sap from leaves, tender shoots and inflorescences of young sapota plants. Severe infestation devitalizes the trees and affects their fruiting potential.

ii) Management: Green shield scale is controlled by the release of *C. motrouzieri* (Mani and Krishnamoorthy, 1997b).

2.3.1.4. Bud Borer, *Anarsia achrasella*

The bud borer was recorded from South Gujarat (Patel *et al.*, 1984).

i) Damage: The larvae bore into flower buds and floral parts making holes on ovary and petals there by reducing the fruit set. A survey carried out in 1979 revealed that, on an average, 36.2% flower buds were infested (Shah *et al.*, 1986).

ii) Biology: The egg stage lasts for 4.91 days and the four larval instars are completed in 2.5, 2.31, 3.25 and 2.25 days, respectively. The pre-pupal and pupal stages are completed in 1.02 and 6.08 days, respectively. Each female moth lays on an average 39.98 eggs in a life span. Eggs are laid singly or in groups of 2-3 on flower buds or bud stalks (Patel, 1981; Patel *et al.*, 1984). The life cycle from egg to adult is completed in 20-31 days.

iii) Management: The pest can be managed by spraying 0.05% malathion, 0.075% endosulfan, 0.01% fenvalerate or 0.01% fenthion.

2.3.1.5. Leaf Miner, *Acrocercops gemoniella*

i) Symptoms: During leaf emergence, the tiny caterpillars mine into the young and tender leaves and feed within. The affected leaves show glistening galleries, get distorted, dry and ultimately fall down. The infestation is quite severe where new flushes are produced throughout the year.

ii) Management

- Remove and destroy the affected leaves in the initial stage of attack to prevent the infestation from spreading.
- Spray with systemic insecticides like dimethoate 30 EC at 2ml/l or oxydemeton methyl 25 EC at 2ml/l.
- Spray 500 g nicotine sulfate (40%) and 2 kg soap mixed in 350 litres of water.

2.3.1.6. Mealy Bugs, *Planococcus citri, P. lilacinus*

i) Damage: Nymphs and adult females congregate on tender leaves, shoots and inflorescences and suck their sap. The excreted honey dew encourages the development of sooty mold. Infested trees are also invaded by ants which feed on honey dew and spread mealy bugs from one plant to another. Severe infestation devitalizes the trees and affects their fruiting potential.

ii) Management: Spray 500 g nicotine sulfate (40%) and 2 kg soap mixed in 350 litres of water. The two encyrtids, *L. dactylopii* and *Coccidoxenoides perigrinus* are able to suppress the population of *P. citri*, while *C. montrouzieri* is highly effective in checking the population of *P. lilacinus* (Mani and Krishnamoorthy, 1997b).

2.3.2. Vertebrate Pests

Jungle crow and House crow caused 12% loss of sapota fruits under Bangalore conditions. The crows are major pests while Redvented Bulbul is a minor pest. In Mudigere, Chichmagalur, Karnataka; crow, monkey, squirrel and bat caused a loss of 18% fruits in a season. In Bangalore, rodents damaging sapota included *B. bengalensis* and *Tatera indica*. Population of the rodents was high October to December coinciding with fruit bearing and harvesting stages. *B. bengalensis* was the dominant species at all stages of crop growth (Anon, 1996).

2.4. PAPAYA

Papaya (*Carica papaya*) is one of the most important fruit crops valued for its rich source of vitamins and nutrient content. It is native to tropical America and its place of origin is said to be in Southern Mexico and Costa Rica. It is believed to have been introduced into India during the 16th century. It is grown both in tropical and sub-tropical countries of the world like Brazil, Australia, USA, Taiwan, Puerto Rico, Peru, Pakistan, Bangla Desh, Malaysia, India and various parts of Central and South Africa. The ripe papaya fruits are used for table purpose, raw fruits are cooked and used as vegetable. Immature fruits are used for the extraction of proteolytic enzyme 'papain', which is mainly used as a drug in digestive disorders and in the treatment of ulcers and diphtheria. It has also varied industrial uses viz., in tanning industry, degumming of silk, etc.

India is the largest producer of papaya in the world. Papaya is grown in 73,000 ha with annual production of 2.317 million tonnes (Table 2.11). Productivity of papaya is the highest (31.7 t/ha) (Table 2.12) among the fruit crops, which has attracted the growers for its commercial cultivation resulting in the spectacular increase in area and production in last few decades. In the last decade, phenomenal increase in area and production is recorded from Southern states owing to demand of fruits for industrial use.

Table 2.11. Major papaya producing countries of the world (2005-05) (Bijay Kumar, 2007)

Country	Area (in '000 ha)	Production (in '000 MT)	Productivity (in MT/ha)	% share of world production
India	73	2317	31.7	28.8
Brazil	33	1574	48.3	19.5
Nigeria	99	834	8.4	10.4
Mexico	19	709	37.5	8.8
Indonesia	24	549	22.8	6.8
Ethiopia	13	259	20.5	3.2
Congo	13	216	17.3	2.7
Peru	14	171	12.5	2.1
Philippines	9	147	15.5	1.8
Colombia	6	138	24.7	1.7
Others	81	1142	14.0	14.2
Total	383	8055		

Table 2.12. Area, production and productivity of leading papaya growing states (2005-06) (Bijay Kumar, 2007)

States	Area (in '000 ha)	Production (in '000 MT)	Productivity (in MT/ha)
Andhra Pradesh	12.9	1009.2	78.2
Gujarat	7.7	323.0	41.8
West Bengal	9.5	263.7	27.7
Karnataka	3.5	230.1	66.1
Assam	8.1	130.5	16.1
Chattisgarh	4.2	119.1	28.2
Kerala	18.9	85.5	4.5
Tamil Nadu	0.3	53.7	182.0
Madhya Pradesh	0.7	19.0	27.0
Others	7.3	83.4	11.4
Total	73.1	2317.2	31.7

On an average, 1000 g of ripe papaya fruit contains 2500 I.U. of Vitamin A, 8 I.U. of Vitamin B, 33 Sherman units of Vitamin B2 and 70 mg of Vitamin C. Besides, it contains papain, a proteolytic enzyme which acts as papsin and is useful in digestive disorders.

2.4.1. Insect Pests

2.4.1.1. White Fly, *Bemisia tabaci*

Besides papaya, this pest also attacks tomato, tobacco, cotton and some winter vegetables.

i) Damage: Small, white nymphs suck sap on the ventral surface of leaves. The affected leaves become yellow and curl downwards. They also wrinkle and results in early shedding. These white flies are important in papaya as they transmit papaya leaf curl disease.

Pear-shaped eggs are laid on the ventral side of leaves. On hatching, the nymphs settle down on the same leaf for feeding. Adults are minute flies covered with white waxy bloom. There are about 12 generations in a year.

ii) Management: Spray lambda cyhalothrin 5 EC at 0.5 ml/l or decamethrin 2.8 EC at 1 ml/l.

2.4.1.2. Aphid, *Aphis gossypii*

i) Damage: Aphids have been reported feeding on papaya leaves. Nymphs and adults inject their saliva in the plant tissues and suck the cell sap. While doing so, these aphids transmit virus diseases (papaya mosaic). The affected trees seldom recover from the disease and gradually die.

ii) Management: Prophylactic sprays with 0.03% dimethoate or 0.05% monocrotophos or 0.05% methyl demeton can protect the plants from the attack of aphids. Virus affected plants should be destroyed.

2.4.2. Mite Pests

2.4.2.1. Red Spider Mite, *Tetranyhus cinnabarinus*

i) Damage: The nymphs and adults suck the cell sap from plant tissues directly. Clusters of yellow spots may be seen on dorsal surface of leaves while minute red mites are seen in silken webs, on the lower surface. The leaves turn yellowish and ultimately dry. In case of severe infestation, fruits are also attacked.

ii) Management

- Removal and burning of the infested leaves.
- Spraying of 0.05% dicofol.

2.4.3. Vertebrate Pests

i) Damage: Ripe fruits of papaya are damaged by birds, squirrels, monkeys and elephants.

ii) Management : Covering fruits with thatch, gunny bag or a festoon of thorny sticks, nylon nets and scaring facilitated management of vertebrate pests.

2.5. PINEAPPLE

Pineapple (*Ananas comosus*) is grown in 85,400 ha with annual production of 1.353 million tonnes (Tables 2.13 and 2.14). There has been phenomenal increase in production and productivity (15.8 MT/ha) due to adoption of improved production technologies especially in Karnataka, West Bengal and Kerala. However, productivity is still low in North Eastern States (8-10 MT/ha) owing to traditional methods of cultivation compared to West Bengal (26.38 MT/ha).

Table 2.13. Major pineapple producing countries of the world (2005-05) (Bijay Kumar, 2007)

Country	Area (in '000 ha)	Production (in '000 MT)	Productivity (in MT/ha)	% share of world production
Brazil	62	2292	37.1	12.7
Thailand	98	2183	22.2	12.1
Philippines	49	1788	36.3	9.9
Coasta Rica	27	1605	59.9	8.9
China	76	1442	19.0	8.0
India	85	1353	15.8	7.5
Indonesia	80	925	11.6	5.1
Nigeria	121	917	7.6	5.1
Kenya	14	600	42.0	3.3
Mexico	15	552	36.2	3.1
Others	268	4386	16.4	24.3
Total	896	18043		

Table 2.14. Area, production and productivity of leading pineapple growing states (2005-06) (Bijay Kumar, 2007)

States	Area (in '000 ha)	Production (in '000 MT)	Productivity (in MT/ha)
West Bengal	13.4	379.2	26.3
Assam	14.6	241.0	16.5
Manipur	2.6	134.5	52.1
Kerala	12.8	109.3	8.6

Bihar	4.2	108.0	25.5
Tripura	5.2	106.4	20.5
Karnataka	10.6	88.6	8.4
Meghalaya	9.5	83.7	8.8
Arunachal Pradesh	7.9	37.0	4.7
Nagaland	2.0	24.8	12.4
Tamil Nadu	0.7	21.7	33.2
Others	2.0	18.9	9.7
Total	85.4	1353.1	15.8

2.5.1. Insect Pests

2.5.1.1. Mealy Bug, *Dyshnicoccus brevipes*

i) Damage: The mealy bug damage is manifested in the form of change in leaf colour from green to red which gradually turns pink and the margins reflex inwards. Finally, the leaf tips start drying. The severely infested plants become stunted and bear small fruits which ripen without developing proper flavour and taste. Normally mealy bugs are found feeding on the roots and leaf axils, but when the population reaches a peak, these are found feeding on fruits.

ii) Management

- Destroy grasses and other monocot weeds which serve as alternate hosts for the pest.
- Control of ants since they spread the mealy bugs from one plant to another (Mathur and Tandon, 1973).
- Healthy suckers should be selected and these should be dipped in 0.05% methyl parathion or 0.05% fenitrothion emulsion for 15 min before planting (Mathur and Tandon, 1973).
- Application of phorate granules at 17.5 kg/ha three months after planting and irrigating the field for quick translocation of the toxicant.

2.5.2. Vertebrate Pests

i) Damage: On an average, pineapple fruit losses due to vertebrate pests (Jungle crow, squirrel and rat) amounted to 22 and 12.5%, respectively in coastal and hill regions of Karnataka.

ii) Management : In a field trial, 70% of fruits uncovered were destroyed by vertebrate pests compared to no damage on fruits covered with either leaf or dry thatch (Chakravarthy, 1993). Fruit losses due to vertebrate pests were reduced by effective watch, timely harvests, covering fruits with thatch and by driving away the animals (Nagarajan *et al.*, 1994).

2.6. JACKFRUIT

2.6.1. Insect Pests

2.6.1.1. Shoot Borer, *Diaphania caesalis*

The shoot borer is one of the major insect pests of jackfruit in India. It has been reported from Assam, Bihar, Karnataka, Maharashtra, Uttar Pradesh and Sikkim (Butani, 1979).

i) Damage: The pest is active during May-October. Eggs are laid on tender shoots and flower buds. On hatching, the reddish brown caterpillars bore into shoots, flower buds and fruits. As a consequence, the affected fruits wilt, dry and drop.

ii) Management

- The affected shoots, flower buds and fruits should be removed and destroyed in the initial stage of attack.
- Fruits should be covered with alkathene bags to prevent from oviposition.

2.6.1.2. Mealy Bugs

Four species of mealy bugs, viz., *Drosicha mangiferae, D. stebbingi, Ferrisia virgata* and *Nipaecoccus viridis* have been reported as pests of jackfruit (Butani, 1978a). These are polyphagous species also recorded on mango, citrus, grapes, guava, ber, fig, etc.

i) Damage: The mealy bugs suck sap from tender leaves and shoots. *D. mangiferae* and *D. stebbingi* gravid females descend the trees during summer and enter the soil in tree basins for egg laying. Eggs remain in diapause till next season (December-January). On hatching, the crawlers emerge from soil and climb the trees.

The white tailed mealy bug, *Ferrisia virgata* is active from August to November. During dry weather, it moves down and inhabits the roots. Prolonged draught may result in a severe outbreak. The *Nipaecoccus viridis* is at the peak activity around November when clusters of the bug are found on leaves and tender shoots.

ii) Management: For effective management, ploughing should be done during summer to expose the eggs to natural enemies and high temperature. The soil in the tree basins should be drenched with 0.3% aldrin emulsion or 5% aldrin dust is applied during December-January to kill the newly hatched nymphs (Tandon and Lal, 1980a). Application of 25 cm wide polythene bands around tree trunks before emergence of crawlers has been found quite effective in checking their migration (Tandon and Lal, 1979). The nymphs congregated below these bands can be killed by spraying 0.1% carbaryl or

0.75% synthetic garlic oil (Tandon and Lal, 1980b). If the mealy bugs have already migrated to shoots, 0.05% chloropyriphos should be sprayed.

2.6.1.3. Bud Weevil, *Ochyromera artocarpi*

The bud weevil is a specific pest of jackfruit. It is quite serious in Assam and coastal region of Karnataka.

i) Damage: The small whitish grubs bore into tender flower buds and fruits and induce premature drop. The grayish brown adult weevils are found nibbling the leaves.

ii) Management: The affected shoots, fallen flower buds and fruits should be removed and destroyed. Carbaryl (0.1%) should be sprayed at the initial stage of infestation.

2.6.1.4. Spittle Bugs

Cosmoscarta relata, Clovia lineaticolis and *Ptyleus* sp. have been reported from South India (Nair, 1975).

i) Damage: These spittle bugs appear in swarms and feed on young shoots and leaves. The affected leaves get curled. The nymphs are found living inside common mass of froth generally on stalks, young shoots and ripening fruits. After the last molt, the nymphs leave the frothy mass and become adults.

ii) Management: To control these bugs, masses of froth should be collected and young nymphs destroyed mechanically. Monocrotophos at 0.05% should be sprayed to kill the left over population.

2.6.1.5. Bark Eating Caterpillar and Borer

i) Damage: Grubs of *Apriona rugicollis* and *Sthenias grisator* have been recorded damaging the jackfruit trees (Nair, 1975). *A. rugicollis* is active during March to October and feeds on bark of living shoots which gradually die. *S. grisator* girdles the stems of the host tree during August-October and the infected trees are often killed.

ii) Management: For the management of borers, hand picking and mechanical destruction of the beetles is advised. To prevent egg laying on the tree trunk, Bordeaux paste with 0.5% aldrin should be applied. If the infestation is detected at late stage, 0.5% dichlorvos should be injected after cleaning the holes which are then sealed with mud.

2.7. MANGO

India is the largest producer of mango in the world with an annual production of 12.538 million tonnes from an area of 2.021 million ha (Tables 2.15 and 2.16). Mango occupies 36.7 per cent of the area under fruits and it contributes 21.3 per cent of the total fruit production in the country. The major mango producing states include Andhra Pradesh, Uttar Pradesh, Karnataka, Bihar, Gujarat, Maharashtra, Tamil Nadu, West Bengal, Kerala and Orissa.

Table. 2.15. Major mango producing countries of the world (2005-06), Bijay Kumar, 2007)

Country	Area (in '000 ha)	Production (in '000 MT)	Productivity (in MT/ha)	% share of world production
India	2021	12538	6.2	40.1
China	434	3676	8.5	11.8
Thailand	285	1800	6.3	5.8
Mexico	183	1679	9.2	5.4
Pakistan	165	1606	9.7	5.1
Indonesia	273	1413	5.2	4.5
Brazil	84	1348	16.0	4.3
Philippines	172	1003	5.8	3.2
Nigeria	138	812	5.9	2.6
Egypt	35	380	10.9	1.2
Others	578	4996	8.6	16.0
Total	4369	31251		

Table. 2.16. Area, production and productivity of leading mango growing states (2005-06) (Bijay Kumar, 2007)

States	Area (in '000 ha)	Production (in '000 MT)	Productivity (in MT/ha)
Andhra Pradesh	399.3	3194.3	8.0
Uttar Pradesh	251.5	2673.3	10.6
Karnataka	124.5	1236.8	9.9
Bihar	140.2	1222.7	8.7
Gujarat	96.0	772.1	8.0
Maharashtra	444.5	638.6	1.4
Tamil Nadu	125.1	537.8	4.3
West Bengal	70.1	513.3	7.3
Kerala	88.0	511.1	5.8
Orissa	125.3	428.8	3.4
Others	156.2	809.1	5.2
Total	2020.6	12537.9	6.2

2.7.1. Insect Pests

2.7.1.1. Fruit Fly, *Dacus dorsalis*

This is one of the most serious pests of mango in India which has created problem in the export of fresh fruits. Its infestation is more in southern states than in northern states.

i) Damage: The well developed adult is stout and measures 14 mm across the wings and 7 mm in maximum length. The female punctures the outer wall of the mature fruits with the help of its pointed ovipositor and inserts eggs just below the fruit epidermis (1-4 mm deep) in small clusters inside the mesocarp. On hatching, the fly is brown or dark brown in colour with transparent wings and yellow legs. The maggots feed on pulp of these fruits (Fig. 2.16) and the infested fruits start rotting and fall down. As a result the brown patch appears around the place of oviposition and the infested fruits start rotting. The maggots come out of affected fruits to pupate in the soil. The flies breed on fruits that are mature and population increases rapidly during summer. High temperature coupled with high humidity prevailing during May-July months are the favourable environmental factors for the development and reproduction of fruit fly. The population declines slowly from August to September. In methyl eugenol traps, it was found that peak trap catch was between 16 and 17 hr (Jayanthi and Verghese, 1998). The pest is active throughout the year in South India, whereas in northern part it hibernates during winter (November to March) in pupal stage (Kapoor, 1993).

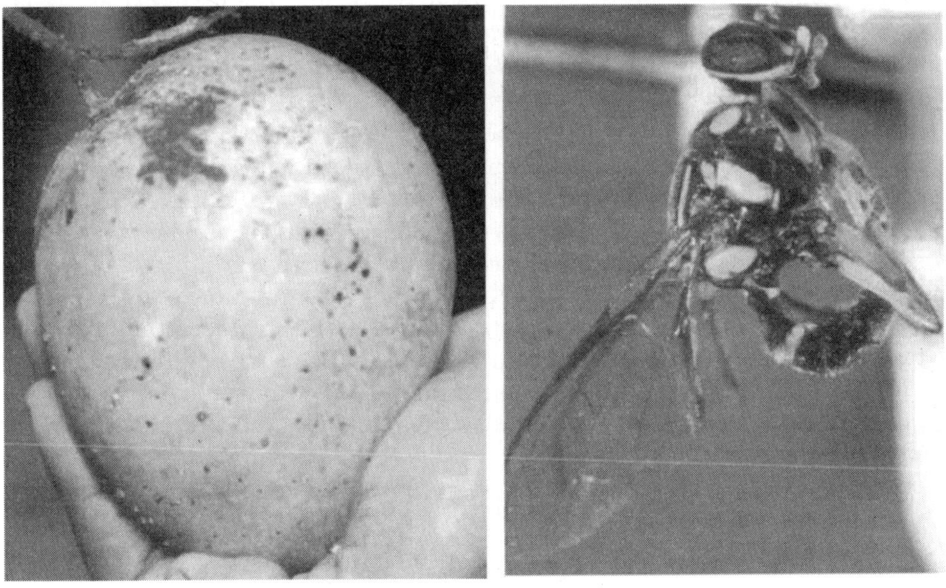

Fig. 2.16. Mango fruit fly and its damage to the fruit (Courtesy: A. Verghese, IIHR, Bangalore).

ii) Management: The following pre-harvest IPM combining sanitation + intercultural ploughing + 3 insecticidal sprays at 15 days interval prior to harvest gives 95 -99% control (Verghese *et al.*, 2000).

Pre-harvest Management: 45 days before harvest the following precautions need to be taken:

- Destroy all fallen fruits at weekly intervals.
- Install at least ten methyl eugenol bottle traps (0.1%) per ha.

- Plough/rake the soil at the tree basin at frequent intervals (1 or 2 times between flowering and harvest) to expose pupae to sun.
- Avoid delay in harvesting.

Three weeks before the harvest, spray decamethrin 2.8 EC at 0.5 ml/l + azadirachtin (0.3%) at 2 ml/l and take up timely harvest.

Post-harvest Management: Within 24 hr after harvest, hot water treatment of fruits at 48°C maintained by thermostat for 1 hr gave 100% fruit fly infestation-free fruits in cvs. Banganapalli and Totapuri. Hot water treatment in 5% salt solution at 55°C for 30 min without thermostat gave 100% control in cvs. Banganapalli, Totapuri and Alphanso.

2.7.1.2. Stone Weevil, *Sternochetus mangiferae*

This is an important pest of mango especially in South India. It is very sensitive to temperature and humidity. Therefore, it is confined to coastal and humid areas. High humidity in coastal areas enhances its infestation. The late varieties suffer more. This pest is very important in view of the major constraint in the export of fresh fruits to foreign countries. It is a specific pest of mango. Sweet varieties like Alphanso, Totapuri, Neelum, etc. are preferred. The weevil causes loss varying from 5 to 80% depending on the place and variety of mango (Verghese, 1998c).

i) Damage: The weevils are about 8 mm in length and about 4 mm in breadth, grayish brown in colour which resembles the background of the bark of mango tree. When the mango fruits are of marble size, the weevils lay eggs singly on the epicarp of partially developed fruits or under the rind of ripening fruits. Freshly hatched grubs bore through the pulp, feed on seed coat (Fig. 2.17) and later damage the cotyledons. Adults have long snout for spotting out a suitable place for egg laying. After emergence the grubs bore through the pulp, feed on seed coat and enter the seeds where they pupate. Pupation takes place inside the seed along the concave side. During the process of adult emergence, the adult drags the faecal matter in the seed into the pulp leading to spoilage of pulp. Adult weevils are inactive for 7-8 months from July-August till next fruiting season. There is only one generation in a year.

ii) Management: Field sanitation including collection and destruction of all fallen fruits at weekly intervals till harvest will destroy adults which act as a source of infestation for the following year. Raking of soil below the tree in October-November and March can contribute partially to weevil management. Spraying main trunk and primary branches prior to flowering (November-December) with 0.05% diazinon or 0.08% fenthion is effective in reducing the number of weevils hiding in the bark. At pea to marble size, spraying of fenthion 0.08% followed by 0.05% monocrotophos or 0.2% carbaryl after 2 weeks gave up to 90% control.

Fig. 2.17. Mango stone weevil and its site of damage (Courtesy: A. Verghese, IIHR, Bangalore).

Integrated Methods

The pest can be managed by following the below mentioned practices:

- Collection and destruction of all fallen fruits at weekly intervals till harvest will destroy adults which are source of infestation for the following year.
- Raking of soil below the tree in October/November and March destroy pupae and contributes partially to weevil management.
- Spraying main trunk, primary branches and junction of branches prior to flowering (November/December) with chlorpyrifos 20 EC at 2.5 ml/l helps to control beetles hiding in the bark.
- Spraying Acephate 75 SP at1.5 g/l should be taken up when the fruits are of lime size (2.5-4.0 cm diameter). This spray should be followed by decamethrin 2.8 EC at 1ml/l after two to three weeks.
- Destruction of all left over seeds in the orchard after harvest and also in the processing industry helps in reducing the pest incidence.

2.7.1.3. Hoppers, *Amritodus* spp., *Idioscopus* spp.

There are several species of hoppers damaging mango crop all over India. Among these *Amritodus atkinsoni, A. brevistylus, A. splendens, Idioscopus niveosparsus, I. cylpealis* and *I. nagpurensis* are important.

i) Damage: Enormous numbers of nymphs are found clustering on the inflorescence during spring, sucking the sap. The infested flowers shrivel, turn brown and ultimately fall off (Fig. 2.18). On attaining maturity, the hoppers leave the blossoms and move on to the leaves. Swarms of adults are commonly seen hovering in mango groves and sitting on all plant parts. The hoppers excrete honeydew which covers the inflorescence, leaves and fruits encouraging growth of fungi, *Capnodium mangiferae* and *Meliola mangiferae* giving rise to sooty mold which affects photosynthetic activity of leaves and market quality of fruits. In case of *I. cylpealis*, Verghese and Rao (1985a) have developed a sequential sampling plan for quick monitoring of field population.

Fig. 2.18. Hoppers on mango leaves and inflorescence (Courtesy: K. Srinivasan, TAFE, Chennai; A. Verghese, IIHR, Bangalore).

ii) Management

a) Chemical Methods: Srivastava and Verghese (1985a) at Lucknow recorded 0% population of hopper species 3-7 days after fogging with 1.5: 8.5 of malathion and diesel mixture. It was cheaper and less time consuming than the conventional spraying.

b) Pheromones: During June, 234 fruit flies (*Bactrocera dorsalis*)/trap were trapped by using methyl eugenol (0.15% + malathion 0.05%) traps tied at 5-6 feet height on mango trees.

c) Integrated Methods

- Orchard sanitation, regular ploughing, removal of weeds, dead, diseased and excess branches to increase sunlight penetration on various sides of the tree are found advantageous in reducing pest damage.
- Successful control of hoppers was achieved by integrating chemical insecticides with botanicals depending upon hopper density (Verghese, 2000a). If the hopper population is low to moderate, two sprays of azadirachtin at 3000 ppm at 2 ml/litre or azadirachtin at 10,000 ppm at 1 ml/litre (first spray at early panicle emergence i.e. before flower opening and second spray at pea-size fruit stage) gave effective control of hoppers. Sprays of neem/pongamia soaps were also highly effective.

- If the number of hoppers is more than 4 per panicle, spray with imidachloprid 0.005% or lambda cyhalothrin 0.0025% at early panicle emergence. Repeat the spray if necessary when the fruits reach pea-size. Further, pollination was found unaffected by imidachloprid, when sprayed at pre-bloom. Generally insecticides affect pollinators of mango such as *Apis florae, Reistalinus arvorum* and certain calliphorids. So it is good to avoid spray during peak bloom.
- Biological suppression of hoppers was achieved by the release of predator, *Mallada boninensis* at 10-12 larvae/tree. The reduviid predator, *Irantha armiper* was found preying voraciously on hoppers. The white halo fungus, *Verticillium lecanii* is observed on mango hoppers in nature resulting in wiping out of hoppers around Lucknow at times (Singh, 1993).

2.7.1.4. Shoot Borer, *Chlumetia transversa*

i) Damage: This pest is usually active from August to October. Eggs are laid on tender leaves. Freshly hatched caterpillars bore into midribs of tender leaves and come out after a couple of days to bore into tender shoots near the growing point tunneling downward, throwing their excreta out of the entrance hole (Fig. 2.19). When full grown, these caterpillars come out and enter into the slits and cracks in the bark of the tree, dried malformed inflorescence or cracks or crevices in the soil for pupation.

Fig. 2.19. Shoot borer on mango (Courtesy: K. Srinivasan, TAFE, Chennai).

Field observation on *C. transversa* showed that the peak percent infection was in September in 1996 (22.26%) and in November 1997 (47.17%). This results in withering, leaf drop and drying of new terminal shoots. Young grafted seedlings are severely affected and may even be killed. Damage symptoms include withering and drying of new terminal shoots due to tunneling. Leaves of affected shoots wither and drop down and affected plant growth is checked. Young grafted seedlings are severely affected and may even be killed.

ii) Management

- Clipping off and destruction of affected shoots.
- Spraying of carbaryl 50 WP at 4 g/l or quinalphos 25 EC at 2ml/l at the time of emergence of new flush.
- The parasitoid, *Bracon greeni* can reduce the infestation of shoot borer.

2.7.1.5. Leaf Webbers, *Orthaga exvinacea, O. eudrusalis*

i) Damage: It is commonly found throughout India. Eggs are laid singly or in clusters within silken webbings or on leaves. On hatching, the caterpillars feed gregariously by scraping the leaf surface. Soon they web together tender shoots and leaves and feed within. Several caterpillars may be found in a single webbed up cluster of leaves and pupation also takes place inside these webs in silken cocoons or in soil (Fig. 2.20). If infestation is serious, panicle emergence will be affected (Verghese, 1998a). Heavy webbing and feeding causes skeletonizing and drying of leaves. Severe infestation makes the tree unproductive and tree bears a burnt look and fails to put forth panicles.

Fig. 2.20. Leaf Webber damage on mango (Courtesy: A. Verghese, IIHR, Bangalore).

This pest is attaining serious proportions. Adult moths are of medium size and somber coloured, eggs are greenish dull in colour. Full grown caterpillar is brownish blue with whitish striations dorsally. Its head is symmetrically variegated with white and brown spots and markings. There are four whitish longitudinal striations dorsally alternated in between by three pale olive green bands. Laterally, there are two brown to black striations and dorsally, two rows of symmetrical dots running longitudinally along the body. The caterpillar is smooth having sparse hair-like white setae. Pupae are dark blackish on maturity. Old orchards with less space between the tree canopies harbour more insects than open orchards. Its infestation starts from April and goes up to December.

High temperature and high humidity prevailing during June-November are favourable environmental conditions for the development and reproduction of leaf webber.

ii) Management: As webber infestation begins by July-August, it is crucial to control it by August (post-harvest phase), to facilitate flushing and flowering in the following year (Verghese, 1998a). Webbed leaves should be removed and burnt wherever possible. Spraying quinalphos 0.05% or carbaryl 0.15% is recommended (Verghese, 1998a). Spraying at this time with the above insecticides will also help in controlling mango psylla, *Apsylla cistellata* which is more predominant in the northern and eastern belts of India like Uttar Pradesh, Bihar and West Bengal.

Bacillus thurungiensis is also recommended at 1 kg/ha for the control of leaf webber. *Aspergillus flavus* and *Beauveria bassiana* are also effective in reducing the infestation of leaf webber.

2.7.1.6. Fruit Borer, *Deonalis albizonalis*

i) Damage: Mango fruit borer is an important pest where the young caterpillars attack fruit from marble size onwards at the distal end. The caterpillars feed within the fruit and may make series of tunnels in the young kernel. Infested fruits fall. Fully grown caterpillars pupate inside the fruits (sometimes pupation takes place outside) and the moths emerge through the exit holes. The characteristic symptom of damage is appearance of small holes (which also serves as exit holes) with dark pale brown ring encircling it.

ii) Management

- Collection and destruction of all fallen fruits.
- Spraying endosulfan 35 EC at 2ml/l at marble size onwards and repeating with decamethrin 2.8 EC at 1ml/l after two weeks in case of heavy infestation.
- No spray should be given a fortnight before harvest.

2.7.1.7. Galls, *Procontarinia matteiana, Amradiplosis* spp.

i) Damage: The gall causing cecidomyiids lay their eggs on ventral surface of leaves. On hatching, the maggots bore inside the leaf tissues and feed within, resulting in the formation of small raised wart-like galls on the upper surface of the leaves. The affected leaves get badly deformed and drop prematurely. Leaf galls adversely affect growth when young plants are infested.

Spatial distribution of the galls has been found to be aggregated (Verghese and Rao, 1985). To some extent fallen leaf sampling can be used as an index of gall population on the tree (Verghese and Sudha Devi, 1998).

ii) Management: To check the loss caused by the leaf gall causing midges, collect and destroy the affected leaves. This will prevent the population build up of the pest.

2.7.1.8. Inflorescence Midge, *Erosomyia indica*

i) Damage: The inflorescence midge attacks panicles which bend and dry. This can be quite serious. The pest lives in association with the mango crop from January to May. It is distributed all over India. The adult is minute yellowish midge with grayish back. The male being larger than female. The wings are broad; claw is slender, strongly arched and dentate. Midge attacks the inflorescence and small fruits in three different stages. The first attack is at floral bud burst stage. The larvae tunnel the axis of inflorescence and thus destroy it completely. Second attack starts at the fruit set as young maggots bore into these tender fruits which slowly turn yellow and finally drop. The third attack is on tender new leaves encircling the inflorescence. The most damaging one is the first attack in which the entire inflorescence is destroyed. The inflorescence finally dries up before flowering and fruit setting. The symptom of attack of this pest is appearance of tiny black spots on inflorescence.

Moderate to high temperature with low humidity prevailing during January-March are ideal environmental factors for the development and reproduction of inflorescence midge.

ii) Management

- Spray of fenitrothion 0.05% at panicle emergence gives good control.
- Raking soil below to destroy pupae also helps to manage this pest (Verghese, 1986).

2.7.1.9. Mealy Bugs

Three pseudococcids, viz. *Rastrococcus iceryoides, Planococcus citri* and *Ferrisia virgata*; and the margoroids viz. *Drosicha mangiferae* and *Perissopneumon ferox* are known to occur on mango. The pest is widely distributed all along the Indo-Gangetic Plains and causes severe damage. This pest is more serious in North India.

i) Damage: Adult male is crimson coloured with brownish black forewings. Female bugs are mealy white in colour, elliptical in shape and covered with numerous minute hairs, while nymphs are densely covered with waxy material. The adult mealy bugs colonize between bark of tree trunk, young shoots and panicles. Eggs which are pink in colour that gradually changes to yellow are usually laid in silken pouches in soil around tree trunk during April-May, which hatch out during November-December. The nymphs ascend the trees and settle on inflorescence causing flower drop affecting fruit set. Nymphs and adults suck the plant sap and reduce the vigour of the plant and destroy inflorescence, causing the fruit drop (Fig. 2.21). The adult female crawls down the tree in April-May and enters the soil for egg laying which hibernates till December.

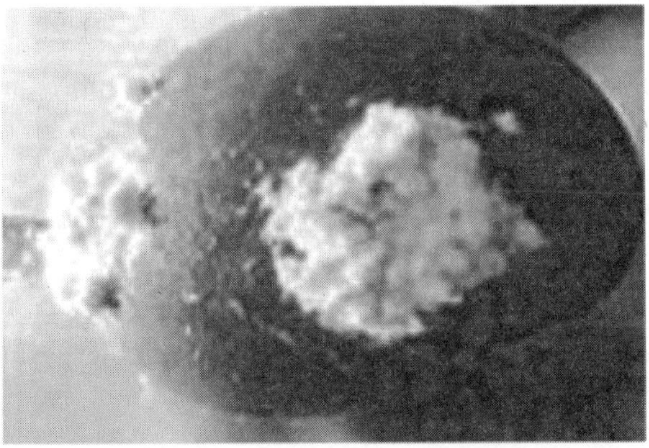

Fig. 2.21. Mealy bug infestation on mango fruits (Courtesy: K. Srinivasan, TAFE, Chennai).

Low temperature prevailing during November-March is optimum environmental conditions for the development and reproduction of mealy bugs. The over-wintering eggs hatch by December-January and ascend the trees and settle on inflorescence causing flower drop affecting fruit set.

ii) Management

- Alkathene sheet (30 cm wide, 400 gauges) should be put around the tree trunk at 0.5 to 1.0 m above the ground level and fastened by twine with a grease barrier at the lower end (Verghese, 1986).
- Sanitary conditions should be maintained near the plants and the soil should be turned regularly up to August-September so that eggs are exposed and destroyed. Raking of basin soil around the tree trunk was found effective.
- Application of karanj based biodynamic preparation resulted in 100% mortality of mealy bugs within 7 days, while neem based biodynamic preparation took 9 days for 100% mortality.
- The parasitoid, *Anagyrus pseudococci* and the predator, *Cacoxenus perspicax* are known to attack *R. iceryoides* (Uma Narasimhan and Chacko, 1988). The release of the coccinellid predator, *C. montrouzieri* at 50 beetles/tree was found very effective in suppressing mealy bug (*R. iceryoides*) population (Mani *et al.*, 1995). *Menochiles sexmacul-*

ata, *Chrysopa scelestes* and *Beauveria bassiana* are also effective against mealy bugs.

- Against crawled nymphs of mango mealy bug (*Drosicha mangiferae*), garlic oil 0.1% was at par with monocrotophos 0.05% and carbaryl 0.1%.

2.7.1.10. Leaf Miner, *Acrocercops syngramma*

i) Damage: Small silvery-gray adult moths lay eggs on the leaves. The tiny caterpillars mine under the dorsal epidermis of tender leaves and feed within. As a result grayish-white round blisters appear on leaves. This pest is usually active from October-November to April-May. Pupation takes place within the tunnels (Butani, 1979).

ii) Management

- Removal and destruction of the affected leaves in the initial stage of attack. This will prevent the pest infestation from spreading.
- Spraying cypermethrin 25 EC at 0.5ml/l or chlorpyriphos 20 EC at 2.5ml/l during new flush emergence.

2.7.1.11. Inflorescence Caterpillars, *Eucrostus* sp. (Geometridae); *Argyroploce aprobola*? (Eucosmidae); *Euproctis fraterna* (Lymantriidae)

i) Damage: The caterpillars attack inflorescence and if not controlled causes heavy loss in fruit bearing (Verghese and Jayanthi, 1999).

ii) Management: Spraying Endosulfan 35 EC at 2 ml/l at early panicle emergence.

2.7.1.12. Termites, *Odontotermes* spp.

i) Symptoms: They usually occur in neglected orchards raised on virgin lands. The severity is more during dry spells in sandy and sandy-loam soils. These termites are white in colour, shy to light and remain underground. They feed on roots or move upwards making the tunnels completely hollow. They construct mud galleries on tree trunks and under the protection of these galleries; they feed on the bark of the trunks. The damage is more severe in nurseries and young and newly planted orchards where the entire seedlings or saplings may dry up.

ii) Management

- Removal of the mud galleries on trunk and swabbing or spraying the trunk with chlorpyrifos 20 EC at 2.5 ml/l.
- After two months, drenching the soil at the base of the tree with chlorpyrifos 20 EC at 4 ml/l.

2.7.1.13. Stem Borer, *Bactrocera rufomaculata*

i) Damage: The stem borer is widely distributed in India. It is a large, prominent, grayish brown beetle. There is also a white triangular patch at the junction of elytra. The general colour being dark brown, covered with grey or yellow grey pubescence. The colour in general resembles that of the mango bark. It has long segmented antennae. During flight they make noise. The freshly hatched larva is creamy white in colour. The grub is fat, fleshy and creamy white in colour. Average length of adult is about 5 cm.

Damage is caused by the grub of this beetle as it feeds inside the stem, boring upward, resulting in drying of branches and in severe cases, the attacked tree is killed. When the stem or any branch is attacked, the sap and masses of frass exude from the bored hole. Often the damage may be visible by falling of the leaves of the attacked branches and sudden collapse of branches. The grub after hatching from the eggs first feed on the bark and makes irregular cavities. As the grub feeds, a harmonious sound is produced by its mandibles. On this account, the insect is popularly known as 'violin beetle'. It makes tunnels which may either be in the peripheral region or may go deep down into base of the tree.

Moderate temperature and moderate humidity prevailing during September-October are ideal weather conditions for the development and reproduction of stem borers.

ii) Management: Borers can be destroyed by inserting a hard wire into the tunnel. They can also be pulled out by using a hook at the end of the wire and then destroyed. The tunnel should be sealed with wet clay after applying petrol/kerosene (10 ml).

2.7.1.14. Scale Insect, *Chloropulvinaria polygonata*

Lately, scale insects are becoming serious pests of mango in certain regions. particularly in Western U.P.

i) Damage: These are immobile, fixed to the plant and survive only on living plants and die if the plant part is wilted. They are about 2.5 cm in size and have a covering of waxy, hairy, glassy, powdery white material (Fig. 2.22). Adult females are wingless while males are winged. Eggs are found in a pouch. Tiny larvae crawl to the tender parts of plants and attack them at a spot. Larvae lose their original form and become a small, footless mass covered over the scale. The main period of activity is summer and reproduces parthenogenetically. The nymphs and adult scales suck the sap of the leaves and other tender parts and reduce the vigour of the plants. They also excrete honeydew which helps in the development of sooty mold on leaves and other tender parts of the tree. Honeydew in due course protects the insects from enemies. In case of severe scale infestation, growth and fruit bearing capacity of the tree is affected adversely.

High temperature and high humidity prevailing during July-August enhances the infestation of scale insects.

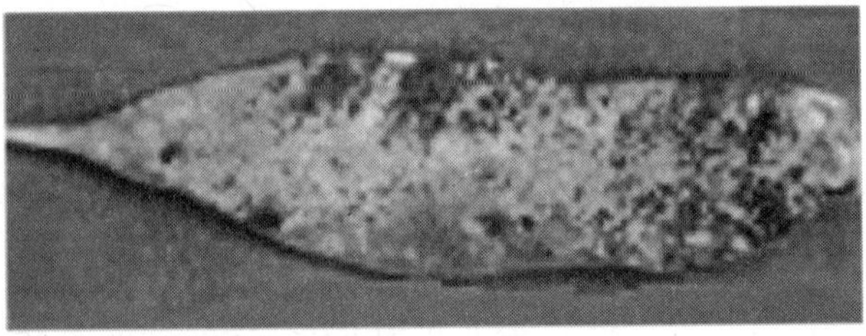

Fig. 2.22. Scales infestation on mango (Courtesy: K. Srinivasan, TAFE, Chennai).

ii) Management: The lycaenid parasitoid, *Spalgis epius* and the green lacewing predator, *Mallada astur* are commonly seen feeding on the ovisacs of *C. polygonata* (Singh, 1993). *C. montrouzieri* which feeds very well on ovisacs is recommended for the control of scale insects (Mani and Krishnamoorthy, 1998a).

IPM

- Ploughing of orchard immediately after harvest and in November-December to expose pupae of fruit fly, midge, leaf webber and eggs of mealy bug to natural enemies.
- Heavy irrigation of orchard in October helps in destruction of eggs of mealy bug, diapause pupae of midge and fruit fly.
- Avoiding dense planting, keeping orchard clean by regular ploughing, removal of weeds and pruning of over crowded and overlapping branches in December to control hoppers.
- Polythene banding (25 cm wide, 400 gauge) of tree trunk 30 cm above the ground level in December-January to prevent migration of freshly hatched first instar nymphs of mealy bugs on tree.
- Spraying 5% NSKE/1% neem soap/*Verticillium lecanii* at bud burst stage (February) to control leaf hoppers, midges and mealy bugs. Repeat spraying during July for controlling next generation of pests.
- Spraying 0.05% imidacloprid during April for the management of hoppers and mealy bugs (adult females).
- Fixing of plastic bottle traps with wooden blocks impregnated with methyl eugenol to capture fruit flies.
- Soil drenching with 5% NSKE/*Beauvaria bassiana* around the tree trunk to prevent migration of mealy bugs on tree.
- Collection and destruction of fruit fly and stone weevil infested fruits from time to time.
- Mechanical removal of leaf webber larvae and webs from April to Sept-Oct by leaf web removing device developed by Central Institute for Sub-tropical Horticulture, Lucknow.
- Spraying of 2% starch to clear sooty mold.
- Early harvesting of matured fruits to avoid fruit fly infestation.

Table. 2.17. Calendar of IPM activities in Mango

Month/s	Target pest	IPM activities
September	Leaf webber	Removal of webs by web removing device and burning them. Pruning over crowded and overlapping branches.
October	Eggs of mealy bug, Pupae of midge, Pupae of fruit fly	Flooding of orchards. Pruning of infected and dried branches, 10 cm below the dried portion and pasting of copper oxychloride.
November	Mealy bug	Deep ploughing of orchards for exposing eggs and pupae of insects. Removal of weeds in orchards which harbour insects.
December	Mealy bug	Fastening of alkathene sheets of 400 gauge thickness, 25 cm wide around the base of tree. Raking of soil around the tree trunk and mixing with neem cake.
January	Inflorescence midge, Mealy bug	Cleaning the alkathene bands at regular interval. Spraying of 0.05% quinalphos or some safer insecticide at bud burst stage.
February	Hopper	Spraying with 5% NSKE at bud burst stage against hoppers.
March	Hopper	Spraying of 0.05% quinalphos.
April	Hopper	Second spraying 0.2% wettable sulphur after fruit setting.
May	Fruit fly, Sooty mold	Hanging of 0.1% methyl eugenol traps + 0.1% malathion.
June	Fruit fly	Continuation of 0.1% methyl eugenol traps + 0.1% malathion. Early harvesting of mature fruits to avoid fruit fly infestation. Collection and destruction of fruit fly infested fruits.
July	Scale insect, Stem borer	Timely picking of fruits. Spraying of 0.04% quinalphos or 0.06% dimethoate for scale insects. Cutting and burning of stem borer affected branches.
August	Shoot gall psylla, Leaf webber	Spraying of 0.05% quinalphos. Removal of leaf webber affected branches.

2.8. GRAPEVINE

Grapevine (*Vitis vinifera*) is one of the important commercial crops grown in different climatic zones of the world for table delicacies and wine preparation. It is fairly good source of minerals like calcium, phosphorus and iron and vitamins like B_1 and B_2. Grapevine is grown in 64,300 ha with annual production of 1.630 million tonnes (Tables 2.18 and 2.19). In grapes, India has recorded the highest productivity (25.4 t/ha) in the world.

Table. 2.18. Major grape producing countries of the world (2005-06) (Bijay Kumar, 2007)

Country	Area (in '000 ha)	Production (in '000 MT)	Productivity (in MT/ha)	% share of world production
Italy	755	8554	11.3	12.6
USA	378	7088	18.7	10.5
France	855	6790	7.9	10.0
China	453	6592	14.5	9.7
Spain	1161	6054	5.2	8.9
Turkey	530	3850	7.3	5.7
Iran	315	2964	9.4	4.4
Argentina	212	2830	13.4	4.2
Chile	178	2250	12.6	3.3
Australia	153	2027	13.2	3.0
South Africa	113	1683	14.9	2.5
India	64	1631	25.4	2.4
Others	2222	15468	7.0	23.0
Total	7385	67696		

Table 2.19. Area, production and productivity of leading grape growing states (2005-06) (Bijay Kumar, 2007)

States	Area (in '000 ha)	Production (in '000 MT)	Productivity (in MT/ha)
Maharashtra	45.1	1275.0	28.3
Karnataka	10.4	193.2	18.5
Tamil Nadu	2.6	84.8	32.6
Andhra Pradesh	1.9	40.3	21.2
Punjab	1.1	30.2	27.0
Haryana	0.1	3.3	33.0
Madhya Pradesh	0.1	2.6	25.0
Mizoram	0.5	0.6	1.2
Jammu & Kashmir	0.2	0.3	1.5
Others	2.2	0.3	0.1
Total	64.3	1630.7	25.4

Currently, about 20,000 tonnes of fresh grapes are exported, accounting for just 2% of the total production. Out of this, about 12,000 tonnes are exported to the Middle East markets and the rest to U.K, Europe and few South-East Asian markets. Among these markets, U.K. is the most paying besides being reliable. Middle East markets are not dependable and are also less paying.

2.8.1. Insect Pests

2.8.1.1. Flea Beetle, *Scelodonta strigicollis*

i) Damage: It is an important pest of grapevine and usually difficult to recognize its significance since the beetles are nocturnal in habit. The grubs feed on roots and pupate in the soil, whereas adults feed on new buds and tender shoots that are formed after pruning of vines and turn it into a 'dried sprout'. Later on, the beetle feeds on mature leaves, causing characteristic longitudinal slits on the lamina of the leaves (Fig. 2.23). During day time, they hide under the loose bark or soil clods.

Fig. 2.23. Flea beetles on grapevine leaf
(Courtesy: K. Srinivasan, TAFE, Chennai; A. Verghese, IIHR, Bangalore).

If not controlled, flea beetle can cause a loss to an extent of 55% (Verghese, 1997). It damages vines from September to December and from March to May in South India, Maharashtra, Gujarat and Rajasthan. In Haryana, Punjab, Uttar Pradesh and Himachal Pradesh, it is active from March to November (Vevai, 1969).

ii) Biology: Adults are 4-5 mm long, copper coloured with three black spots on each elytran. Eggs are cigar-shaped, creamy white and about 1 mm long. Female lays 200-600 eggs in its life span of about eight months in clusters of 20-40 eggs. Incubation, larval and pupal periods lasted 4-8, 34-45 and 7-11 days, respectively (Kulkarni, 1971). At Hyderabad, Rao *et al.* (1984) observed that the population of beetles reached a peak in late September just prior to pruning, and most larvae and pupae occurred in top 10 cm soil.

iii) Management

- Remove all loose bark to prevent beetles from hiding.
- Rake the soil in the basin after vine pruning to expose grubs and pupae to sunlight.
- Seven days after pruning (at early bud sprout), spray carbaryl 50 WP at 4 g/l or quinalphos 25 EC at 2ml/l or chlorpyrifos 20 EC at 2.5 ml/l.
- Rao *et al.* (1983) reported that 0.05% monocrotophos and 0.15% carbaryl caused 90% mortality of beetles four days after the spray and these remained toxic for 33 and 32 days, respectively.

- Repeat any of the above sprays after 15 days.
- Mulching frequently in July-September reduces the population of grubs and pupae in soil.
- Neem and castor cakes both at 5 to 7.5 MT/ha and tobacco waste at 2.5 MT/ha were effective against flea beetle and all were at par.

2.8.1.2. Thrips, *Rhipiphorothrips cruentatus, Scirtothrips dorsalis*

i) Damage: These are important rasping and sucking insects, which emerge after pruning when tender leaves are ready as ovipositional sites (Fig. 2.24). Both nymphs and adults feed on the ventral surface of tender leaves, flower stalks and mustard sized berries and suck the oozing cell sap. Their injury manifests as 'thrips scab' on full-grown berries, reducing the market and export value. So, it is always better to manage thrips before flowering.

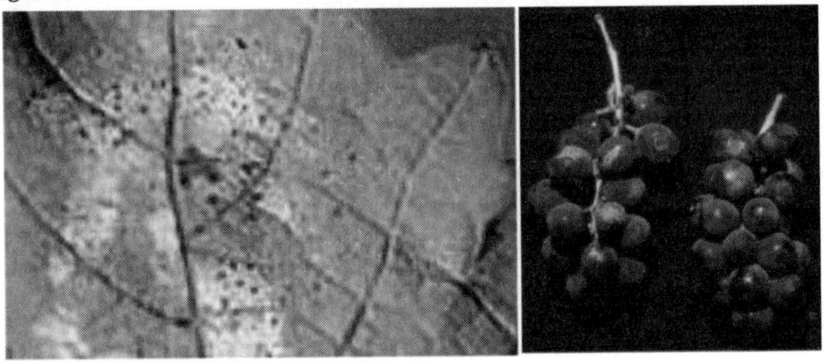

Fig. 2.24. Thrips damage on grapevine leaf and fruits (Courtesy: K. Srinivasan, TAFE, Chennai)

ii) Management

- Remove pruned material from main field and burn.
- Rake the soil periodically.
- Spray acephate 75 SP at 1 g/l before flowering followed by endosulfan 35 EC at 2ml/l or carbaryl 50 WP at 4 g/l at post berry formation.
- Spray spinosad 45% EC (chitin inhibitor) at 0.25 ml/l for the effective management of thrips especially in exportable grapes because of shorter pre-harvest interval and safety to natural enemies especially to *Cryptolaemus montrouzieri*.

Note : Repeated use of same pesticide for sprays may result in development of resistance to pesticides by the thrips and hence avoid their indiscriminate use.

2.8.1.3. Mealy Bug, *Maconellicoccus hirsutus*

i) Damage: This is the most serious pest on grape particularly in Peninsular India. Mani and Thontadarya (1987) recorded 80% damage to bunches at Bangalore. Manjunath (1985) recorded serious outbreak of *M.*

hirsutus in some parts of Karnataka and Andhra Pradesh with peak infestation in February-March. The bunches were shriveled and dropped up to 90% (Fig. 2.25). Due to the concealed habitat and waxy coating, it is very difficult to manage these bugs. Soon after hatching, majority of nymphs start crawling up the vines and colonies of nymphs and adults are seen on young shoots sucking the cell sap. Overlapping generations and co-occurrence of other mealy bug species makes their control difficult.

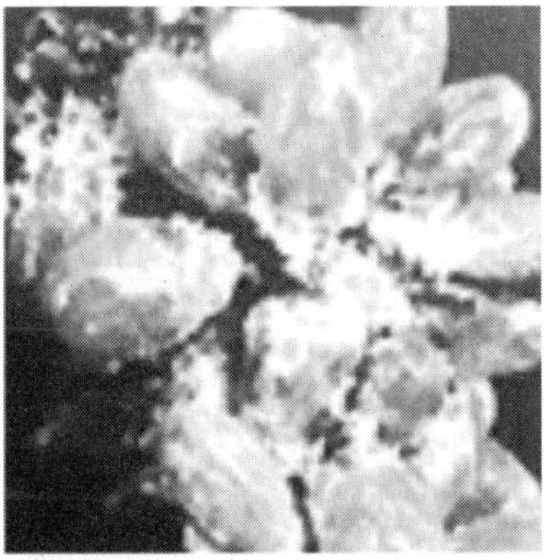

Fig. 2.25. Mealy bug infestation on grapevine berries (Courtesy: K. Srinivasan, TAFE, Chennai).

ii) Biology: Duration of one life cycle was 24 days at 31°C and 48 days in November at 25°C (Babu and Azam, 1987). The female passed through three and male four nymphal instars. The female lays about 317 eggs over a period of 46 days.

iii) Management

a) Biological Control: Release of 100 day-old adult beetles of *C. montrouzieri* at 3000/ha (10/vine) resulted in significant reduction of mealy bug population 30 days after release. A peak of 12.7 grubs/vine was recorded on 60th day after release. The predator has given good control of mealy bugs in Karnataka, Maharashtra, Andhra Pradesh and Tamil Nadu. Tree like mango and hibiscus hedges were found ideal for multiplication and relaying predatory ladybirds (*C. montrouzieri* and *Scymnus coccivora*) to the new areas. The mealy bug indicator plant like okra may be planted and explored as trapping source for conservation of predatory ladybirds. Predators (*C. montrouzieri* and *S. coccivora*) and parasitic wasp, *Anagyrus kamali* are effective against mealy bug. Growing short duration crops like onion, okra, corn, cucumber, capsicum, chilli, brinjal, yam, sweet potato, pumpkin and long duration crops like sugarcane and banana are possible in the mealy bug-hot spot area provided ladybirds have established and the farm area is surrounded by forage plants like glyricidia and ornamentals like oleander alongside a few coconut trees. Since mealy bug attacks graminaceous weeds including congress grass (*Parthenium hysterophorus*), the residue of mealy bugs remain either on glyricidia or congress grass which sustain *C. montrouzieri* population. The okra (the most susceptible plant) grown in vicinity gets infestation so late, by the time its harvesting is over and the plant residues are able to create balance between mealy bug and predator population. Okra acts as a trap crop which may be used even for the field multiplication of ladybirds if grown at different intervals.

Anagyrus dactylopii was the principal parasitoid causing a mean parasitization of 62.5% in nature. Higher percentage of parasitization and more female progeny production were observed when 3rd instar and adult female mealy bugs were exposed to the parasitoid. The release of adult beetles (10 days old) of *C. motrouzieri* at 5000/ha in the evening hours is recommended for the management of grape mealy bug.

Release of predatory beetle, *C. motrouzieri*, keeping the surroundings of the trees neat and clean and turning the soil around the trees up to September helps in reducing the pest incidence.

b) Integrated Methods (Verghese and Tandon, 1989)

- Remove all loose bark after pruning from main trunk and primary arms.
- Swab the trunk with IIHR swab mixture (carbaryl-6 g + copper oxychloride-10 g + neem oil-1 ml + kerosene-1 ml + sticker-1 ml per litre of water).
- Repeat the above treatment after six months coinciding with pruning.
- Spray phenthoate 50 EC at 1ml/l, two weeks after pruning which can be repeated 15 days later.
- At fruit formation stage, spray dichlorvos 76 EC at 1 ml/l.
- As an alternate for the above spray, introduce *Cryptolaemus montrouzieri* predator at 10 grubs/vine at the site of infestation.
- The encyrtid, *Anagyrus dactylopii* was parasitizing up to 60-70% in March-April despite regular insecticidal sprays.
- Stop all sprays three weeks prior to harvest to prevent pesticide residue problem in fruits.

2.8.1.4. Shot-Hole Borer, *Xyleborus crassiusculus*

i) Damage: Shot-hole borer is a serious pest on grapes where the beetle attacks the main trunk, starting from the base of the plant. It belongs to the group of insects popularly known as 'ambrosia beetles'. In their shot-hole nests, these beetles cultivate 'Ambrosia' fungus that serves as a feed for the grubs. This nest building activity of the beetle on the main trunk damages the vine. Pinholes are observed on the main trunk (Fig. 2.26) with a powdery waste coming out. Later, gummy exudates are seen on the trunk and the plants show wilting and yellowing. If proper management of the pest is not taken, total mortality of the vine occurs in 15-20 months.

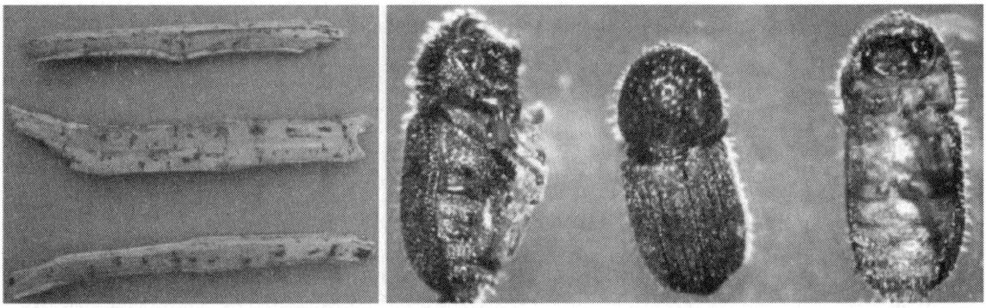

Fig. 2.26. Shot hole borer damage on grapevine main trunk and adult beetles (Courtesy: A. Verghese, IIHR, Bangalore).

ii) Management

Integrated Methods (Verghese and Tandon, 1995)

a) Prevention

- Remove all loose bark after pruning from main trunk and primary arms.
- Treat the trunk with IIHR swab mixture (as mentioned under mealy bug).
- Repeat the above treatment after six months coinciding with pruning.

b) Management

- Remove loose bark and scrape gently gummy exudates using a knife.
- Spray dichlorvos 100 EC at 2.5 ml/l on main trunk (not on the leaves and vines).
- Inject dichlorvos 100 EC at 2.5 ml/l, if the holes are only few in numbers and band the trunks with plastic/polythene/synthetic gunny bags for 3-5 days.
- Swab with the IIHR mixture after removing the bands.
- If serious, repeat the above after a month.
- If necessary, remove the vine and take up fresh planting.

2.8.1.5. Lepidopterous Caterpillars, *Helicoverpa armigera, Spodoptcra litura*

i) Damage: These polyphagous caterpillar pests cause severe damage to the leaves and berries of grapes. These can be managed by following IPM with minimum insecticides.

ii) Management

- A combination of pheromone traps (5/ha) + NPV at 500 LE/ha + sprays of endosulfan 35 EC at 2ml/l or cypermethrin 25 EC at 0.5ml/l (depending on severity) + hand picking and destroying.

- Wherever possible, spray *Bacillus thurungiensis.*
- All sprays should be avoided a month prior to harvesting.
- Application of *Sl* NPV and *Ha* NPV gave effective control of the above pests.

2.8.1.6. Leaf and Berry Webber, *Adoxophyes privatana*

i) Damage: Leaf and berry webber is a new pest of grape (Verghese, 1999b). The larva initially rolls a single or two leaves into webs and feeds on the lamina leaving behind the dried skeletonized veins. Later it migrates on to the berries and webs and feeds within causing direct yield loss. A single caterpillar can destroy the entire bunch.

ii) Management

- Spray fenvalerate 20 EC at 0.25 ml/litre 45 days prior to harvest.
- In case a second spray is required nearer to harvest, use 4% neem cake extract (Verghese, 1999b).

2.8.1.7. Defoliating Beetles, *Holotrichia serrata; Adoretus bengalensis*

i) Damage: The adult beetles defoliate leaves, especially after the first few rains.

ii) Management: They can be managed by erecting light traps and a single spray of fenvalerate 20 EC at 0.25 ml/l or decamethrin 2.8 EC at 1ml/l, during July/August (Verghese and Tandon, 1994).

2.8.1.8. Lantana Scale, *Hemiberlesia lataniae*

i) Damage: Scale insects occur on leaves, bunches and the trunk.

ii) Management: The parasitoid, *Signiphora* sp. is found causing up to 30% parasitism in nature. Release of the predator, *Chilocorous nigrita* helps in bringing down the scale population in grapevine.

2.8.2. Vertebrate Pests

i) Damage: Bank mynas (*Acridotheres ginginianus*) and Indian mynas (*Acridotheres tristis*) damage both immature and mature grape berries. In Ludhiana, Bank mynas caused a fruit loss of 19% and damage was severe during May-June (Toor, 1982), when alternate food sources are meagre. In Punjab, more damage (54.6%) occurred in grapevines trained with the Head System than in Bower System (35%). Birds preferred Beauty Seedless variety of grapes having purple coloured berries. In Bangalore, Jungle crow (*Corvus macrorhynchos*) and House crow (*Corvus splendens*) are the major pests on grapes. The Barbet (*Megalaima viridis*) and three species of bats also incur heavy losses (30-36%) in grape cvs. Arka Hans, Arka Shyam, Arka Kanchan and Bangalore Blue (Verghese, 1993).

Fig. 2.27. Bat damage on grape berries.

The bat (*Cynopterus sphinx*) is a serious pest of grapes in Bangalore, India (Fig. 2.27). Bat foraging was maximum at maturity of fruits and vines adjacent to open space had more bat damage.

ii) Management: In home vineyards, bunches may be covered with muslin bags. The vineyards are also covered with nylon netting or electric fencing grid (1.5 volts). Bird scaring with traditional methods is also adopted.

Fig. 2.28. Covering the canopy gap and erecting nylon netting to control bat damage.

Erecting nylon netting around bower and covering the canopy gap on the bower with briar and twigs gave almost complete control of bat damage (Verghese, 1993, 1998) (Fig. 2.28).

2.9. GAUVA

Guava rank fifth in area and production with an estimated production of 1.823 million tonnes from an area of 0.182 million hectares (Table 2.20). Guava occupies 3.3 per cent of the area under fruits and contributes 3.1 per cent of the total fruit production in the country.

Table. 2.20. Area, production and productivity of leading guava growing states (2005-06) (Bijay Kumar, 2007)

States	Area (in '000 ha)	Production (in '000 MT)	Productivity (in MT/ha)
Maharashtra	31.3	223.6	7.1
Bihar	27.7	199.0	7.2
Uttar Pradesh	15.5	157.8	10.2
Karnataka	7.3	154.1	21.2
West Bengal	9.9	153.0	15.5
Punjab	7.8	135.7	17.5
Andhra Pradesh	10.1	131.6	13.1
Uttarakhand	14.3	99.6	7.0
Gujarat	7.0	98.7	14.1
Tamil Nadu	8.5	92.2	10.9
Orissa	13.8	89.4	6.5
Others	28.7	288.5	10.0
Total	181.7	1823.3	10.0

2.9.1. Insect Pests

2.9.1.1. Green Shield Scale, *Chloropulvinaria psidii*

i) Damage: It is a serious pest of guava in India. Small greenish-yellow, sessile, oval shaped scales feed by sucking on the leaves or tender twigs and shoots. As the pest excretes honeydew, ants will be attracted. Secondary fungal infections results in black sooty mold on the leaves and twigs. The infestation is more severe during summer months.

ii) Management

- During non-fruiting stage, prune affected parts and burn.
- In case of serious infestation, prune affected parts and spray with diazinon 40 EC at 2ml/l or quinalphos 25 EC at 2ml/l or monocrotophos 36 SL at 1.5 ml/l.
- Dust methyl parathion 2% DP at 25 g around the base of the tree to prevent ascent of ants.
- Prune all adjoining branches of infested tree to prevent the easy spread of pests to neighboring trees.
- Spray 2% starch to remove the sooty mold.
- The predator, *Cryptolaemus montrouzieri* and the ant, *Camponotus compressus* are found associated with the scale (Verghese and Ramachander, 1998).
- Release *Cryptolaemus montrouzieri* grubs in January/February at 20 grubs/tree (Mani and Krishnamoorthy, 1990a).

- Application of non-drying adhesive on tree trunk prevents ascent of ants.

Three parasitoids namely *Coccophagus cowperii, C. bogoriensis* and *Aneristis* sp. and the predator, *Scymnus coccivora* are known to attack *C. psidii. Cryptolaemus montrouzieri* is known to feed about 4000 eggs during its development.

2.9.1.2. Bark Eating Caterpillars, *Indarbela quadrinotata, I. tetraonis*

i) Damage: The caterpillars are polyphagous pests which cause problem in the neglected orchards where crop sanitation is lacking. *I. tetraonis* causes more damage in Punjab, Uttar Pradesh and in Southern States.

ii) Biology: Full grown larvae are dirty brown and 38-45 mm long. The moths are stout with wing span of 35-38 mm in males and 46-50 mm in females, pale brown with brown spots and streaks on the fore wings and whitish hind wings.

Eggs are laid in cuts and crevices in the bark. On hatching, the larvae bore into the bark or main stems. The larvae remain within the bored holes during day and come out at night to feed on the bark (Fig. 2.29). They also bore into the trunk, main stems and thick branches of trees. They feed and shelter under the silken galleries full of frass and faecal matter on the bark surface. Pupation takes place within the galleries. Incubation, larval and pupal periods varied from 8-10 days, 10-11 months and 15-25 days, respectively.

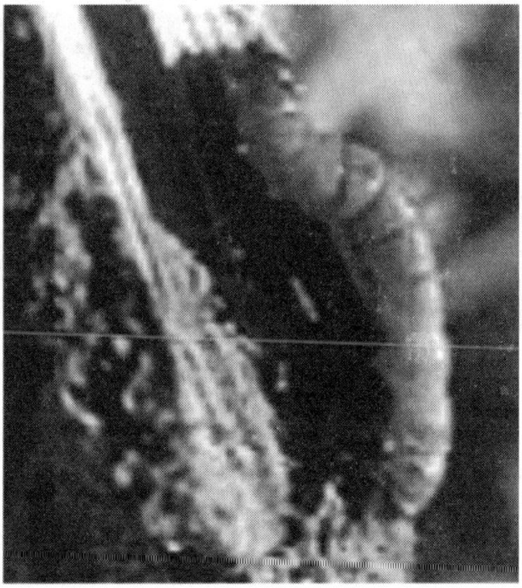

Fig. 2.29. Bark eating caterpillar damage on grapevine (Courtesy: K. Srinivasan, TAFE, Chennai).

iii) Management

- To prevent infestation of these borers keep the orchards clean.
- If and when any infestation is noticed, kill the caterpillars mechanically by inserting an iron spike into the holes.

- In case of severe infestation, clean the affected portion of the trunk or main stem and insert into the holes swab of cotton wool soaked in dichlorvos 100 EC at 2ml/l.
- Spraying the trunk and branches with 0.1% aldrin is effective.

2.9.1.3. Aphid, *Aphis gossypii*

i) Damage: The aphids occur in large numbers on tender leaves and suck the cell sap. They excrete honeydew resulting in development of sooty mold on leaves and shoots.

ii) Management : The coccinellid predators, *Chilomenes sexmaculata, Scymnus* sp., *Pseudospidimerus circumflexa, Chilocorus nigrita, Cryptolaemus montrouzieri, Ishiodon scutellaris* and *Paragus serratus* were quite effective in controlling aphids within 45 days of their appearance (Mani and Krishnamoorthy, 1989a).

2.9.1.4. Mealy Bugs

i) Damage: Five mealy bugs namely, *Ferrisia virgata, R. iceryoides, Planococcus citri, P. lilacinus* and *Maconellicoccus hirsutus* cause serious damage to many guava orchards in South India. They occur on shoots and fruits causing heavy economic losses (Fig. 2.30). Fruits infected with mealy bugs and covered with sooty mold lose the market value.

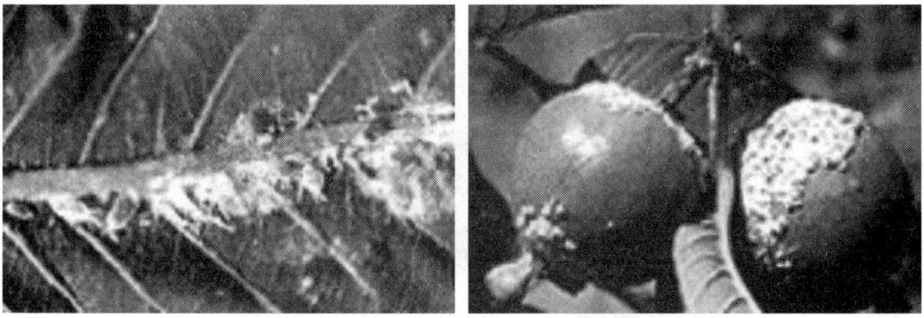

Fig. 2.30. Mealy bug infestation on leaves and fruits of guava (Courtesy: K. Srinivasan, TAFE, Chennai).

ii) Management: The parasitoids, *Aenasius advena, Blepyrus insularis* and the predators, *S. coccivora, Mallada boninensis, Brumus suturalis* and *Spalgis epius* play a significant role in reducing the mealy bug population. The encyrtid, *A. advena* is a very important parasitoid of *F. virgata* causing up to 50% parasitism in the field. *Blepyrus insularis* also causes up to 32% parasitism. Release of 380 adult beetles of *C. motrouzieri*/ha (10/tree) resulted in reduction of mealy bug population to 2.8/plant as against 145/plant in control on 40th day and complete control on 50th day of release. Release of *C. motrouzieri* also helps in clearing *P. lilacinus* and *M. hirsutus* (Mani *et al.*, 1989).

In guava orchards, parasitoids like *C. perigrinus* and *Allotropa citri* and predators, *S. epius* and *C. montrouzieri* regulate *P. citri* population in nature.

The release of exotic parasitoid, *Leptomastix dactylopii* also gives effective and permanent control of *P. citri* in guava orchards. *P. citri* disappear subsequently due to the activity of *L. dactylopii* (Mani, 1994a).

2.9.1.5. Fruit Fly, *Dacus dorsalis, D. diversus, D. zonatus*

i) Symptoms: The fruit flies feed on guava from September to May when fruits are available, which later feed on loquat and mango. They can fly one to two km in search of a suitable host.

ii) Biology: Adults of *D. diversus* are smoky brown with greenish black thorax having yellow markings unlike those of *D. dorsalis* which have brownish black thorax without any yellow middle stripe. Shukla and Prasad (1985) using methyl eugenol, observed peaks in the population of *D. dorsalis* in guava orchards at Bangalore in March-April, May-June and September-October coinciding with the ripening of fruits.

iii) Management : Avoid taking rainy season crop, as the incidence is very high. Collect fruit fly infested fruits and destroy them. Plough during summer to expose and kill pupae. Spraying with 0.03% phophamidon or oxydemeton methyl is effective against the pest.

2.9.1.6. Fruit Borers

i) Damage: *Virachola isocrates* and *Dichorosis punctiferalis* have been recorded boring into guava fruits (Fig. 2.31). The pomegranate butterfly, *V. isocrates* has been found damaging guava, amla, citrus, etc. Its infestation on guava was up to 10%. *D. punctiferalis* damage the shoots, buds and fruits.

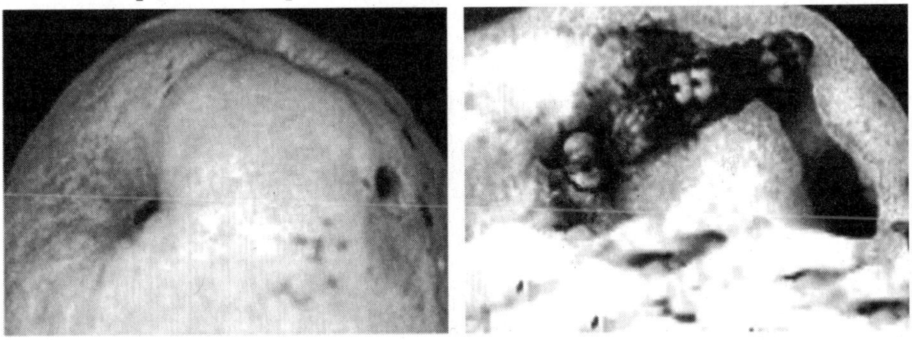

Fig. 2.31. Fruit borer damage on guava (Courtesy: K. Srinivasan, TAFE, Chennai).

ii) Management : Spraying with 0.2% carbaryl at 10-12 days interval in June-July can provide good control if damage exceeds 5%.

2.9.1.7. Tea Mosquito Bug, *Helopeltis antonii*

i) Damage: Guava cvs. Lucknow-46, Lucknow-47, Bangalore and Red Fleshed were found susceptible to tea mosquito bug (Fig. 2.32).

ii) Management : Guava cvs. Safeda and Seedless were rarely infected, while Smooth Green was less affected.

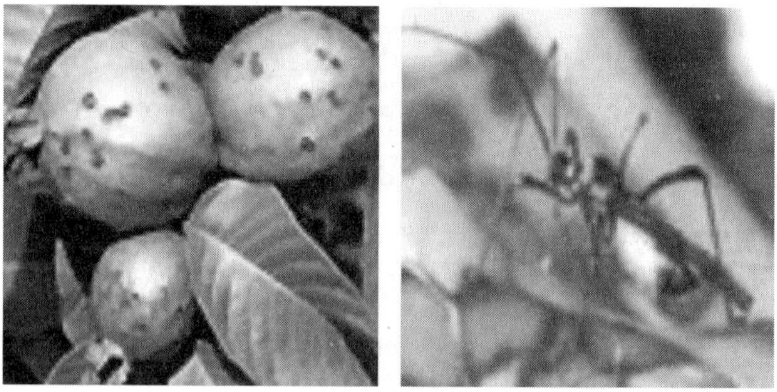

Fig. 2.32. Tea mosquito bugs on fruits and leaf of guava (Courtesy: K. Srinivasan, TAFE, Chennai).

2.9.1.8. Waxy Scale, *Drepanococcus chiton*

i) Damage: The scale insect has become very serious in some orchards around Bangalore, India.

ii) Management : The parasitoids, *Anicetus ceylonensis* and *Cephaleta brunniventris* are highly effective in suppressing the scale population. The local predator, *Chilocorous nigrita* is also observed in large numbers helping in bringing down the scale population. In subsequent years, the scale population does not appear at all (Mani, 1995).

2.9.1.9. Spiralling Whitefly, *Aleurodicus dispersus*

i) Damage: The exotic whitefly is causing serious damage to several fruit crops including guava (Fig. 2.33), papaya, banana, etc. in Peninsular India.

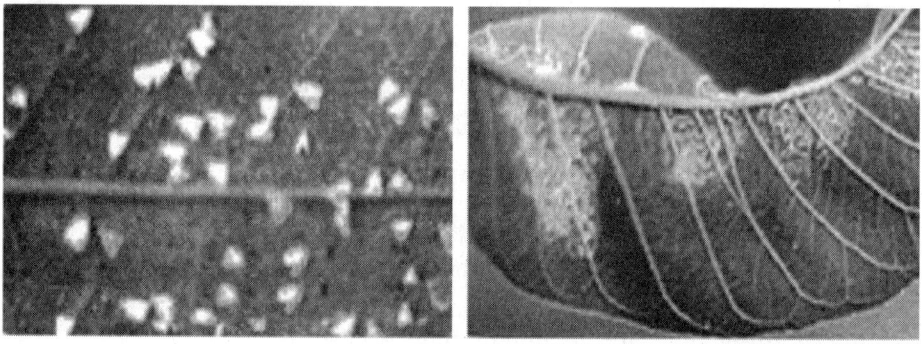

Fig. 2.33. Spiralling whitefly on guava leaf (Courtesy: K. Srinivasan, TAFE, Chennai)

ii) Management: The predators, *C. montrouzieri, Axinoscymnus puttarudriahi* and *Mallada astur* are commonly associated with *Aleurodicus dispersus* (Mani and Krishnamoorthy, 1999d). Two aphelinid parasitoids, *Encarsia haitiensis* and *E. guadeloupae* have appeared recently on the spiralling whitefly in Peninsular India. The build up and dispersal of these parasitoids would bring down the population of spiralling whitefly (Mani *et al.*, 2000).

2.9.2. Vertebrate Pests

i) Damage: Rose-ringed parakeets (*Psittacula krameri*) feed on seeds, berries, fruits, blossoms and nectar and are serious pests on guava (Bhatnagar *et al.*, 1982). Small green barbet (*Megalaima viridis*) and Jungle crow (*Corvus macrorhynchos*) damaged guava fruits in hill region of Karnataka, where birds on an average incurred 14% unripe, 24% partially ripe and 33% ripened fruit loss (Chakravarthy, 1993). Guava suffered 20-30% fruit losses due to birds in Ludhiana (Toor, 1982). Indian flying fox (*Pteropus giganteus*) in Rajasthan desert and squirrel (*Funambulus pennanti*) are major pests of guava in north Rajasthan (Tripathi *et al.*, 1992).

In guava orchards, birds particularly parakeets caused on an average 23.5% fruit loss in Mudigere, Karnataka, India, and hence protection is warranted (Chakravarthy, 1993).

ii) Management: Shooting, trapping, scaring, destruction of bird roosts and nests and encouragement of natural enemies like owls help in the management of bird damage. Covering vulnerable portions of canopy during fruit bearing period with thatch (locally available materials) saved 20% fruits (Chakravarthy, 1993).

2.10. LITCHI (*Litchi chinensis*)

Table 2.21. Area, production and productivity of leading litchi growing states (2005-06) (Bijay Kumar, 2007)

States	Area (in '000 ha)	Production (in '000 MT)	Productivity (in MT/ha)
Bihar	28.4	200.1	7.0
West Bengal	8.1	74.9	9.3
Assam	4.7	24.4	5.2
Jharkhand	1.4	16.5	12.0
Tripura	2.3	14.3	6.2
Punjab	1.4	13.8	10.0
Uttarakhand	7.9	13.6	1.7
Orissa	3.9	11.9	3.0
Chattisgarh	1.0	6.7	6.8
Himachal Pradesh	3.5	3.7	1.0
Haryana	0.2	0.7	3.5
Mizoram	0.2	0.3	1.5
Uttar Pradesh	0.2	0.2	0.9
Others	0.1	0.2	2.0
Total	63.3	381.4	6.0

2.10.1. Insect Pests

2.10.1.1. Fruit Borer and Leaf Miner, *Acrocercops cramerella*

i) Damage: Fruit borer and leaf miner has been reported as one of the major pests of litchi in Bihar. Its larvae mine the leaves during August-February and migrate to *Eugenia jambolina* or *Cassava tora* during March-April and return to litchi in May when the larvae bore into the fruits and feed below the petiole. The highest population density was observed in September and the lowest in December. Damaged shoots and leaves eventually wither and fall off. Full grown caterpillars were 4 to 6 mm long and green in colour. These pupate either in mines or in yellowish brown cocoons on the leaves.

ii) Management: The pest can be controlled by spraying 0.025% methyl demeton, 0.045% dimethoate or 0.08% thiometon. Spraying of 0.5% quinalphos, 0.03% dicrotophos or fenthion or 0.05% carbaryl were also found effective. Sharma and Lall (1983) found that 0.025% demeton methyl and 0.05% phosphamidon gave the highest mortality 3 and 7 days after spraying.

2.10.2. Mite Pests

2.10.2.1. Leaf Curl Mite, *Acerya litchi*

i) Damage: In most parts of India, leaf curl mite is one of the major pests of litchi. The erinose mites lacerate the tissues of the leaves and suck the cell sap (Lall and Rahman, 1975). Erinose, the chocolate velvety brown growth on the ventral surface of the leaves, is the characteristic symptom of this mite. It is also responsible for the leaf curl disease of litchi (Ghai, 1976).

ii) Biology: The white eggs are 0.04 mm in diameter and round in shape. These are laid singly on the ventral surface of the leaves at the base of the hair. Incubation period is 2-3 days, the nymphal period lasts 8-12 days and the adult longevity is 2-3 days. Adults have sexual dimorphism, are 0.15 to 0.2 mm long with greatly enlarged abdomen having about 55 ring-like segments. The mites over winter in adult stage.

Sharma *et al.* (1986) reported that the temperature and relative humidity significantly influenced mite population and the largest number existed in April-May and the lowest in November-December.

iii) Management: Sharma and Rahman (1982) obtained the best control by spraying 0.05% dicofol before and after flowering followed by 0.05% phosphamidon and dimethoate. Spraying of kelthane at 1.5 ml/litre before and after flowering was found to be most effective in reducing leaf curl and increasing fruit yield (Sharma, 1985).

2.11. LOQUAT

2.11.1. Insect Pests

2.11.1.1. Fruit Fly, *Dacus dorsalis*

i) Damage: Loquat fruits are known to serve as the food for the fruit flies during April-May. Damage may be quite considerable on some varieties.

At Pantnagar, fruit flies were observed causing fruit damage up to 80% in May (Singh, 1988).

ii) Management: Raking of soil below the plants and mixing 5% aldrin dust would kill the pupae or pupating larvae. Poison baiting with yeast hydrolysate + sugar + malathion was effective. Spraying 0.1% carbaryl or 0.05% fenthion at least a fortnight before harvesting of fruits will reduce the fruit fly infestation.

2.12. APPLE

Apple (*Malus pumila*) is the most important temperate fruit of the North-Western Himalayan region. It is predominantly grown in Jammu and Kashmir, Himachal Pradesh and hills of Uttarakhand accounting for about 90% of the total production. Its cultivation has also been extended to Arunachal Pradesh, Sikkim, Nagaland and Meghalaya in North-Eastern region and Nilgiri hills in Tamil Nadu. Apple is grown in an area of 2, 34,700 ha with annual production of 1.755 million tonnes (Tables 2.22 and 2.23).

Table 2.22. Major apple producing countries of the world (2005-06) (Bijay Kumar, 2007)

Country	Area (in '000 ha)	Production (in '000 MT)	Productivity (in MT/ha)	% share of world production
China	1851	24020	13.0	38.7
USA	153	4409	18.8	7.1
Iran	201	2662	13.2	4.3
Turkey	121	2570	21.2	4.1
Italy	57	2192	38.4	3.5
Poland	170	2075	12.2	3.3
France	47	1857	39.1	3.0
Russian Federation	384	1773	4.6	2.9
India	235	1756	7.5	2.8
Chile	37	1350	37.0	2.2
Argentina	45	1272	28.3	2.0
Others	1531	16109	10.5	26.0
Total	4832	62044		

Table 2.23. Area, production and productivity of leading apple growing states (2005- 06) (Bijay Kumar, 2007)

States	Area (in '000 ha)	Production (in '000 MT)	Productivity (in MT/ha)
Jammu & Kashmir	107.9	1093.3	10.1
Himachal Pradesh	88.6	540.4	6.1
Uttarakhand	28.8	112.3	3.9

Arunachal Pradesh	9.1	9.6	1.1
Others	0.3	0.0	0.0
Total	234.7	1755.6	7.5

2.12.1. Insect Pests

2.12.1.1. San Jose Scale, *Quadraspidiotus pernicious*

i) Damage: San Jose scale is a highly destructive and polyphagous pest and is widely distributed in all the apple growing regions of the world. These are soft-bodied scale insects, covered with a waxy shell, which protects them from predators and heavy wind. Females remain fixed to the host. Males are elongated with pale yellow bodies, dark eyes and a long projection from the base of abdomen. Females are dark yellow with an oval shape tapering to the point at one end. Males are oval and 1 mm long. It emerges from the scale as a small, two-winged insect. Heavy infestation of scale insects particularly in young trees can seriously reduce tree vigour and even death of limbs. Injury from San Jose scale can be confirmed by cutting away the bark to reveal a reddish discolouration (Fig. 2.34). Injury to fruit is more often encountered and more economically significant. The second generation of crawlers that infests the fruits cause direct economic loss. Bud bursting on infected twigs is delayed in the spring. Affected fruits get deformed and lose market value.

Pest is active from April to November. Population development is very rapid from April to November depending upon temperature, rainfall and relative humidity.

Fig. 2.34. San Jose Scale infestation on apple.

ii) Management

a) Cultural Methods: Apple nursery should not be raised in infested field. Orchard sanitation *viz.* removal of alternate hosts, weed hosts, volunteer plants and crop residues greatly reduce the infestation potential. Heavily infested branches should be pruned and burnt.

b) Chemical Methods: Soil should be treated with carbofuran at 6 g/m^2 before raising the nursery. Select healthy plants from nursery and treat them with 0.05% chlorpyriphos before planting.

Dormant spray (after leaf fall during winter) with 3% tree spray oil or diesel oil emulsion prepared locally with fish oil soap (potash base) at 6.33% (stock solution and water in 1:5 to 1:7 ratio). This will also give adequate control of aphids during the years of light infestations. Severe scale infestation in neglected orchard management requires additional summer sprays with 0.04% methyl demeton to be synchronized with crawler emergence (Bhardwaj, 1988).

c) Biological Methods: Release of Aphelinid endoparasitoid, *Encarsia perniciosi* at 2000 adults/infested tree and coccinellid predator, *Chilocorus bijugus* once in spring is effective in bringing down the population of San Jose scale. In endemic areas, repeat release, if necessary.

The predators, *Chilocorus bijugus* and *Pharoscymnus flexibilis* introduced from Jammu and Kashmir have not only established in Thanedhar area of Himachal Pradesh but are also providing reasonable control of San Jose Scale (Rawat *et al.*, 1988b).

The Russian strain of the parasitoid, *Encarsia periniciosi* gave 89% parasitism in Himachal Pradesh. *Aphytis diaspidis* in combination with *E. periniciosi* gave 86.5% parasitism. American and Chinese strains of *E. periniciosi* were also released in the Kumaon hills of Uttar Pradesh. The population of the pest was reduced by about 95% (Sahai and Joshi, 1965).

2.12.1.2. Woolly Aphid, *Eriosoma lanigerum*

The woolly aphid is a serious pest of apple and is reported to occur throughout the temperate regions including the North-Eastern hill states where apple cultivation has recently been started. It is an exotic but a specific pest of apple in India.

i) Damage: It sustains both on aerial parts and root systems and frequently moves up and down. The insect can multiply and spread to growing twigs and the leaf axils of water sprouts. The woolly aphid has a distinctive white waxy covering (Fig. 2.35). The aphid colonies resemble small tufts of wool or cotton batting. Beneath the waxy covering, the aphids are reddish brown to purple in colour. Woolly aphids do not feed directly and damage apple leaves. Their feeding forms knots or gals on twigs or roots. Galls are more visible on water sprouts than on the tree wounds. Underground colonies also form galls on the roots. Aphid colonies feed on healing tissues of limbs and trunk wounds. Where a canker disease had established, aphid feeding may spread the pathogen. Woolly aphids also secrete sticky material called honeydew that drips on fruits and leaves. Honeydew can cause russet spots on the fruit and a black sooty mold fungus may establish thereby downgrading fruit quality. This is also a nuisance to harvesters because of stickiness to clothing on contact.

Population multiplies rapidly during onset of spring and warming up of weather. High temperature and high rainfall are detrimental to its growth. Autumn period is also favourable for its flare up.

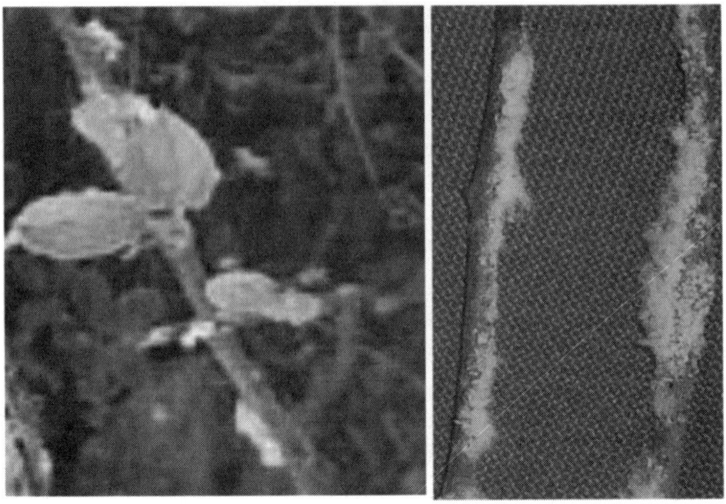

Fig. 2.35. Woolly aphid infestation on apple.

ii) Management

a) Cultural Methods: Cracks, crevices and wounds caused by pruning should be covered with Bordeaux paste. Remove suckers/water shoots. Keep the tree basin free from weeds.

b) Chemical Methods: Since infested nursery plants are main source for spread of the woolly apple aphid, nursery beds should be treated with granules of aldicarb, carbofuran, disulfoton or phorate at 0.5 g a.i./plant. The infested plants should be treated with 0.05% chloropyrifos or fenitrothion before planting in the orchard.

The aphid remains covered with flocculent waxy mass which creates a barrier for penetration of the insecticides. If thorough application of insecticide is made with high volume sprayer, aerial population can be checked by 0.05% vamidothion or 0.03% phosphamidon or 0.05% phosalone which are comparatively less harmful to the endoparasitoid, *Aphelinus mali* (Gaffar *et al.*, 1989). For the control of root infestation, drenching the tree basin with any of these insecticides or application of granular insecticides, viz., aldicarb, carbofuran, dimethoate, thiometon, phorate, etc. would be satisfactory (Thakur and Dogra, 1980). Chander and Dogra (1977) recommended application of aldicarb or phorate at 2.5-3.0 g a.i./tree around the base of tree trunk in April and again in October-November to check migration of the aphid.

c) Biological Methods: The introduced parasitoid, *Aphelinus mali* has already been successfully colonized in all the apple growing areas of India. In Himachal Pradesh, it is more effective in suppressing woolly aphid in valleys rather than on hills slopes. The parasitoid is released at 1000 adults or mummies/infested tree once, as soon as infestation is noticed. The parasitoid is effective against aerial population of woolly aphid. Among the mummies collected, 63.3% were found parasitized by *A. mali*. Prepupal and pupal parasitization ranged from 58.4 to 78.5% (Rawat *et al.*, 1989).

Coccinella septumpunctata is the most important predator of woolly aphid. Each grub consumed 426 to 450 aphids. In areas where natural population of *C. septumpunctata* is high during summer months (April-June), application of toxic chemicals should as far as possible be avoided.

Hippodamia variegata is the most important predator of *E. lanigerum*. The beetles are capable of clearing the aphid colonies within 10-15 days. At Mashobra (Himachal Pradesh) during spring and summer, entire woolly aphid population was wiped out within 24 days due to heavy predation by *C. septumpunctata* and *H. variegata*.

Syrphid fly, *Syrphid confrator*, *Chrysoperla* sp. and lady bird beetle are also effective against woolly aphid.

d) Host Resistance: Use resistant rootstocks of M-21, M-25 and MM series (Thakur and Dogra, 1980; Rai and Tripathi, 1984). Among rootstocks, Merton Stokes 778, 779, 789 and 793 were tolerant. Golden Delicious suffers less from woolly aphid infestation.

2.12.1.3. Stem Borer, *Apriona cinerea*

i) Damage: Pest is widespread in northern India. Adult beetles are of ash grey to grayish yellow, long horn coloured with conspicuous blackish tubercles at the base of elytra (Fig. 2.36), and appear in July-August. The eggs are laid in chambers in bark. In one trunk six grubs may be seen. Larvae are creamy white, elongate and 60 x 12 mm in size. Chamber inside the trunk is covered with coarse fibres for over-wintering of larvae. In August, borer chews out its way through the emergence hole. The larvae bore into branches, stem and tree trunk just under the bark and, sometimes, feed on heartwood. The larvae cut 8-9 circular holes on the bark of each branch at a distance of 10-15 cm and tunnel through and effuse frass. The trunk is hollowed inside at several places. The affected trees are fragile and often get broken even with moderate wind. Adults feed on bark and leaves and girdle young shoots which may die. They fed actively from August to November.

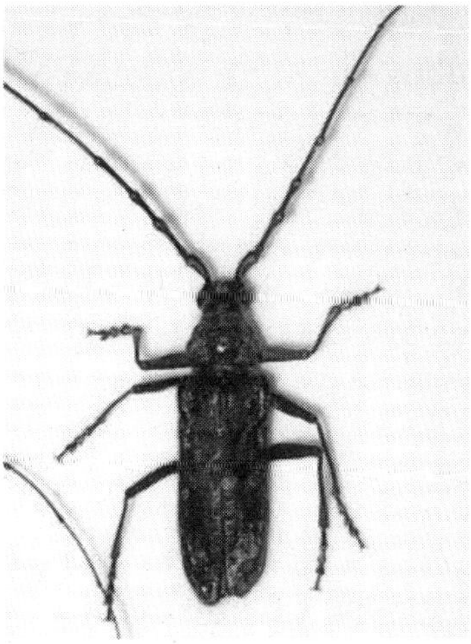

Fig. 2.36. Stem borer on apple.

ii) Management: The stem borer can be controlled by pruning and destruction of newly attacked branches and killing the older larvae by inserting flexible wire into the live tunnel. If the attack is heavy, satisfactory control can be achieved by inserting cotton wick soaked in petrol or kerosene or 0.1% dichlorvos or 0.1% methyl parathion (10 ml) into the holes and plugging the holes with mud. The holes which get reopened should be treated again.

2.12.1.4. Shot Hole Borer, *Scolytus* spp., *Xyleborus* spp.

i) Damage: Shot hole borers are bark feeders and often attack weak, declining, decaying and poorly managed trees. They also transmit pathogenic fungi. Damage is initiated by small (2-6 mm) dark coloured cylindrical beetles with deflexed head. They burrow the bark making small perporations exuding dark fluid. Each female excavates a serpentine egg gallery in upward direction towards the bask making tiny pin hole-like depressions, in each of which a small, shining, creamish, spherical egg is deposited. A female lays 50-75 eggs. The hatching grubs bore into the bast and sap wood at right angle or in upward direction away from the egg gallery. Larvae become full fed in about a month, pupate at the end of larval gallery and adults emerge within two weeks of pupation. There may be up to four overlapping generations, the last over wintering as larva or sometimes as adult, if adult emergence occurs by November.

ii) Management: Clip off the terminal shoots with unshed cluster of dry leaves in winter. Shot hole borers are the pests of diseased and weak trees. The best way to prevent the attack is to follow good horticultural practices and grow healthy trees. All badly infested trees or dead and dying wood should be cut and burnt before leaf drop. Collect and kill grubs of root borer while preparing basins. Install light traps near the orchard to collect and kill the beetles in kerosinized water.

2.12.1.5. Codling Moth, *Cydia pomonella*

i) Damage: The codling moth is a serious pest of apple and other fruits in Ladakh district of Jammu and Kashmir. About 30-70% of the apple fruits are rendered unmarketable by this pest (Singh, 1986) (Fig. 2.37). The larvae of *C. pomonella* hibernate under the bark of the tree as well as in fruits and soil. Some birds like *Parus caeruleus, P. minor* and *Passer domesticus* prey upon the over wintering larvae and a number of other natural enemies including granulosis virus disease of the larvae cause considerable mortality. In spite of this, the pest continues to cause serious damage in apple growing areas of Ladakh (Khaltse, Nurla, Nemo, Sapol, Basgo, Leh, Saboo, Kargil, etc.).

ii) Management: Scrape loose bark from trunks to eliminate shelter for over-wintering caterpillars of codling moth. Promptly remove and destroy all fallen fruits which could contain caterpillars that will help in suppression of pest population.

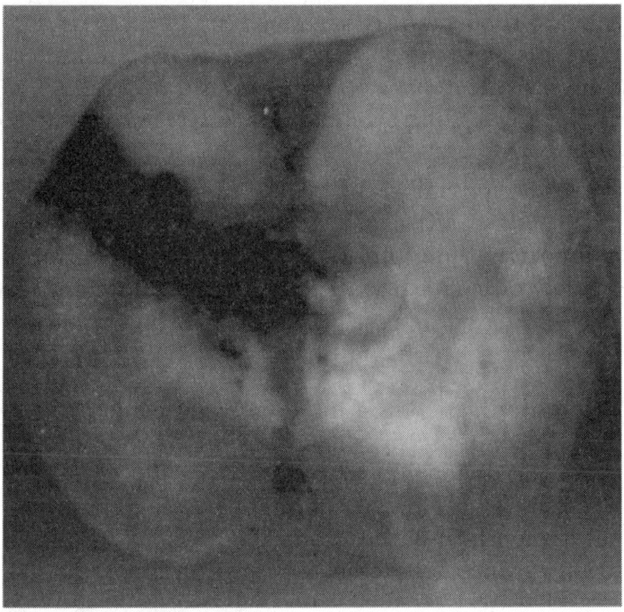

Fig. 2.37. Codling moth damage on apple.

There is a need to initiate control measures by monitoring of moth by large scale use of pheromones, conservation of natural enemies, synchronization of first release of egg parasitoid, *Trichogramma embryophagum* at 2000 adults/tree, with the first appearance of the moth, subsequent releases at weekly intervals and large scale production and spray of granulosis virus at 8 x 10 virus capsids/ml at fortnightly interval.

2.12.1.6. Collar Zone Weevil, *Dyscerus fletcheri*

The collar zone weevil was first observed during July 1979 on apple orchard at Chaubattia (Gupta *et al.*, 1984). This pest has emerged as a major bottleneck to apple cultivation in the hills of Uttar Pradesh.

i) Damage: The grub and weevil feeds on internal tissues of the stem at collar zone. The damaged stem gets hollow and weak. In severely infected apple tree, the tunnels are filled with frass and excreta and rotting takes place around the feeding site thereby reducing their vigour and yield. One to two tunnels and 6-10 weevils are usually recorded in a tree. But in a few instances, as many as five tunnels and 25-39 weevils per tree are recorded in which case the growth of the tree is markedly affected. The grubs are creamy white with stout dark brown head. The adult weevils are dark brown in colour (Singh *et al.*, 2001).

ii) Management

- Keep the tree basin clean.
- Collect and destroy the grubs and adults at the time of digging.

- Interculturing under the trees helps in killing the grubs and weevils.
- Expose the collar zone of the infested plants by removing the soil and spray with 0.07% endosulfan or 0.04% monocrotophos. Repeat the spray after 15 days according to need.

2.12.1.7. Root Borer, *Dorysthenes hugelii*

The pest is more serious menace in hilly areas of North India and in Almora and Nainital districts of Uttarakhand and in loamy and sandy loam soils (Verma and Chander, 1988). Occasionally, it also attacks the roots of apricot, cherry, peach, pear and walnut.

i) Damage: White grubs measure about 7.5-10.0 cm. Adults are of chestnut red colour with their white grubs surviving in soil. It has a life cycle of four years. Eggs are elongated, 3 x 1 mm in size and are laid either singly or in clusters. Adults emerge after rainfall. These affect new orchards. The young larvae after hatching bore into the roots and girdle them. Consequently, roots get severed and branches wither and die. Such trees become shaky, get easily uprooted with wind and usually die within three years. Young and small trees with one or few main roots suffer the most as the roots in the older orchards are well developed. Lower leaves usually become yellow in colour due to its attack and cracks can be seen in stem. Its incidence and population increases rapidly during rainy season.

ii) Biology: Adult beetles emerge from the soil with the advent of monsoon, around the end of June up to the middle of July. Immediately after emergence, they start mating and the males die soon after, while females live for 10-12 days. A female lays, on an average, 300 ovoid-shaped, yellow white eggs, 8-12 mm below the soil surface. The eggs hatch in 30-40 days. After hatching, the grubs go down into the soil and feed on organic matter and on the roots of trees. The grub takes up to 3.5 years to mature and become full grown. When full grown, they pupate inside the earthen cocoons and then emerge as beetles after the first monsoon shower.

iii) Management: Avoid dry sandy soils for establishing new apple orchards as they provide congenial conditions for survival and multiplication of the pest. Interculturing under the trees helps in killing the grubs. Marking infested trees in July and digging their root system in winter for collection and mechanical destruction of grubs. Treat the tree basin with 0.05% chloropyriphos or monocrotophos in July against neonate grubs.

Verma and Chander (1988) suggested chemical treatment at three stages: i) tree basin treatment with 5% aldrin dust at 150-200 g in July against neonate grubs, ii) marking infested trees in July and digging their root system in winter for collection and mechanical destruction of grubs, and further drenching the basin with 0.1% aldrin or 0.05% methyl parathion, iii) again drenching the basin with either of the two insecticides or 0.05% chloropyriphos in March. It is best to treat the basin in July and opening the tree basin for collection and killing of grubs during dormancy.

2.12.1.8. Blossom Thrips

i) Damage: Some blossom infesting thrips of significance are *Frankliniella dampfi, Taeniothrips* sp., *T. rhopalantennalis, Thrips flavus, T. florum* and *Haplothrips tenuipennis*. These are small insects having narrow, fringed wings with greatly reduced venation and long marginal setae. Nymphs are elongate white or yellow and have three larval instars. Full grown nymphs fall, grow and pupate in soil. After hatching, nymphs are initially white in colour, turn yellow orange at later stages. These are associated with the blossom and their population increases with the onset of the spring and flowering. Both nymphs and adults lacerate all floral parts and delicate and unfolding leaves of vegetative buds. Consequently, brown spots develop. The lesions on androecia and gynoecia deteriorate the fruit quality and reduce fruit set and yield. The set fruits may drop. Heavily infested flowers bear sticky and faded appearance with indication of early senescence.

Thrips population increases rapidly in spring with the flowering. In rainy season, its population decreases. Population increases again in October-November and decreases in summer and rainy season.

ii) Management: Coccinellids, anthocorid (*Orius* spp.) and lygaeid (*Goecoris* spp.) bugs feed on blossom thrips (Kumar and Ananthakrishnan, 1984) and curb the population build up. If the population of blossom thrips is too high, spraying 0.05% fenitrothion, 0.1% isofenphos or 0.05% methamidophos could be done at pink bud stage when predator activity is very low (Mishra *et al.*, 1987).

2.12.1.9. Leaf Folder, *Archips pomivora*

i) Damage: The black headed green caterpillar commonly occur in leaf folds or in leaves stitched together in apple orchards (Chander and Dogra, 1983; Verma, 1985). The intensity of attack is more on foliage but the injury is also inflicted by scraping the fruit skin and its subsequent rotting causing 11-41% fruit loss during storage (Verma, 1985).

The pest over winters as larva in fallen leaves or cracks and crevices of the trees as well as in wrappers of the stored fruit for 227-258 days (Verma, 1985). The caterpillars shift to foliage and blossoms in April. Adult emergence occurs in May and until September. Two generations are completed and larvae of third generation over winter. Maximum damage is done in June and on late maturing varieties of apple.

Eggs are dorsoventrally flattened and are laid in irregular clusters of 35-178 each on dorsal surface of the leaf and depressed fruit surface. Eggs hatch in 8-16 days. Caterpillars feed on young buds, blossom and terminal delicate leaves after binding them with silken threads which act as hiding sites for feeding. They molt six times and become full-fed in 20-37 days in first two generations. Prepupal and pupal periods last for 2-10 and 6-15 days, respectively. Adult emergence occurs during the day and they mate in the evening hours after 1-2 days. Preoviposition period is 3-5 days.

ii) Management: Bhardwaj (1987) demonstrated that *Archips pomivora* males are attracted to the oblique banded leaf roller pheromone (OBLR) traps which were used for monitoring population of the moths in May. Carbaryl at 0.1%, malathion at 0.05% or endosulfan at 0.05% sprayed at three weeks prior to the harvest, when the moth activity is maximum, gives satisfactory control of the pest infestation in storage (Verma, 1985).

2.12.2. Mite Pests

2.12.2.1. European Red Mite, *Panonychus ulmi*

i) Damage: The feeding by mite on the lower surface of foliage (Fig. 2.38) results in off colour foliage which in severe cases becomes bronzed as compared to uninfested foliage. The leaf efficiency and productivity is indirectly affected and their response to micronutrients and growth regulators is affected. Heavy mite feeding early in the season can not only reduce the tree growth and yield but also drastically affect fruit bud formation, thereby reducing yields in the following year.

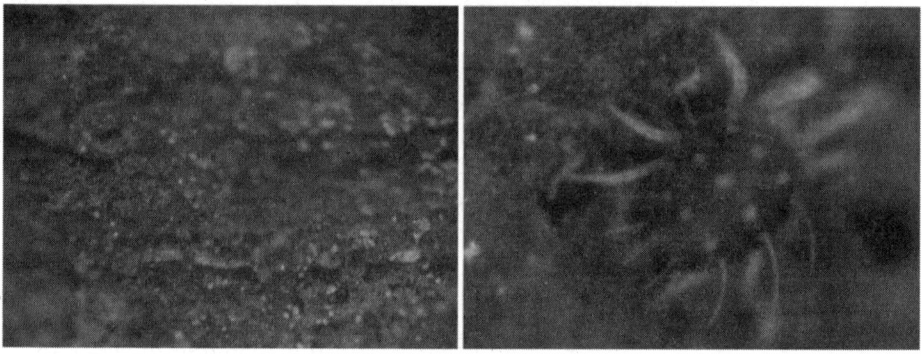

Fig. 2.38. Red mite infestation on apple.

High temperature and low humidity prevalent during summer periods in May-June are highly favourable for rapid multiplication of mites.

ii) Management: Natural enemies such as coccinellid, *Stethorus punctum*; anthocorid bug, *Orius* sp.; *Chrysoperla* sp. and *Typhlodromus pyri* are effective in reducing the population of mites.

2.12.3. Vertebrate Pests

i) Damage: Narang and Chandel (1995) assessed the extent of damage caused by seven species of birds to apple orchards at Nauni, Solan, Himachal Pradesh. The damage was maximum during 6 to 9 AM mostly confined to the upper and middle canopy of the tree. Maximum damage to fruits (80%) was caused by Blossom headed parakeet (*Psittacula cyanocephala*). The damaged fruits become unfit for marketing. The early maturing compared to late maturing cvs. suffered more bird damage.

Flying foxes (*Pteropus giganteus*) caused damage to apple, citrus, peach, guava and other fruits in India. In Himachal Pradesh, apple orchards suffered 6.7 to 21.5% damage at nursery and 4.4% damage during storage due to rodents.

Twenty percent of the fruits in apple, pear, peach and apricot orchards in Himachal Pradesh are damaged by birds like Redbilled Blue Magpie (*Cissa erythrorhyncha*), Redvented Bulbul (*Pycnonotus cafer*), White Cheeked Bulbul (*Pycnonotus leucogenys*) and Slatyheaded Parakeets (*Psittacula himalayana*).

ii) Management: The orchards were successfully protected from depredations of birds by repeated sprays of landarin at 1.5 kg a.i./ha. However, the fruits become unfit for consumption because of accumulation of toxic residues (Kakar *et al.*, 1986).

Branches of thorny trees and old fish nets or wire netting protected the fruit bunches from flying foxes. Bats may be deterred to roost near orchards. Zinc phosphide fumigation resulted in 86% control of rodents in apple orchards. Burrow baiting proved superior to surface baiting (Anon, 1996).

IPM

- Spraying 5% NSKE/Neemarin at blossom stage to control blossom thrips.
- Spraying of horticultural mineral oil at 2 litres/200 litres of water in June-July *i.e.* just before breaking of dormancy to control San Jose scale, European red spider mite and over wintering eggs of red spider mite during pink stage.
- Releasing of chrysopids like *Chrysoperla carnea* and *C. scelestes* for aphid control.
- Spraying dormant oil at 4 litres/200 litres of water to manage San Jose scale and aphids and to prevent hatching of European red spider mite.
- Release of predatory mites like *Amblysius fallacies* and *Zitzellia mali* against European red spider mite.
- Soil application of neem and mustard cake at 250 kg/ha for controlling the soil insect pests, collar rot and root rot pathogens.
- Spraying of kerosene oil in water during 15th Nov-15th Mar to reduce root borer population.
- Spraying of *B. thuringiensis* (Bilep) at 1 kg/ha against mites and fruit scrapper pests.

Table. 2.24. Calendar of IPM activities in Apple

Crop stage	Target pests	IPM activities
Dormant season module (October-November)		
Tight cluster stage	San Jose scale and over wintering eggs of mite	Apply dormant oil at 4 litres/tree

Spring season module (April-June)		
Pink bud, petal fall	San Jose scale, Woolly aphid, Blossom thrips, Mites, Scales	Spray dormant oil in first week of May for San Jose scale and European mite when 1st stage crawlers emerge. Release of coccinellid, *Chilocorus bijugus* for San Jose scale. Release of *Chrysoperla scelestes* and *C. carnea* for woolly aphids. Release of predatory mites, *Amblysius fallacies* and *Zitzellia mali* for mites. Spray 0.2% chlorpyriphos.
Summer season module (July-September)		
Fruit development (Walnut-sized fruits)	Red spider mite, Caterpillar	Spray horticultural mineral oil for mites when ETL > 6-8 mites/leaf and temperature is < 24°C. Encourage predatory mites. Spray 0.15% endosulfan for caterpillars.
Fruit development	Mite	Encourage predatory mites. Spray horticultural mineral oil for mites.
Fruit development (20 days after earlier spray)	Mite	Spray acaricide after monitoring for mites ETL. Encourage predatory mites. Spray horticultural mineral oil for mites.

2.13. PEACH & PLUM

Peach (*Prunus persica*) is a temperate fruit rich in proteins, sugar, minerals and vitamins. In India, peach is grown in Himachal Pradesh, Punjab, Haryana, Delhi, West Uttar Pradesh, Jammu & Kashmir and Assam.

2.13.1. Insect Pests

2.13.1.1. Fruit fly, *Dacus dorsalis*

i) Damage: When the fruit is about to ripen, the fly punctures it by its oviscapt (false ovipositor) and lays eggs inside the fruit. Maggots feed on pulp. Fruit becomes soft, ferments and drops. The attack is more on late maturing varieties. The incidence reduces yield and quality of the fruit.

ii) Management

- Use early maturing varieties like 'World's Earliest', 'Early White Giant', 'Pratap', 'Shan-e-Punjab' and 'Florida Sun' in infested localities.
- Do not delay harvesting of matured fruits.
- Collect and destroy the fallen fruits.
- Deep ploughing near the plants during May-June and December-January to expose the pupae to the biotic and abiotic factors.

2.13.1.2. Leaf Curl Aphid, *Brachycaudus helichrysi*

Peach leaf curl aphid is destructive pest of peach, apricot, almond and plum in order of preference. In India, this species occurs in the plains as well as hills and is extremely polyphagous pest (Ghosh and Verma, 1988).

i) Damage: The aphid damages floral and vegetative buds by sucking sap due to which unfolding leaves curls up, remain smaller, then get much distorted and turn pale. Numerous aphids and nymphs remaining inside the flowers suck the sap and thus they appear sickly and dull. Consequently, poor set, premature drop and subnormal formation of fruit occur. The badly curled leaf whorls usually dry and shed.

ii) Management: Removal of weed hosts which acts as secondary hosts is essential. Regular pruning of current vegetative growth which usually carries eggs during December is recommended (Gupta *et al.*, 1988).

Application of systemic insecticides, 0.025% methyl demeton and 0.03% dimethoate at pink bud stage has been suggested. Gupta *et al.* (1986) obtained best control of the aphid on mature peach trees by application of aldicarb at 15 g a.i./tree to the soil below the tree canopy before bud burst stage. During the months of March-April, avoid chemical sprays to encourage biotic agents like Coccinellids, Syrphids and Anthocorid bugs which are very active.

Coccinella septempunctata, Hippodamia sp. (*Adonia variegata*) and *Anthocoris minki* are reported as suitable predators in Himachal Pradesh. Cvs. Stark Early Giant, Early White Giant and Flavour Crest are found moderately susceptible (10-30% leaf whorl infestation) to leaf curl aphid (Gupta *et al.*, 1988).

2.13.1.3. Stem Borer, *Apriona cinerea*

i) Damage: Pest is widespread in northern India and destructive to peach. Adult beetles are of ash grey to grayish yellow, long horn coloured with conspicuous blackish tubercles at the base of elytra, and appear in July-August. The eggs are laid in chambers in bark. In one trunk six grubs may be seen. Larvae are creamy white, elongate and 60 x 12 mm in size. Chamber inside the trunk is covered with coarse fibres for over-wintering of larvae. In August, borer chews out its way through the emergence hole. The larvae bore into branches, stem and tree trunk just under the bark and, sometimes, feed on heartwood. The larvae cut 8-9 circular holes on the bark of each branch at a distance of 10-15 cm and tunnel through and effuse frass. The trunk is hollowed inside at several places. The affected trees are fragile and often get broken even with moderate wind. Adults feed on bark and leaves and girdle young shoots which may die. They fed actively from August to November.

ii) Management: The stem borer can be controlled by pruning and destruction of newly attacked branches and killing the older larvae by inserting flexible wire into the live tunnel. If the attack is heavy, satisfactory control can be achieved by inserting cotton wick soaked in petrol or kerosene or 0.1% dichlorvos or 0.1% methyl parathion (10 ml) into the holes and plugging the holes with mud. The holes which get reopened should be treated again.

2.13.1.4. Flat Headed Stem Borer, *Sphenoptera lafertei*

Flat headed stem borer is of frequent occurrence on stone fruits, mainly peach and plum. Sharma (1986) found 53.4% infestation on peach and 37.8% on plum. At Hisar, Lakra *et al.* (1980) observed its damage on peach trees to the extent of 45%.

i) Damage: The pest attacks trees of all sizes but the worst affected are older, weak, injured, unhealthy and sun-burnt trees. The orchards located in southern aspect are severely damaged, particularly the portion of the tree exposed to sun. The larvae feed on fresh bast and then bore into the sapwood (phloem and cambium), making it powdery and sometimes girdle the branches and stem killing them outright. Affected portion shows small oval holes about 1 cm long and 6 mm broad. Small coppery black adult beetles (10-13 mm) feed on foliage.

Beetles from overwintering/prepupae emerge by mid March, feed on foliage and lay white spherical eggs singly in crevices all over the tree trunk and main branches. Larvae hatch out in three weeks in spring. The neonate white larva with a small yellow head and greatly enlarged prothoracic segment bores into the bark and from the entry hole oozes out a lump of resinous gum. As larva grows, it penetrates deep into the sapwood. The full grown grub (18-24 mm long) drills a small pupation chamber (4-6 x 2-4 mm and 1 cm deep) in the wood and pupates in it. Larval duration varies from two months in summer to six months in winter. Prepupal and pupal periods last 1-2 and 8-12 days, respectively. Depending on altitude, there may be 2-4 overlapping generations in a year.

ii) Management: Proper training and pruning of the tree to avoid direct exposure of the stem to sun and covering the trunk with craft paper or muslin cloth provide some protection from the pest. Maintaining the tree vigour is essential. The dead or heavily infested branches oozing out gum should be cut and destroyed. Sharma *et al.* (1963) suggested 2-4 swabbings, first with 0.5% and subsequently with 0.25% dieldrin, each treatment synchronizing with laying of broods. Sharma (1986) noticed that spraying the infested branches with 0.1% methyl parathion, 0.3% aldrin, 0.1% fenitrothion or 0.1% DDVP was more satisfactory than swabbing.

2.13.1.5. Twig Borer, *Anarsia lineeatella*

i) Damage: The caterpillar damages twigs of peach, plum, apricot and almond in Kashmir. Tiny grayish moths (1.2-1.4 cm wing span) active at dusk, lay about 170 eggs singly or in small groups of 2-3 on shoots, twigs, fruits or leaves during the life span of about a week. Egg, larval, prepupal and pupal period is of 6-11, 18-35, 3-9 and 6-11 days, respectively thereby completing the life cycle in 40-70 days. There are three generations in a year and third instar larvae of last generation overwinter from September end to

March end under flimsy silken webs or in silk-lined cracks on the bark. Larvae resume feeding by the end of March and bore into the new growth (twigs or terminal buds) making small galleries. Pupation takes place on the bark in cocoon. The second generation larvae feed mostly on fruits.

ii) Management: Spraying of 0.1% carbaryl, 0.05% malathion or 0.05% endosulfan three weeks prior to the harvest, when the moth activity is maximum, gave satisfactory control of pest infestation (Verma, 1985).

2.13.1.6. Bark Eating Caterpillar, *Indarbela quadrinotata*

i) Damage: The incidence of this pest has been recorded up to 1500 m altitude in Himachal Pradesh on plum (Verma, 1985a). Its attack has also been observed on peach in mid hills. Verma (1985a) recorded an incidence of 90% on 12-15 years old plum trees each bearing about 13 holes confined to forking of branches or to depression of more than 4 cm on the stem and branches. Old and weak trees in the neglected orchard are more susceptible to the attack.

ii) Biology: Eggs are laid under loose bark or in cracks and crevices in clusters of 15-25 from April-June. They hatch in 8-11 days and after 2-3 days, larvae bore into the trunk or main branches. During day time larva seeks shelter in the undivided tunnel and comes out at night to feed on the bark under silken web interspersed with fine chips of wood and faecal pellets. The infested trees often suffer from canker. Full fed larvae (46-52 mm) are with dark brown head and dirty brown body. They may feed up to April (larval duration of 9-10 months) when pupation occurs. Pupal period is of 3-4 weeks. Thus, there is one generation in a year (Butani, 1979).

iii) Management: Injection of 5 ml dichlorvos or methyl parathion at 0.0125% into the live borer holes (Verma *et al.*, 1974) or painting of the bark around the borer holes with 0.05% monocrotophos (Verma and Singh, 1975) are the effective control measures.

2.13.1.7. Plum Fruit Moth, *Cydia* sp.

i) Damage: Infestation of a torticid caterpillar as fruit borer has been observed on plum in Himachal Pradesh at an altitude of 1500-2100 m (Chander, 1984). It causes 35-65% infestation on plum. The larva makes gallery of variable shape and size in the mesocarp by feeding on the flesh. The damage is conspicuous around the stone. A fruit may harbour up to 7 larvae depending upon size and stage of its development.

The moth has two generations. Moths are seen on wings at dusk during April end to May. They usually lay eggs singly on fruit which hatch in 8-14 days and larva enters the fruit. Depending upon extent of damage, infested fruits may drop. Larvae become full grown in 21-24 days and come out of fruit to construct cocoons on the tree, near its base or any suitable site on the ground, and pupate. Moths emerge in June. Developing larvae of second generation are more destructive. They become full fed by August end, overwinter as prepupae in silken web and pupate in April.

ii) Management: Collection and destruction of fallen fruits and debris harbouring cocooned larvae reduced the inoculum of the pest for ensuing year. In orchards sprayed with dormant spray of 2% oil mixed with 0.05% fenitrothion, there was negligible attack of the pest. In hot spots, spraying 0.05% fenitrothion, 0.05% quinalphos, 0.1% malathion or 0.1% carbaryl in first week of May followed by a second spray in first fortnight of June provides effective control.

2.13.1.8. Plum Scale, *Eulecanium teliae*

The plum scale is a serious pest of apple and plum in Himachal Pradesh. Unarmoured coccids are of concern which infests plum in Kashmir (Zaka-ur-Rab, 1982), plum and peach in Uttar Pradesh (Gupta and Ray, 1971), and plum in Himachal Pradesh (Sharma and Dogra, 1982).

i) Damage: The stages of development include eggs, two nymphal instars and adult female (Sharma and Dogra, 1982). Adult females (1.16 x 0.75 mm) emerge during February-March, swell on voracious sap feeding and become dome-shaped measuring 3.3 mm long, 2.5 mm wide and 3.1 mm high. They secrete copious amount of honeydew from the anal end which attracts a variety of flies, ants and other insects. Eggs are laid beneath the body throughout April. The female dies after egg laying and its Venter shrinks until held against hardened dorsal skin. Eggs remain protected underneath the dead female hatching out during May-July. Crawlers leave the dome shaped body of the mother through anal slit and often settle on underside of leaves along midrib and main veins. They molt to second instar in August and migrate to 1-2 year old twigs during September-October. The insect overwinters in this stage and adult female is formed in March.

ii) Management: Mishra and Bhalla (1974) recorded high parasitism of 73% by the aphelinid *Coccophagus ishii* in Himachal Pradesh. Sharma and Dogra (1986) observed parasitization of second instar nymphs and adult females by the *Coccophagus* sp. and *Blastothrix sericea* (Encytidae), respectively, to the extent of 7.2 and 40.7% and the fungus *Rhinocladiella* sp. caused 13.6% mortality of second instar resulting in a total mortality of 89.3% on leaves and 65.8% on twigs. *Harmonia eucharis* was a good predator of *E. teliae* in Kashmir (Zaka-ur-Rab, 1982).

Although pre-bloom application of dormant spray oil is effective, yet post-bloom applications of 0.04% monocrotophos, 0.025% oxydemeton methyl, 0.03% dimethoate, 0.05% quinalphos, 0.05% fenitrothion or 0.03% diazinon are more effective to kill the crawlers (Sharma and Dogra, 1986).

2.13.2. Vertebrate Pests

i) Damage: Rose-ringed parakeets, House crow and Jungle crow together caused 21.2% fruit loss in peach in Punjab (Toor, 1982). In Punjab, peach cv. Floridasun attracted flocks of House sparrow (*Passer domesticus*) during second week of January as birds and new flush begin growing. As soon as the greenish-white flower buds appear, sparrows feed on flower buds to the extent that no flowers are produced. The average yield of Floridasun cv. is 75 kg/tree. But the bird affected trees yield 0-2 kg only (Mann, 1986).

Field mice (*Mus booduga*), Brown-spiny field mice (*Mus platythrix*), Common house rat (*Rattus rattus*) and Indian bush rat (*Golunda ellioti*) constitute rodent pest species on plum, peach, apple and other orchards in Himachal Pradesh.

ii) Management: Regular live trapping as a part of non-toxic management approach in plum orchards in Himachal Pradesh yielded 64.29, 62.50 and 50.00% reduction in population of *Bandicota bengalensis* (Anon, 1996).

2.14. POMEGRANATE

Pomegranate (*Punica granatum*) is an ancient and favourable table fruit in tropical and sub-tropical regions of the world. It is commercially grown for its sweet-acidic fruits which provide a cool refreshing juice and is valued for its medicinal properties. Its popularity is also due to the ornamental nature of the plant especially when bearing bright red flowers throughout the year.

In India, it is considered as a crop of the arid and semi-arid regions because it can be grown even under conditions of severe drought and frost. Pomegranate is grown in an area of 1,16,400 ha with annual production of 0.849 million tonnes (Table 2.25). The fruit rind, juice, leaf and roots are used in the preparation of various Ayurvedic medicines. Fruit is good for stomach inflammations and heart pain, diarrhea and dysentery. Juice is useful for leprosy patients. The stem, root and bark are useful in destroying parasitic worms and cure of fever. The dry rind is useful in teeth and gum disorders. The fruit juice contains glucose and fructose. The juice and seed contain large quantities of tannin and agolic acid which are essential in curing several diseases.

Table 2.25. Area, production and productivity of leading pomegranate growing states (2005-06) (Bijay Kumar, 2007)

States	Area (in '000 ha)	Production (in '000 MT)	Productivity (in MT/ha)
Maharashtra	91.0	593.6	6.5
Karnataka	12.4	138.6	11.2
Andhra Pradesh	7.7	65.0	8.4
Gujarat	3.8	38.2	10.1
Tamil Nadu	0.4	11.1	25.0
Rajasthan	0.3	1.8	6.0
Chhattisgarh	0.1	0.6	6.7
Himachal Pradesh	0.6	0.2	0.4
Total	116.4	849.1	7.3

2.14.1. Insect Pests

2.14.1.1. Anar Butterfly, *Deudorix isocrates, D. epijarbas*

i) Damage: The larvae bore into the fruits of pomegranate and can destroy up to 50% of fruits (Fig. 2.39). The female butterfly lays eggs singly on calyx of flowers or small fruits. On hatching, the caterpillars bore inside the developing fruits and feed on pulp and seeds. This will allow the entry of fungi and bacteria causing fruit rot. The conspicuous symptoms of damage are offensive smell and excreta of caterpillars coming out of the entry holes. The affected fruits rot, get dried and ultimately fall down. Besides pomegranate, the pest also damages guava, amla, annona, apple, ber, citrus, litchi, loquat, sapota, mulberry, peach, pear, plum *etc.*

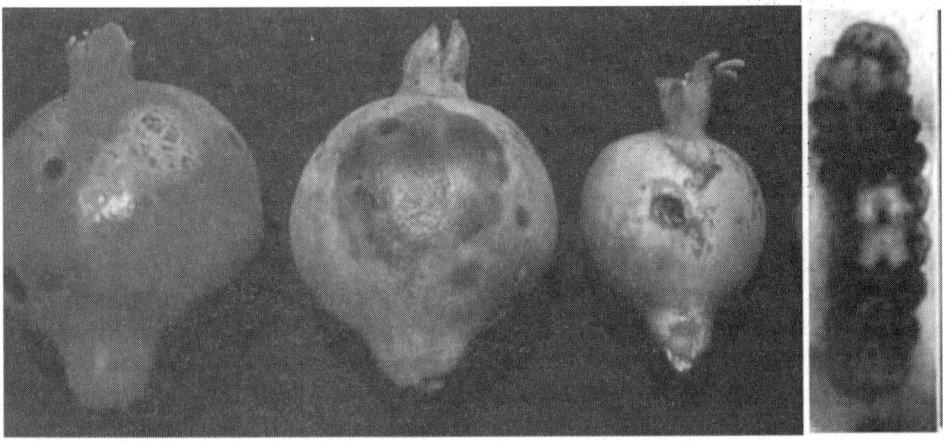

Fig. 2.39. Anar butterfly on pomegranate fruit (Courtesy: K. Srinivasan, TAFE, Chennai).

ii) Biology: Female butterflies lay eggs in the day time singly on the calyx of flowers or on fruits which are shiny white and round in shape and hatch in 7 to 10 days depending upon the season. Caterpillars feed on the seed and fill the fruit with their excreta. Full grown caterpillars are dark brown with short hairs and white patches all over the body and measure 17-20 mm in length. Larval period extends from 18 to 47 days. Pupation takes place outside the fruits and lasts 7 to 34 days. The pest breeds throughout the year on various host plants. Four generations have been reported in a year. Adult butterflies have 40-50 mm wing span. The males are slightly smaller, glossy, bluish violet and the females are brownish violet and have conspicuous patches on fore wings (Butani, 1976).

iii) Management

- Removal and destruction of all the affected fruits (fruits with exit holes).
- Spray decamethrin 2.8 EC at 1ml/l at the time when more than 50% of fruits have set.

- Repeat after two weeks with carbaryl 50 WP at 4g/l or fenvalerate 20 EC at 0.25ml/l or quinalphos 25 EC at 2ml/l in non-rainy season.
- Nanjan and Kumar (1983) reported that spraying of 0.1% phosalone in December after pruning followed by two sprays of 0.12% monocrotophos in February and April resulted in increased number of healthy fruits to 81% per tree from 32% in control.
- Kakar *et al.* (1987) found that cypermethrin at 50 g a.i./ha, fenvalerate at 50 g a.i./ha and deltamethrin at 7.5 g a.i./ha were more effective than methyl parathion.
- Two sprays of fenvalerate, cypermethrin or deltamethrin in July followed by a cover spray after 45 days gave effective and profitable control.
- Remove flowering weeds especially of Compositae family.
- After fruit setting, cover the fruits with paper or cloth bags.
- The release of egg parasitoid, *T. chilonis* at 2.5 lakhs/ha 4 times at 10 days interval reduces the fruit borer infestation up to 50% (Kakar and Sharma, 1988).
- The egg parasitoids namely, *Telenomus* sp., *Ooencyrtis papilionis* and *Trichogramma chilotraeae* are known to cause up to 60% parasitism on *D. isocrates* in Peninsular India (Mani and Krishnamoorthy, 1996a, 1999c).
- Weekly spraying of *B. thuringiensis* at 1 g/litre is effective.
- Spray application of 5% NSKE gave satisfactory control of fruit borer.

2.14.1.2. Shot Hole Borer, *Xyleborus* sp.

i) Damage: This has become a major pest on pomegranate in Peninsular India. The adult beetles bore holes on the roots and later on the lower parts of main trunk. The holes are used as nest. The early stage of infestation in an orchard begins with a mild yellowing of a lateral branch on one or more trees, usually in a contiguous patch. Within a week, the whole tree show yellowing followed by drying of branches. Infested trees may show heavy bearing with reduction in fruit size, which ripen prematurely. The main trunk from a foot above the soil shows small pinholes. These holes cut through xylem and phloem, resulting in the death of the tree. The adults migrate within a month, to the nearest healthy trees. If care is not taken at this time the rate of spread will be rapid and the whole orchard may be infested in 3-6 months.

ii) Management: Immediately after noticing the pest in the orchard the following management practices may be taken up:

- Drench the soil around main trunk with a mixture of chlorpyrifos at 2.5 ml + tridemorph 1 ml/litre. Use 2-3 litres of mixture/tree.
- After three weeks repeat with dimethoate 30 EC at 2ml/l + carbendazim at 1g/litre.

- If pest is severe, repeat the above drenching after a month.
- If infestation is low, drench with 0.15% azadirachtin (3 ml/litre) around main trunk with 2-3 litres of mixture/tree along with either of the above along fungicides.
- Avoid water logging and keep soil raked and aerated.
- Infested trees should be uprooted and burnt, especially the root zone.
- Pits of uprooted trees should be drenched with chlorpyrifos at 2.5 ml/litre.

Note : As a prophylactic measure, drench soil with chlorpyriphos 20 EC at 2.5ml/l around all trees in the orchard once in six months, followed by a spray on trees with quinalphos 25 EC at 2ml/l, followed by azadirachtin 1500 ppm at 3 ml/litre.

2.14.1.3. Stem Boring Beetles, *Coelosterna spinator*

i) Damage: The beetles keep emerging from first summer shower in May/June, till October/November. These beetles are active during night and lay eggs on the stem especially during the onset of monsoon. The eggs hatch in a fortnight time and the grubs bore into the trunk, primary and secondary branches. The leaves of infested stem will show yellowing, drying leading to dieback symptoms. Secondary infection with pathogens leads to slow death of the tree.

ii) Management

- Periodically inspect the orchards for dry branches.
- If holes/excreta/gummosis is detected on the branches, inject 5-10 ml of dichlorvos 76 EC at 4 ml/l using a disposable syringe (without needle) and seal the hole with clay.
- Cut out the drying portion of the branch and swab copper-oxy chloride 50% WP on cut end.
- Spray the surrounding trees with quinalphos 25 EC at 2ml/l or chlorpyrifos 20 EC at 2.5 ml/l.
- Give a prophylactic swab of IIHR mixture (mentioned under grape mealy bug) during May/June on the main trunk.

2.14.1.4. Thrips, *Rhipiphorothrips cruentatus, Retithrips syriacus, Anaphothrips oligochaetus, Ramaswamiahiella subnudula, Scirtothrips dorsalis*

i) Damage: These rasping and sucking insects infests leaves and flowers causing leaf tip curl and shedding of flowers. When they feed on tender fruits typically causes scab and reduces the market value (Fig. 2.40).

Fig. 2.40. Thrips damage on leaves and fruit of pomegranate (Courtesy: K. Srinivasan, TAFE, Chennai).

ii) Management

- Spraying dimethoate 30 EC at 2ml/l prior to flowering.
- If incidence is more, spray methyl oxy-demeton 25 EC at 2ml/l or dimethoate 30 EC at 2ml/l after fruit set.
- Keep the trunk basins clean.
- If necessary spray multineem at 0.5%.

2.14.1.5. Aphid, *Aphis punicae*

i) Damage: These are small green plant bugs feeding on young leaves (Fig. 2.41) and flowers and causes shoot shriveling. Accumulation of honey-dew and sooty mould development is common when their incidence is severe.

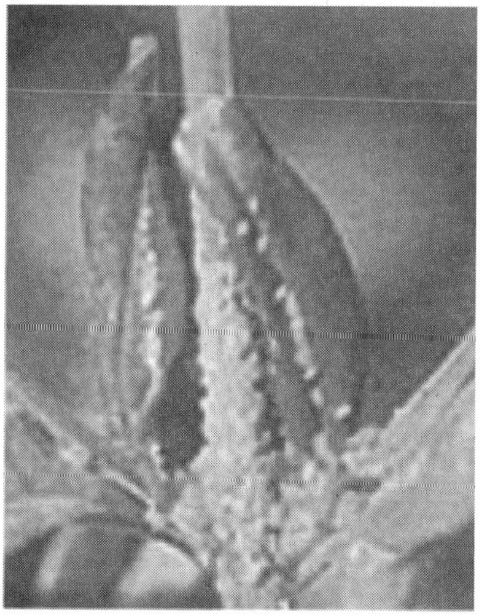

Fig. 2.41. Aphid infestation on pomegranate leaves (Courtesy: K. Srinivasan, TAFE, Chennai).

ii) Management

- Spray dimethoate 30 EC at 2ml/l as new shoots emerge.
- If predators like syrphids and coccinellids are found, delay spraying and in some cases, natural enemies can sufficiently suppress the aphids.
- NSKE 5% is found to be effective in reducing the aphid population without affecting the population of natural enemies (Verghese, 2000d).
- The coccinellids, *Scymnus castaneus, S. latemaculatus* and *C. sexmaculata*; and syrphid, *P. serratus* keep the aphid population under check (Mani and Krishnamoorthy, 1995).

2.14.1.6. Mealy Bugs

Many mealy bugs like *Icerya purchasi*, *Planococcus* sp., *Pinnaspis* sp., *Maconellicoccus hirsutus, etc.* have been reported on pomegranate.

i) Management

- Prune the affected parts of the plant and destroy.
- Spray chlorpyrifos 20 EC at 2.5 m/l + dichlorvos 76 EC at 0.75 ml/l after pruning infested parts.
- The predators *S. epius* and *C. montrouzieri* are found efficient in clearing the mealy bugs (Mani and Krishnamoorthy, 1990b).
- Spray *Verticillium lecanii* at 2 g/l.
- The parasitoid, *Leptomastix dactylopii* is recommended for suppression of *P. citri* and *C. motrouzieri* for all other mealy bugs (Mani and Krishnamoorthy, 2001).

2.14.1.7. Ash Whitefly, *Siphoninus phillyreae*

i) Damage: The whiteflies occur in serious form on the leaves at times. Entire plant becomes sick due to the development of black sooty mold on the honeydew excreted by the whiteflies.

ii) Management: The aphelenid, *Encarsia azimi* cause a high level of parasitism up to 90% in nature (Mani and Krishnamoorthy, 1994, 1998c).

2.14.2. Vertebrate Pests

i) Damage: Squirrels and species of *Funambulus* are the principal pests feeding on pomegranate fruits throughout India. The damage starts when the fruits are ripe (Patel, 1993). The fruit loss is 100% during summer (February to May) when alternate food sources are scarce.

ii) Management: Snap trapping, sticky bands, covering fruits with polybags, cloth bags laced with waste oils, etc. were the methods adopted to protect the fruits under Bangalore conditions in small orchards (less than 1 ha) and backyards of houses. Ripening fruits of pomegranate and custard apple laced with oils of neem and castor protected the fruits from vertebrate depredation for a longer period (Chakravarthy, 1993).

2.15. BER

Ber, *Ziziphus mauritiana* (Syn. *Z. zuzuba*) is a native of Indian Subcontinent. It grows in tropical and sub-tropical regions, found all over India. Fruits are used as an ingredient in the preparation of 'Joshanda" (an Ayurvedic medicine used in chest trouble). The fruits are laxative and invigorating. The leaves are used to treat conjunctivitis. They are antihelminthic and good in stomatitis, gum bleeding, asthma and liver complaints. Bark is used for the treatment of diarrhea. The root is used as decoction in fever and its powder is applied to ulcers and old wounds. The kernels are sedative and as soporific which are used to stop vomiting. It is also used for abdominal pain in pregnancy.

2.15.1. Insect Pests

2.15.1.1. Fruit Fly, *Carpomyia vesuviana*

i) Damage: The fruit fly is responsible for 77 to 80% damage rendering the ber cultivation unprofitable. Lakra and Singh (1984) found that the losses caused in Haryana on *Z. nummularia* fruits varied from 5.9% during off season (June-September) to 51.0% during the main season (September-December) and on *Z. mauritiana* from 6.7% to 58.6% during the main season (November to April). It remains active from autumn through the winter and spring. Maximum damage is, however, caused during February-March and as many as 18 maggots have been found in a single fruit. An increase in daily maximum temperature over 40°C together with low humidity (20-30%) was unfavourable for the pest development (Lakra and Singh, 1985). A temperature of 5°C prolonged the duration of immature stages. Intermittent light rainfall of 20-40 mm per week during July-August enhanced adult activity while moderate to heavy rainfall of 50-120 mm per week curtailed it. The mean weekly minimum temperature of 10-15°C and the maximum of 25-40°C and relative humidity of 25-90% were ideal for pest activity. Larval incidence in fruits of *Zizyphus nummularia* reached the peak (76.4%) during August-September and those of *Z. mauritiana* during July, January and April (47.4%).

ii) Biology: The fruit fly lays dull creamy white eggs which are opaque and 0.9 to 1.0 mm long. Full grown maggots are creamy white and about 6 mm long. Adult flies are small, brownish yellow with brown longitudinal stripes on the thorax surrounded by black spots. Wings are hyaline with four yellow cross bands.

Each female lays 12-18 eggs in groups of 2-4. Incubation, larval and pupal periods last 1-4, 9-12 and 11-31 days, respectively in February-March. There are 3-4 generations in a year.

iii) Management

a) Cultural Methods: Regular summer ploughing/digging of the soil around trees will expose pupae to the sun's heat and birds. Collection and destruction of fallen infested fruits and wild jujube bushes from the vicinity of the orchard helps to reduce the pest population. Preventing the fruits to ripe on the trees to escape egg laying on fruits and harvesting of green and firm fruit helps in reducing the pest incidence.

b) Chemical Methods: Spraying 0.1% gamma BHC in early or late November and again after three weeks followed by 1-2 applications 0.6% malathion reduced the damage to 3.7% as compared to 46.9% in control.

c) Host Resistance: Singh and Vashistha (1985) reported that ber cv. Tikadi was resistant, while Illaichi was moderately resistant. Susceptibility to fruitfly varied in decreasing order in cvs. Seb, Jogia, Gola and Mundia-Murhera (Sachan, 1984).

2.15.1.2. Leaf Defoliating Beetles

i) Damage: The adults of cockchafer beetles such as *Holotrichia* spp., *Anomala* spp., *Adoretus* spp. feed on leaves and defoliate ber trees in rainy season. They cut the tender shoot tips affecting growth of the young plants. In case of severe infestation, the entire foliage may disappear and such trees do not bear any fruits. The attack starts early in the spring and continue up to August. White eggs of these beetles are laid in soil in July and cream coloured larvae feed in soil on rotten material, roots, etc. The larvae make an earthen cocoon and hibernate throughout the whole winter. Pupation takes place in next April. These have only one generation in a year.

ii) Management: Ploughing or digging of soil would expose eggs, larvae and pupae to adverse weather conditions and natural enemies. Flooding of fields before rains in summer will stimulate the adults to come out (Baketia *et al.*, 1986). Collection of beetles by jerking the trees on a sheet of cloth and killing them in kerosenized water is very effective. They are also attracted to light traps.

Spraying endosulfan in the evening as soon as the damage is noticed helps in reducing the incidence of the pest.

2.15.1.3. Leaf Webber, *Pagyda traducalis*

i) Damage: Leaf webber cause damage with a population of 17 to 63 larvae/pupae per 100 leaves (Mann and Singh, 1981). These population peaks were observed in August-September, January and April-May. The population in May was higher than in other months.

ii) Management: Batra *et al.* (1980) and Mann (1981) found that 0.05% dichlorvos, fenvalerate, monocrotophos or quinalphos gave good control of leaf webber in Punjab.

2.15.1.4. Hairy Caterpillar, *Thiacidas postica*

i) Damage: It occurs in serious form at times causing extensive defoliation of the plant.

ii) Management: The braconid, *Apanteles creatonoti*, the ichneumonid, *Charops obtusus*, the tachnids, *Chaetoxorista* sp. and *Exorista zanthaspis* and NPV cause mortality up to 40% in nature on *T. postica* in ber orchards.

2.15.1.5. Mealy Bugs, *Ferrisia virgata, Planococcus citri, P. lilacinus, Maconellicoccus hirsutus, Nipaecoccus viridis*

i) Damage: They cause extensive damage to the shoots and fruits. Severe economic losses occur when the fruits are infested with mealy bugs. In severe infestation, the growth of tree and fruiting capacity are adversely affected. The bugs also exude honeydew which falls on dorsal surface of lower leaves, stems and even on fruits on which sooty mold grows covering the affected areas with a superficial black coating.

Damage is caused by nymphs and wingless females which are oval, flattened and have body covered with a white mealy powder. Adult female is wrinkled and opterous covered with waxy secretion.

ii) Biology: The female lay 100-300 eggs in soil about 150 mm deep during September-October. The egg over winter and hatch in spring (April and May). Nymphal period lasts from 40 to 62 days. There is only one generation in a year.

iii) Management: Frequent raking of soil under the tree during winter helps in destroying the eggs. Tying of alkathene sheet (400 gauge) around the tree trunk will prevent the nymphs from climbing the trees.

Soil treatment with phorate at 300-500 g/tree helps to reduce the population of newly hatched nymphs. In case of severe infestation, spraying of monocrotophos is recommended.

The local parasitoids and predator, *S. epius* play a great role in reducing the pest population. Release of *C. motrouzieri* is also found to be effective against *N. viridis*, and *P. citri*. *L. dactylopii* is used against *P. citri* in ber orchards (Mani, 1993b).

2.15.1.6. Waxy Scale, *Drepanococcus chiton*

i) Damage: This insect appears in serious form on the shoots and sometimes on leaves. Severe scale infestation leads to drying of the affected shoots.

ii) Management: Besides the two predators, *C. motrouzieri* and *Scymnus* sp. feeding on the crawlers of scale insects, two key parasitoids,

Cephaleta brunniventris and *Anicetus ceylonensis* cause heavy parasitism. The scales disappear subsequently indicating check by the natural enemies (Mani, 1995).

2.16. CUSTARD APPLE

The custard apple, also known as *Sitaphal*, Sugar apple, Sweet sop and *Sharifa*, is hardy and tolerant to drought. It grows very well even in shallow soils and its flowering coincides with maximum moisture availability period. It also sheds off leaves during stress periods to evade moisture loss from plant tissues through transpiration and thus a most appropriate fruit crop for arid region. Custard apple fruits are very popular in Deccan plateau and are also grown commercially in smaller scale in Assam, Andhra Pradesh, Gujarat, Bihar, Karnataka, Madhya Pradesh, Maharashtra, Orissa, Uttar Pradesh and Rajasthan. Large area is reported from Andhra Pradesh.

The fruit is used for table purpose. The edible portion or pulp is creamy or custard-like, granular with good blend of sweetness and acidity. The pleasant flavour and wild aroma have a universal liking. The ripe fruits are rich in carbohydrates and a good source of Vitamins A and C. Glucose is the predominant sugar. Fruit pulp can be used in ice creams, confectionary and certain milk products. The seeds contain 27-30% oil which can be used in soap and paint industry. The seed cake containing 4% nitrogen is beneficial as manure. Powdered seed when mixed with mung seed at 100 ppm is effective against storage pests for 100 days. Seed extract is effective for the control of houseflies. Crushed leaves have insecticidal properties. Roots are reported to have purgative properties.

2.16.1. Insect Pests

2.16.1.1. Mealy Bugs

i) Damage: The common species of mealy bugs observed damaging custard apple fruits include *Planococcus citri, P. pacificus, P. lilacinus, Perissopneumon ferox, Ferrisia virgata* and *Maconellicoccus hirsutus* (Mani and Krishnamoorthy, 1989). Among these *M. hirsutus* and *F. virgata* are the two major species observed infesting 60 and 40% fruits, respectively. They infest leaves, shoots and fruits and suck the sap there from. The affected plant parts get deformed and remain small in size. *P. lilacinus* feeds on roots as well as leaves. The infested roots develop spongy tissue. Honey dew excreted by mealy bugs encourages the growth of sooty mold on leaves and fruits (Fig. 2.42) and also attracts black ants which help in the spread of these mealy bugs. The infestation starts in May and continues up to November.

ii) Management: The affected shoots and fruits should be removed and destroyed. Aldrin dust (5%) should be sprinkled around the base of tree trunk to check the activity of ants. Spraying with 0.04% diazinon or 0.05% monocrotophos or 0.05% chloropyriphos at 10 days interval is effective for the control of mealy bugs. Spraying should be stopped 15 days before harvest of fruits.

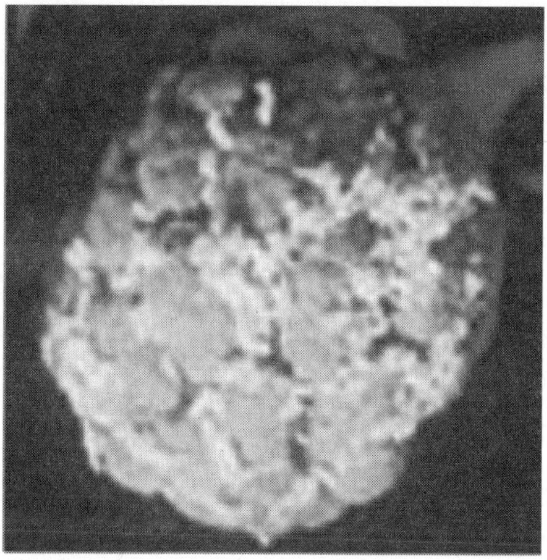

Fig. 2.42. Mealy bug infestation on custard apple fruit (Courtesy: K. Srinivasan, TAFE, Chennai).

Encyrtid parasitoids, *Anagyrus dactylopii* and *Aenasius advena* were observed parasitizing *M. hirsutus* and *F. virgata*. Besides, five species of predators, namely, *Spalgis epius, Scymnus coccivora, Cryptolaemus montrouzieri, Triommata coccidivora* and *Cacoxenus perspicaux* were observed predating on mealy bugs. These natural enemies partially check the population of mealy bugs on custard apple. Release of *C. montrouzieri* helps in clearing the mealy bugs quickly on the fruits (Mani and Krishnamoorthy, 1989b).

2.16.1.2. Scale Insects, *Aonidiella orientalis, Parasaissetia nigra*

i) Management : *Aphytis* sp. is known to attack the scale causing up to 30% parasitism on *Aonidiella orientalis*. A pteromalid, *Scutellista cyanea* causes up to 42% parasitism on *Parasaissetia nigra* (Mani and Krishnamoorthy, 1999e). Release of *Chilocorus nigrita* also helps in bringing down the scale population.

2.16.1.3. Fruitfly, *Dacus zonatus*

i) Damage: The peach fruitfly, *D. zonatus* has been reported attacking custard apple fruits. The males are active throughout the year except in winter months when pupae hibernate in the soil or within fallen infested fruits. The female fruitflies oviposit in the epicarp. The infested fruits rot due to the feeding of maggots and exhibit brown patches externally.

ii) Management: To prevent the attack of fruitflies, fruit should be harvested before ripening. All fallen fruits should be collected and destroyed. Bait spray consisting of 0.1% carbaryl or 0.05% fenthion or 20 ppm decamethrin + yeast hydrolysate at 1 g/litre + sugar at 2 g/litre is recommended. The spray may be repeated twice at 15 days interval. Spraying should be done after harvesting the mature fruits and next harvest should be done 10-15 days after the spray.

2.16.1.4. Fruit Borer, *Anonaepestis bengalella*

i) Damage: The fruit borer is another sporadic pest of custard apple in South India. The caterpillars bore inside the fruits, making irregular tunnels and damaging mesocarp. The growth of affected fruits is arrested and these ultimately drop. The main symptom of the attack is the presence of excreta of the caterpillars near the entry holes on the affected fruits.

ii) Management: Spraying with 0.1% carbaryl or 0.05% phosphamidon or 0.05% endosulfan is recommended for the control of fruit borer. Spraying should coincide with egg laying period when the fruits are immature.

2.17. PHALSA

Phalsa, *Grewia subinaequalis* (Syn. *Grewia asiatica*) belonging to Family Tiliaceae is of Indian origin. It can be grown in all parts of north India except at higher elevations. Its plantation is mainly confined to the states of Haryana, Punjab, Gujarat, Rajasthan, Uttar Pradesh and Maharashtra. It can come up very well as filler crop. The tiny fruits have to be picked from bush several times during the fruiting season. It comes to the market during April-May.

Phalsa fruits are used as table fruit and in preparation of squash and juice. Jam and pickle can also be prepared from phalsa fruits. Ripe fruits are sub acidic and good source of Vitamins A and C.

2.17.1. Insect Pests

2.17.1.1. Aphids, *Aphis craccivora*

i) Damage: *Aphis craccivora* appears on the sprouts after pruning.

ii) Management: The predators, *C. sexmaculata* and *Scymnus* sp. reduce the population of phalsa aphids (Mani and Krishnamoorthy, 1996c).

2.17.1.2. Mealy Bugs, *Maconellicoccus hirsutus, Drosicha mangiferae*

i) Damage: The pink mealy bug (*M. hirsutus*) occurs on the leaves, flowers and fruits. *D. mangiferae* causes severe damage to fruits.

ii) Management: The pest can be controlled by spraying with 0.04% diazinon or monocrotophos.

The coccinellid, *C. montrouzieri* and the lycaenid, *S. epius* help to suppress the mealy bugs on phalsa (Mani and Krishnamoorthy, 1996c).

2.17.1.3. Bark Eating Caterpillar, *Indarbela tetraonis*

i) Damage: It is a polyphagous pest which damages the plant by making tunnels in the main branches or trunk.

ii) Management: Injecting petrol or kerosene in the holes and plugging the mouth with mud controls the pest.

2.17.1.4. Leaf Eating Caterpillar, *Euproctis fraternae*

i) Damage: In case of severe infestation, the entire tree may be defoliated.

ii) Management: Spraying with 0.1% carbaryl keep the pest under control.

2.18. AMLA

Amla (*Emblica officinalis*) is an important fruit crop owing to its hardy nature, suitability to various waste lands, high productivity/unit area (15-20 tonnes/ha), nutritive and therapeutic value. It can be cultivated in salt-affected and ravenous areas and semi-arid regions of Karnataka. There has been a steady increase in area and production of amla as a result of identification and development of suitable varieties [Kanchan (NA 4), NA 6 and NA 7] and development of improved production technologies.

A budded/grafted amla tree starts bearing third year onwards after planting, attain full bearing within 10-12 years and continue to bear for 60-75 years under well managed conditions. An amla tree may bear 100-300 kg fruits/tree, giving 15-20 MT/ha.

Amla being a deep-rooted, deciduous tree with sparse foliage, is an ideal plant amicable for 2- and 3-tier cropping system. Fruits, vegetables, flowers and a few medicinal and aromatic plants are well suited for intercropping in amla orchards. Some models are: amla + ber, amla + guava, amla + ber + phalsa, amla + dhaincha + wheat/barley, amla + dhaincha + onion/garlic/fenugreek/brinjal and amla + dhaincha + German chamomile.

Amla fruits are rich in Vitamin C (ascorbic acid content-500-625 mg/100g pulp). Amla fruits are highly valued among indigenous medicines. It is acrid, refrigerant, diuretic and laxative. Dried fruits have been reported to be useful in hemorrhages, diarrhea, dysentery, anemia, jaundice, dyspepsia and cough. 'Trifla' and 'Chavanprash' are well known indigenous medicines in Ayurvedic system prepared from amla.

2.18.1. Insect Pests

2.18.1.1. Bark Eating Caterpillar, *Indarbela* spp.

i) Damage: The bark eating caterpillar causes significant damage to amla trees requiring control measures. At Pantnagar, up to 80% plants were affected.

ii) Management: Cleaning of holes, spraying with 0.1% gamma BHC or aldrin and fumigation of the borer holes gives effective control of the pest.

2.18.1.2. Shoot Gall Insect, *Betonsa stylophora*

i) Damage: The apical gall maker is one of the major pests of amla. On almost all varieties, galls are caused by caterpillars. A blackish caterpillar inhabits the galls on growing shoots. The young caterpillars bore into the shoot and reaches the pith during August-September where it feeds and causes gall formation (Fig. 2.43). The growth of heavily affected branches may be checked.

Fig. 2.43. Shoot gall insect on stem of amla (Courtesy: K. Srinivasan, TAFE, Chennai).

ii) Management: Cutting and destroying the twigs having galls helps in reducing the incidence of the pest.

2.18.1.3. Mealy Bug, *Ferrisia virgata*

i) Damage: Innumerable nymphs and adults are seen congregating on shoot, twigs and tender leaves during summer (Fig. 2.44). As a result of desapping, there is loss of vitality. In case of severe infestation, the growth of tree and its fruiting capacity are adversely affected.

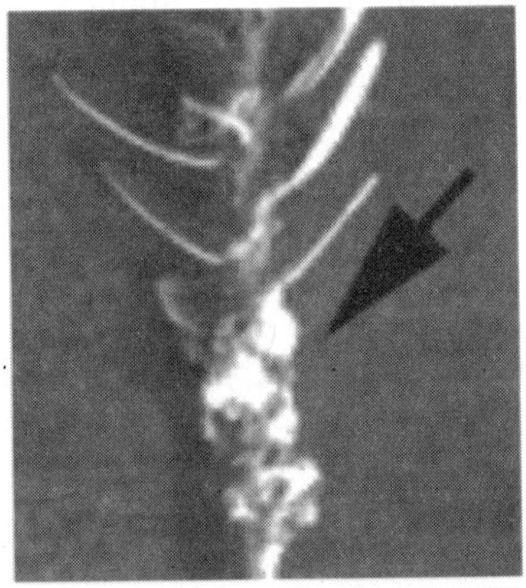

Fig. 2.44. Mealy bug on stem of amla (Courtesy: K. Srinivasan, TAFE, Chennai).

The bugs also exude honey dew which falls on dorsal surface of lower leaves, stems and even fruits on which sooty mold grows covering the affected areas with a superficial black coating. Damage is caused by nymphs and wingless females which are oval, flattened and have body covered with a white mealy powder. Adult female is wrinkled and opterous covered with waxy secretion.

In the month of September-October, the female lays 100-300 eggs in soil about 50-150 mm deep. The eggs over winter and hatch in spring (April-May). Nymphal period lasts from 40 to 62 days. There is only one generation in a year.

ii) Management

- Frequent raking of soil under the tree during winter helps in destroying the eggs.
- Soil treatment with neem cake at 300-500 g/tree reduces the population of newly hatched nymphs.
- Tying of alkathene sheet (400 gauge) around the tree trunk prevents the nymphs from ascending the trees.
- In case of severe infestation, spraying with monocrotophos helps in reducing the pest population.

2.19. FIG

The fruits are valued for their laxative property and are rich in protein digestive enzymes. It is also effective in the treatment of piles, constipation, asthma, sexual weakness, corns and chronic cough. It also removes physical and mental exertion and endows the body with renewed vigour and strength. It is an excellent tonic for weak people who suffers cracks in lips, tongue and mouth.

2.19.1. Insect Pests

2.19.1.1. Stem Borer, *Bactrocera rufomaculata*

i) Damage: The stem borer is a major pest of fig in India. The damage is done by the grubs which feed on the portion of bark and inner wood (xylem) by following zig zag path and filling the tunnel with frass consisting of fibrous matter and excreta (Fig. 2.45). The accumulation of frass on the ground on the side of the trunk indicates stem borer damage. The attacked branches of the tree get weakened and fall down if there is strong wind. The fruiting capacity of the tree is greatly impaired.

ii) Biology: The beetles emerge during rainy season (June-August). The female makes a cut in the bark and deposits a dirty white egg in it. A female lays about 100-200 eggs during June-October. The incubation period ranges from 7 to 14 days. The larval stages last 3-6 months. Before pupation, the grubs enter diapause during winter. Pupation takes place in an elliptical chamber within the stem 5 cm below the surface. The life cycle is completed in 4-7 months. There is only one generation in a year.

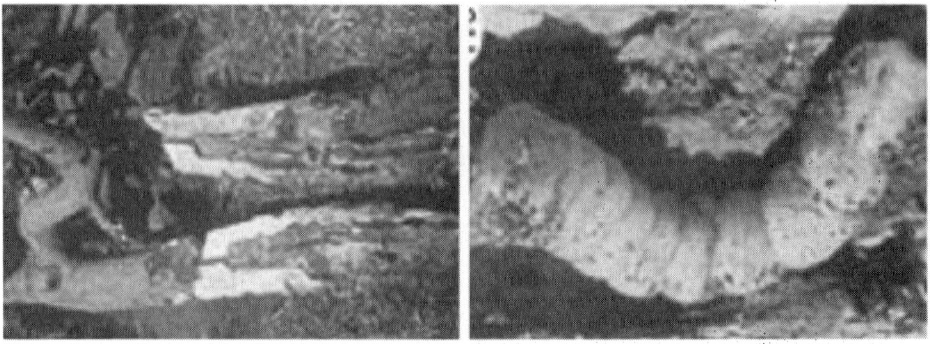

Fig. 2.45. Stem borer on fig (Courtesy: K. Srinivasan, TAFE, Chennai).

iii) Management

- Cut the infested dry twigs and burn them.
- Put kerosene or petrol or dichlorvos soaked cotton wad in the holes and seal them with mud.
- A mixture of ethylene dichloride and carbon tetrachloride (3: 1) may be injected in the hole and sealed.
- The stem of fig tree can be protected from oviposition by spraying with 0.1% chloropyriphos or by application of coal tar painted paper.

2.19.1.2. Cockchafer Beetles

Five species of defoliating beetles, viz., *Adoretus duvauceli, A. horticola, A. lasiopygus, A. verusutus* and *Brahmina coriacea* have been reported damaging foliage of fig trees. All these are polyphagous pests.

i) Damage: Eggs are laid in soil and on hatching; the grubs feed on roots of plants. Pupation takes place deep in the soil. During winter, pupae enter into diapause. After emergence, the adult beetles continue to live in the soil till rain starts. Adults of these beetles are nocturnal in habit and damage the foliage from dusk to dawn. They defoliate fig trees in June-July. These have only one generation in a year. *Holotrichia consanguinea* causes defoliation of fig plants.

ii) Management

- Collection of adults at the time of their emergence is advised.
- Dusting with 10% BHC or spraying 0.2% carbaryl was recommended against adults of *Adoretus* spp. (Butani, 1975).
- Spraying of 0.05% fenitrothion and 0.1% carbaryl was reported to give best control of *H. consanguinea* (Bindra and Singh, 1971).
- The use of bioagents, *Beauveria brongniartii* and *Metarrhizium anisopliae*, have been recommended against the grubs.

2.19.1.3. Gall Midges

i) Damage: The midge, *Anjeerodiplosis peshawarensis* caused major damage to raw fig fruits, which become elongated, wrinkled and soft. Midge adults appeared in May and damaged up to 25% fruits. Oviposition is done by female midges in raw and tender fruits. On hatching, the maggots feed on the pulp. The infested fruits wither and drop prematurely. Larval period was one and a half month in May and June. Pupation occurred in soil for 4-12 days.

Udumbarie nainiensis damages the flower buds by ovipositing in them. On hatching, the maggots feed on the interior contents of inflorescences and cause discolouration, wrinkling and small protuberances. One female of the midge lays 200-250 eggs in cavities on flower buds during March to May and again in August to October.

ii) Management

- Infested flowers and fruits should be destroyed.
- The damage can be checked by ploughing or digging the soil below the trees (to expose and kill the pupae) and mixing 5% aldrin dust in soil.
- Spraying 0.05% phosphomidon or monocrotophos or fenitrothion at flower bud stage and when the fruits are of pea size was also found effective.

2.19.1.4. Leaf Rollers, *Phycodes minor, P. radiata*

i) Damage: Leaf rollers cause significant damage to leaves which may fall. The pest is active during spring. Incubation, larval and pupal periods for *P. minor* were 5-6, 4-5 and 8-11 days, respectively (Simwat and Sidhu, 1974). Moths of both the species may be seen during March-May sucking nectar from the blossoms. Eggs are laid on tender leaves. On hatching, the larvae feed on epidermis of leaves during spring. Later on, they fold the leaves and feed within. Affected leaves ultimately fall. Pupae hibernate in cracks and crevices of stem bark.

ii) Management

- Infested rolled leaves should be collected and destroyed.
- Spraying with 0.05% monocrotophos, carbaryl, fenitrothion or endosulfan was recommended.

2.20. JAMUN

The jamun, *Syzygium cuminii* (*Eugenia jambolana*) belonging to Family Myrtaceae is considered to be native to India or East Indies. Fully ripe fruits are eaten fresh and can be processed into beverages, jelly, jam, squash, wine, vinegar and pickles. It is a good source of iron. The powdered seed is useful in the treatment of diabetes.

It is widely grown in the larger part of India from Indo-Gangetic plains in the north to Tamil Nadu in the south. It is one of the most hardy fruit crop and can be easily grown in neglected and marshy areas where other fruit plants can not be grown successfully.

2.20.1. Insect Pests

2.20.1.1. Fruit Fly

i) Damage: Affected fruits gets wormy due to infestation of fruitfly.

ii) Management: Collection and destruction of infested fruits. Digging the soil around the tree trunk so that the maggots and pupae hibernating in the soil are destroyed.

2.20.1.2. Leaf Eating Caterpillar, *Carea subtilis*

i) Damage: Caterpillars attack the leaves and the tree become defoliated.

ii) Management: The pest can be controlled by spraying dimethoate or malathion both at 1 ml/l of water.

2.21. DATE PALM

2.21.1. Insect Pests

2.21.1.1. Black Palm Beetle, *Oryctes rhinoceros*

i) Damage: This is a serious pest and may cause considerable damage to date palm. Adult beetle is the only destructive stage which feeds on the crowns by boring into unopened tender fronds. Young trees may die due to the attack while growth and yield of older palms suffer. Damage is maximum during the monsoon.

White oval eggs (3-4 mm long) are laid in heaps of decaying organic matter, cow dung, manure pits, etc. and the grub feeds there. Pupation also occurs in soil. Incubation, grub, pupal and pre-ovipositional periods were 8-14, 47-191, 14-29 and 5-26 days, respectively.

ii) Management: Control can be achieved by keeping the orchard clean. Compost heaps should be treated with 10% BHC dust. Treating the crowns with a mixture of saw dust and lindane (9 : 1) is also recommended. Fungus *Metarrhizium anisopliae* and parasitoid *Sarcophaga fuscicunda* were reported on its larvae.

2.21.1.2. Red Palm Weevil, *Rhynchoforus ferrugineus*

i) Damage: The red palm weevil is a major pest of date palm. The pest breeds throughout the year and the grub is the chief damaging stage which bores into the tissues of the stem mostly from the top. Growing tips may dry and the trees may bend.

Eggs are laid singly in soft tissues at the base of leaf sheaths or in the cuts or wounds. Incubation, grub and pupal periods were 2-5, 24-61 and 18-34 days, respectively.

ii) Management: Reduction in damage is possible by prompt removal of the infested trees. Application of lindane + saw dust (1: 1) in the leaf axils three times in a year would kill weevils. All cuts or wounds on the trunk should be treated with lindane or gamma BHC paste at collar.

2.21.1.3. Scale Insect, *Parlatoria blanchardi*

i) Damage: Scale insect is reported to be serious on date palms, especially on the young palms which have been grown under shade. *P blanchardi* lays rose-coloured eggs. Young crawlers are also rose-coloured which turn light brown. Female nymphs become adults in 30-40 days, while winged males pass through prepupal and pupal stages.

ii) Management: Removal and burning of the infested fronds is effective. Monocrotophos at 0.05% was most effective followed by dimethoate and phosphamidon.

2.22. TAMARIND

The major breakthrough in recent years has been the identification of tamarind types with less acidic pulp, commonly referred as sweet tamarinds. In Thailand, more than 50 sweet tamarind cultivars are grown, while in Philippines, 8 selected cultivars are popular among the farmers. Presently, sweet tamarind plantations are interplanted with annual crops in Thailand. Small scale plantations of sweet tamarind can be raised for the fresh fruit market.

India, the largest producers and exporters of tamarind, continue to grow sour types. Tamarind is most suitable for growing in regions that have extended spells of dry weather. Those that farm under adverse climatic conditions will benefit most due to the draught resistant nature of this species. Due to ease of cultivation, particularly in marginal areas which are often unsuitable for other crops and the potential for domestic use and export, tamarind could be considered a "Tree for the future". Tamarind attains maximum production during tenth year which yields 4 tonnes/ha (40 kg/tree) with a net income of Rs. 38,000 annually.

The leaves are an important source of food and herbal medicine and the edible pulp of ripe fruits is used as flavouring agent in cooking, soups, jams, chutneys, sauces, juices, etc. The fruit pulp of tamarind is the richest natural source of tartaric acid (8 to 18%) and is the chief acidulant used in the preparation of foods. The major industrial product of tamarind is the seed kernel powder which is an important material used in sizing of textiles, paper and jute. Tamarind seed protein contains highly favourable amino acids and could complement methionine and cystine in the diet, to alleviate protein malnutrition which is widely prevalent in developing nations.

Table 2.26. The major tamarind growing countries and its production (Gunasena & Hughes, 2000)

Countries	Production (in MT)
India	2,50,000 (1964)
Thailand	1,40,000 (1995)
Coata Rica	221 (1997)
Mexico	37
Puerto Rico	23 (1977)

During 1993, exports reached 11,145 tonnes valued at Rs. 10.8 million (US$ 2,40,300), and were in the form of fresh fruit, pulp, seed powder and paste. The importing countries are mainly Pakistan followed by UAE, Japan and Yemen. Over 75% of the imports are of sweet type, having a premium demand over sour types.

2.22.1. Insect Pests

The major pests which attack tamarind include shot hole borers, toy beetles, leaf feeding caterpillars, bag worms, mealy bugs and scale insects (Coronel, 1991). Among the worst are the sap suckers, which affect the tender shoots and fresh foliage. The most destructive among these are white flies, thrips, coccids and aphids.

2.22.1.1. Scale Insects

i) Damage: Some of the most serious pests in India are hard scale insects (*Aonidiella orientalis, Aspidiotus destructor* and *Saisetia oleae*) (Timyan, 1996). The oriental yellow scale (*Aonidiella orientalis*) (Hemiptera: Diaphididae) is a polyphagous pest which attacks tamarind. The eggs are laid under a protected covering such as cracks in the bark of the trees. After hatching, the nymphs crawl in search of suitable succulent parts of the plant.

ii) Management: The scale insects can be controlled by removing the affected parts in the initial stages. When the infestation is serious, chemical spraying with 0.1% diazinon or carbosulfan or 0.3% dimethoate provide effective control (Butani, 1979).

2.22.1.2. Mealy Bugs

i) Damage: Mealy bugs (*Nipaecoccus viridis* and *Planococcus lilacinus*) (Hemiptera: Pseudococcidae) also attack tamarind trees. Nymphs and the females suck the sap on the ventral surface of the leaflets, the base of leaf petioles, tender shoots and even the mature shoots. *P. lilacinus* also attacks the roots. The leaflets become chlorotic and defoliation is common. Immature fruit fall is sometimes observed.

ii) Management: Mealy bugs can be controlled by mechanical removal of infected shoots on the small scale. In serious infestations, chemicals such as 0.1% diazinon or carbosulfan can be sprayed. *P. lilacinus* populations living in the soil can be displaced by raking or be killed using 5% aldrin at 500 g/tree (Butani, 1979).

2.22.1.3. Thrips

i) Damage: Some thrips are reported to attack tamarind flowers. *Scirtothrips dorsalis* (Thysonoptera: Thripidae) is a polyphagous thrip and the adults live for 10 or 15 days and complete 25 overlapping generations per year. *Ramaswamiahiella subnudula* is another small yellowish thrip which attacks tamarind.

ii) Management: Thrips can be controlled by spraying with dimethoate at 20-30 ml/10 litres of water or fenthion at 10-15 ml/10 litres of water.

2.22.1.4. Bruchid Beetle

i) Damage: The bruchid beetle [*Pachymerus (Coryoborus) gonogra* (Coleoptera: Bruchidae)] is the most serious pest of tamarind in India and Pakistan (Chaturvedi, 1985). The eggs of this insect are laid on the pods or on seeds and the contents are eaten by the larvae. The infestation frequently continues in storage.

Larvae of tamarind weevil (*Caryedon serratus*) are reported as a serious pest which attacks the pods, seeds and fruits in the field and fruits and roasted seeds in storage in India (Butani, 1979), Colombia and Puerto Rico (Timyan, 1996).

The beetle, *Calandra linearis* also attacks fruits and is reported as a common pest in Florida and Hawaii (Anon, 1979).

ii) Management: Neem oil at 5-20 ml/kg of seed, has been found effective to control bruchid beetles attacking tamarind pods and seeds. The use of neem oil has pronounced adulticidal and ovicidal effects, which are maintained over six months, without any adverse effect on seed viability. This low cost bio-pesticide is very cost effective and efficient in protecting seeds of tamarind from bruchid damage. This method is widely used in the Sahel (Cardet *et al.*, 1998).

The bruchid beetles can also be controlled by fumigation with phosphine from aluminium phosphide tablets (1.5 g/tonne of pods/seeds) or ethylene bromide or carbon tetrachloride or acrylnitrile (Mital and Khanna, 1978).

2.22.1.5. Lac Insect, *Kerria lacca* (Hemiptera: Lacciferidae)

i) Damage: The lac insect is widely distributed polyphagous pest in India which attacks tamarind stems, twigs and leaves.

ii) Management: Removing the affected parts in early stages, as commonly practiced by farmers can control the lac insect. In serious infestations, spraying 1% diazinon or carbosulfan is effective.

2.22.1.6. *Achaea janata* (Lepidoptera: Noctuidae)

i) Damage: Larvae of *Achaea janata* are reported to cause heavy losses when epidemics of the moth infest flowers in tamarind plantations in Tamil Nadu (Ahmed, 1990). In such epidemics, over 400-650 larvae can be found on a single tree causing serious growth reduction. It is also considered as a serious pest causing damage to flowers and defoliation in the Salem district of Tamil Nadu, India.

ii) Management: Spraying 0.5% permethrin can control this pest.

2.22.1.7. Aphids, *Toxoptera aurantii* (Hemiptera: Aphididae)

i) Damage: Aphids are serious pests that attack tamarind. It is a major pest which sucks sap of tender shoots and leaflets causing them to become distorted and covered with molds growing on the secreted honeydew.

ii) Management: Aphids can be controlled by spraying endosulfan at 1 ml/l of water or dimethoate at 2-4 ml/litre of water, repeated every two weeks until the pest is no longer observed.

2.22.2. Vertebrate Pests

Monkeys are reported to devour the fruits greedily while animal browsing causes damage to seedlings unless effectively protected in the early stages (Hocking, 1993).

Chapter 3

VEGETABLE CROPS

China is the largest producer of vegetables contributing 47.3% of the total world production (Table 3.1). Vegetables occupy an area of 22.518 million ha in China with a production of 434.332 million MT. Productivity of vegetable crops is highest (33.8 MT/ha) in Spain.

India is the second largest producer of vegetables next only to China contributing 11.9% of the total world production. Vegetables occupy an area of 7.164 million ha with a production of 109.050 million tonnes (Bijay Kumar, 2007) (Tables 3.1 and 3.2). Vegetable crops contribute 58.93% share to the total production of horticultural crops. India produces 14.7% of the world onions. In vegetables, India occupies the first position in peas; second in cabbage, cauliflower, onion, brinjal; third position in potato and fourth in tomato. The per capita availability of vegetables has increased from 76 g in 1951 to 256 g in 2001 as against 280 g/day recommended by ICMR for a balanced diet. Increase in production is mainly due to increase in productivity. Productivity is still lower in India (15 t/ha) than the world average productivity (15.7 t/ha). There is scope for increasing the productivity of vegetables in India further. The vegetable production target for 2020 has been fixed at 150 million tonnes.

Table 3.1. Major vegetable producing countries of the world

Country	Area (in '000 ha)	Production (in '000 MT)	Productivity (in MT/ha)	% share of world production
China	22518	434332	19.3	47.3
India	7164	109050	15.2	11.9
USA	1307	36845	28.2	4.0
Turkey	1060	26290	24.8	2.9
Egypt	600	16165	26.9	1.8
Italy	593	15994	27.0	1.7
Iran	631	15760	25.0	1.7
Russia	928	15404	16.6	1.7

Spain	395	13356	33.8	1.5
Japan	424	11745	27.7	1.3
Mexico	687	11241	16.4	1.2
Korea	329	10965	33.3	1.2
Others	15984	200416	12.5	22.0
Total	52620	917563	15.7	---

Source: FAO (except Indian data) (Indian data source: Indian Horticulture Database 2006)

Table 3.2. Area, production and productivity of vegetable crops in India (2005-06)

Vegetable crops	Area ('000 ha)	Production ('000 MT)	Productivity (MT/ha)	% share of India production
Brinjal	553.9	9136.3	16.5	8.4
Cabbage	267.4	6921.6	22.1	5.4
Cauliflower	291.8	5260.1	18.0	4.8
Okra	376.1	3684.0	9.8	3.4
Onion	695.1	9248.4	13.1	8.5
Peas	314.0	2298.7	7.3	2.1
Potato	1550.2	29093.7	18.8	26.7
Sweet potato	129.4	1148.6	8.9	1.1
Tapioca	252.7	7924.9	31.4	7.3
Tomato	534.5	9361.8	17.5	8.6
Others	2199.5	25971.4	11.8	23.8
Total	7163.9	109049.5	15.2	

(Source: Indian Horticulture Database 2006)

Potato, tomato, brinjal, chilli, cabbage, cauliflower, onion, peas, gourds, okra and melons are now grown on commercial scale with increasing demand, better output, change in production system and development of suitable cvs. These crops have high potential for foreign exchange earning. Today the emphasis on vegetable breeding is on value addition, disease resistance and better quality.

The important vegetables exported include onions-7.1%, cabbage-4%, cauliflower-4%, okra-3%, peas-3%. Traditional vegetables like onion, potato, okra, bottle gourd and chilli are exported to Malaysia, Singapore, Gulf, Sri Lanka, Bangladesh, Pakistan and Nepal. Non-traditional vegetables like asparagus, cabbage, capsicum, sweet corn, baby corn, green peas, French bean, cucumber, gherkin, cherry tomato, Brussels sprout, broccoli, processed and frozen French bean and lima bean are exported to European countries, Australia, Gulf and S.E. Asian countries. French bean, lima bean, sweet corn and capsicum are exported to Gulf. Some of the new products introduced in export market are tomato paste in bulk aseptic packs, processed button mushrooms, freeze dried and instant frozen vegetables. Dehydrated onion and tomato products have great potential as export commodities.

Pest Scenario: Successful cultivation of vegetables is hampered due to the incidence of several insect pests. Given the sub-tropical climate, intensive cultivation of high yielding varieties/hybrids under high fertigation, off-season production, lack of adequate knowledge on plant nutrition and protection, economic constraints to maximize qualitative returns and inability to afford risk, insect pest management in vegetables in South Asia is a very complex issue. Diamondback moth (*Plutella xylostella*) in cabbage and cauliflower; fruit borer (*Helicoverpa armigera*) in tomato; shoot and fruit borer (*Leucinodes orbonalis*) in brinjal; serpentine leaf miner (*Liriomyza trifolii*), thrips (*Thrips tabaci*) and red spider mites (*Tetranychus urticae*) in many crops have become difficult to control. Bean fly (*Ophiomyia phaseoli*) and pod borers in leguminous vegetables, and fruit fly in cucurbits are also becoming important. Insect control in vegetables is usually through a mixture/cocktail of insecticides that either can be under or over dosed. Pesticides have offered a simple solution to farmers as they are cheap, readily available, easy to apply and non-specific. Thus, it is not surprising that with 3% of the area; nearly 17% of pesticide uptake is on vegetables in India.

However, lack of adequate control, insecticide resistance, resurgence of secondary pests and out-break of new pests have necessitated many growers to seek scientific information on management of insect pests and re-evaluate the role of pesticides in vegetable production. Many IPM technologies using trap crop, botanicals and bioagents have been developed in some vegetables and a few are validated in farmer's field. Ecologically sound and economically viable IPM is possible only through dissemination of scientific and up to date information to farmers either by the extension agencies or those who train them.

3.1. POTATO

Potato (*Solanum tuberosum*) is one of the important vegetable crops grown in India. It is being cultivated in 1.550 million hectares producing 29.094 million tonnes of potatoes with an average yield of 18.8 tonnes per hectare (Bijay Kumar, 2007) (Tables 3.3 and 3.4). It thrives best in cool climate. Therefore it is a summer crop in the hills and a winter crop in the plains.

Table 3.3. Major potato producing countries of the world

Country	Area (in '000 ha)	Production (in '000 MT)	Productivity (in MT/ha)	% Share of world production
China	4884	73460	15.0	22.4
Russia	3071	37280	12.1	11.4
India	1550	29094	18.8	8.9
Ukraine	1516	19462	12.8	5.9

USA	439	19091	43.5	5.8
Germany	277	11624	42.0	3.6
Poland	588	10369	17.6	3.2
Belarus	462	8185	17.7	2.5
Netherlands	156	6777	43.4	2.1
France	156	6605	42.2	2.0
Others	6102	105368	17.3	32.2
Total	19200	327314	---	---

Source: FAO (except Indian data) (Indian data source: Indian Horticulture Database 2006)

Table 3.4. Area, production and productivity of leading potato growing states in India (2005-06)

State	Area (in'000 ha)	Production (in '000 MT)	Productivity (MT/ha)
Uttar Pradesh	443.2	9986.8	22.5
West Bengal	320.6	7106.6	22.2
Bihar	308.9	5702.5	18.5
Punjab	75.0	1223.5	16.3
Gujarat	41.0	1088.5	26.6
Madhya Pradesh	53.3	799.7	15.0
Assam	90.1	629.1	7.0
Uttarakhand	22.2	450.6	20.3
Haryana	19.0	374.4	19.7
Karnataka	57.5	368.4	6.4
Others	119.5	1363.6	11.4
Total	1550.2	29093.7	18.8

(Source: Indian Horticulture Database 2006)

3.1.1. Insect Pests

3.1.1.1. Aphid, *Myzus persicae*

i) Damage: Wingless green peach aphids tend to be pear shaped. They are straw-coloured or light-green (Fig. 3.1). Hind cornicles slant so they converge. A key characteristic is tubercles at the base of the antennae. Potato aphids are more elongate and larger than the green peach aphid. They often are bright green, but pinkish forms occur. Both potato and green peach aphids produce winged forms. Winged aphids are most important in spread of viruses to plants since they move readily within and between fields. Wingless green peach aphids tend to occur on lower, older potato leaves. Potato aphids more commonly colonize the upper part of the potato plant. During the summer, aphids reproduce asexually (no males) and give live birth to young. Under optimal conditions, a generation may be completed in 10 days.

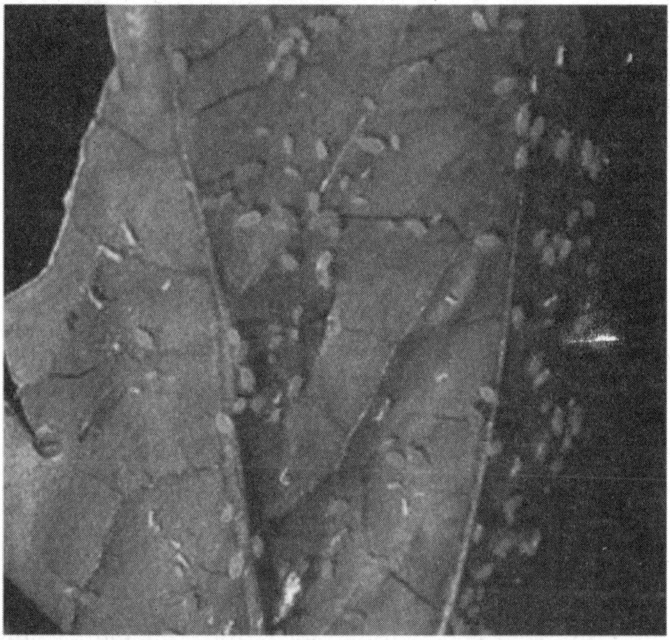

Fig 3.1. Potato leaf infested with aphids

ii) Management: Cropping in aphid-free periods or when their population is low by adjusting the planting dates in the Indo-Gangetic Plains, where about 90% of seed crop is grown i.e. up to 15th October in North Western Plains, up to 5th November in North Eastern Plains. Remove all weed hosts susceptible to viruses and for aphids (especially those having yellow flowers) from within and around the vicinity of seed potato plots. The haulm cutting of seed crop should be done as soon as the aphid number crosses the critical level i.e. 20 aphids/100 leaves. Care must be taken to prevent any regrowth of stumps before harvest.

Monitor flights of winged aphids by using yellow traps (yellow dish pans with water and a bit of bleach). Sweep fields at least weekly to monitor aphids. Make sure to get the sweep net deep in the canopy since a lot of aphids are found on lower leaves. Pull up yellowish plants to check for aphid colonies. Apply pesticides only when necessary. Use of some insecticides can exacerbate aphid problems, because the control has more effect on natural enemies than on aphids. Insecticides can only provide limited control of aphid transmitted virus diseases. The green peach aphid can be highly resistant to insecticides. Therefore, choose insecticides carefully for effective control. Protect natural predator populations that are important in aphid management. Trap crops around the edges of fields can help aphid control. Direct damage from aphid feeding in a strictly commercial field needs high populations. If aphid levels are approaching 50 aphids per plant, treat the field.

Several predators and parasites help control aphids. Among the more important insects are lady beetles, damsel bugs, green lacewings, minute pirate, and parasitic wasps.

3.1.1.2. Cutworm, *Agrotis ipsilon, A. flammatra*

i) Damage: Cutworms feed at night. During the day they can be found under clods of soil or in cracks in the ground near injured plants (Fig. 3.2). Cutworms have a smooth appearance, three pairs of legs and five pairs of prolegs. Damage early in the season includes stems cut off at or below ground level. Later, chewed foliage is the most common symptom. Tubers that are exposed on the surface or in cracks in the soil can be chewed by some cutworms. Cutworms over winter as immature larvae or eggs in the soil.

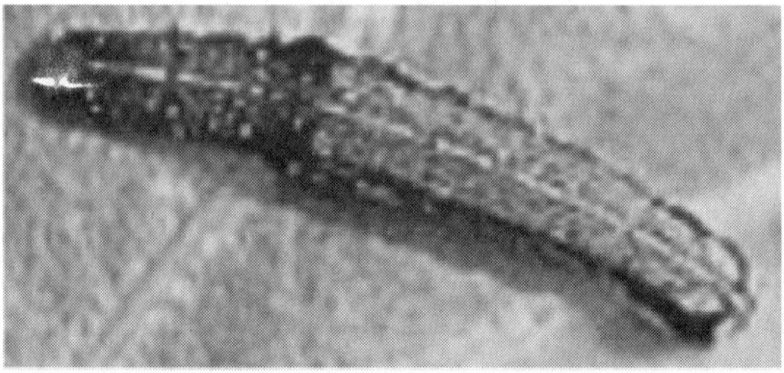

Fig 3.2. Cut worm of potato

ii) Management: Hot weather ploughing in plains and autumn ploughing in hills reduce the population of immature stages. A number of birds such as crow, mynah, starling, etc. feed on insects that get exposed upon ploughing. Planting of potato on 25th November recorded least foliage damage (1.25%).

Some of the natural enemies which play important role in the management of cut worms are *Broscus punctatus, Liogryllus bimaculatus, Anpopus hypsipylae, Ichneumon* sp. and *Turanogonia chinensis.*

Bacillus thurigiensis is quite effective against cut worms. The percentage foliage damage in *Bacillus thurigiensis* treated plots ranged between 0.0 to 2.0% compared to 1.6 to 6.4% in control. Similarly, tuber damage recorded at harvest was also less in *B. thurigiensis* treated plots. Further, significantly higher yield of potatoes was recorded from *B. thurigiensis* treated plots.

3.1.1.3. White Grub, *Holotrichia consanguinea*

i) Damage: White grubs can severely damage seed pieces and tubers (Fig. 3.3). These insects feed on grass roots and should be considered as a serious threat when potatoes are to be grown in ground immediately following sod.

ii) Management: Mechanical control by collection of beetles should be organized on community basis and started in late May or early June after the first shower by vigorous shaking of host trees/shrubs and then killed. Repeated ploughing before monsoon (April-May) exposes the grubs and pupae for predation by natural enemies (birds) or may be hand collected and destroyed. Flooding of the fields for 7-10 days, wherever possible may be undertaken. Adopt suitable crop rotations. Apply only well rotten FYM/ compost in the fields.

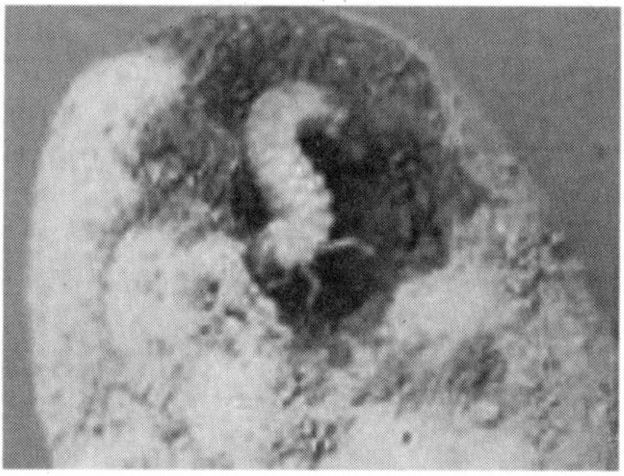

Fig 3.3. White grub damage to potato tuber

A majority of beetles are attracted to the light source. Hence, light traps may be operated for mass collection.

3.1.1.4. Tuber Moth, *Phthorimaea operculella*

i) Damage: Tuber moth is one of the most serious pests of potato in the field as well as in storage. On an average 105 eggs are laid by each female at an average temperature of 28°C with an incubation period of 4.93 days. The larval, pupal and adult stages last for 8.87, 5.10 and 24 days, respectively. The pest is observed during December to March in the field and April onwards in stored potatoes.

The tuber worm is a small caterpillar, the larva of an inconspicuous grey and black moth called the potato tuber moth. The larva is usually dull white with a brown head. The tuber worm can have several generations a year in warm areas. Tuber worms also feed on other Solanaceous species like nightshade, tomato and eggplant. On potato tubers, the larva feed just below the surface and sometimes deep into the flesh. The tubers can appear black since the tuber worm tunnels are filled with larval feces (Fig. 3.4). In any case, tuber worm feeding leaves the tuber un-marketable. Varieties that have tubers well protected by soil can tolerate higher populations of tuber worms.

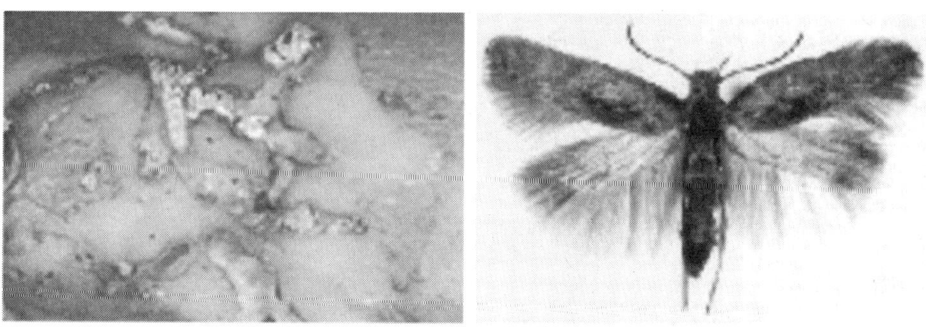

Fig 3.4. Tuber moth damage on potato

ii) Management: The pest is controlled through deep planting, high ridging and cold storage of seed. Larvae failed to enter the potato tubers through wet soil layers. Keep the tubers in storage before sun set covered with 5 cm sand layer to prevent the adult moths from laying eggs on the eyes of the tubers (Patel, 1980). Use 2.5 cm thick layers of dried lantana or eucalyptus leaves below and on top of the potato heaps which checks tuber moth infestation. Planting varieties that set tubers deep in the soil and hilling the plants up well to protect the tubers from tuber worm can help to reduce the damage.

In country stores, use of sex pheromones on water traps at 4 traps/100 m^3 store space are effective for mass trapping of males. Pheromones are available to monitor the tuber moth populations. The trap consists of a dishpan of water with a little soap to break the surface tension of the water. Place a tent over the trap to help catch adult moths and place the pheromone inside to attract them. If you catch 10-20 per day you should treat the field.

Foliar application of carbaryl at 0.15% and chlorodane 5% dust at 20 kg/ha is found effective in controlling the pest under field conditions. Pesticide treatment must occur before vine kill.

Use of biocontrol agents like *Bacillus thuringiensis* and Granulosis Virus are also advocated.

3.1.1.5. Epilachna beetle, *Epilachna vigintioctopunctata*

i) Damage: Grubs scrape the chlorophyll from the leaves leaving only veins (Fig. 3.5).

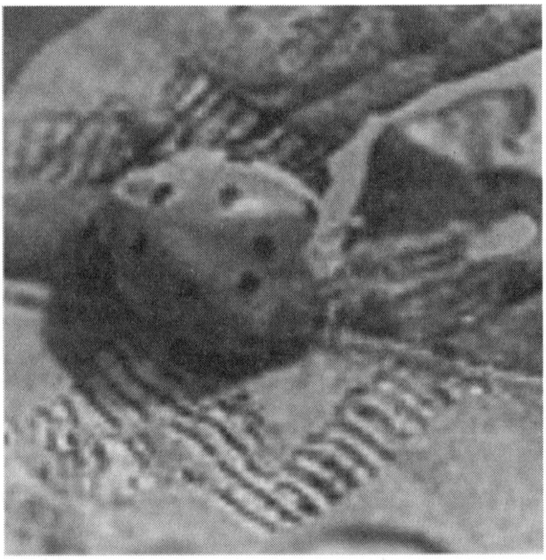

Fig 3.5. Epilachna beetles on potato leaves.

ii) Management : Apply 5% carbaryl dust at 25 kg/ha.

3.2. TOMATO

Tomato (*Lycopersicon esculentum*) is the most important and remunerative vegetable crop in India. It is being cultivated in 0.535 million hectares producing 9.362 million tonnes of tomatoes with an average yield of 17.5 tonnes per hectare (Bijay Kumar, 2007) (Tables 3.5 and 3.6). A rich source of minerals, vitamins and organic acids, tomato fruits provides 3-4% total solids, 15-30 mg/100 g ascorbic acid, 7.5-10.0 mg/100 ml titratable acidity and 20-50 mg/100 g fruit weight of lycopene. Tomato fruits contain 0.4% protein, 2.5% carbohydrates, 1.5 mg iron and 2 mg vitamin C in 100g of fresh weight. Fruits are good for people suffering from constipation, jaundice and indigestion.

Table 3.5. Major tomato producing countries of the world

Country	Area (in '000 ha)	Production (in '000 MT)	Productivity (in MT/ha)	% Share of world production
China	1305	31626	24.2	24.7
USA	167	11043	66.3	8.6
Turkey	260	10050	38.7	7.9
India	535	9362	17.5	7.3
Egypt	195	7600	39.0	5.9
Italy	139	7187	51.8	5.6
Spain	72	4810	66.5	3.8
Iran	139	4781	34.4	3.7
Brazil	61	3453	57.0	2.7
Mexico	119	2800	23.6	2.2
Others	1624	35281	21.7	27.6
Total	4615	127993		

Source: FAO (except Indian data) (Indian data source: Indian Horticulture Database 2006)

Table 3.6. Area, production and productivity of leading tomato growing states in India (2005-06)

State	Area (in'000 ha)	Production (in '000 MT)	Productivity (MT/ha)
Andhra Pradesh	76.5	1453.5	19.0
Orissa	100.4	1332.2	13.3
Karnataka	44.5	1188.1	26.7
Maharashtra	35.0	987.0	28.2
West Bengal	50.0	857.2	17.2
Bihar	46.0	727.2	15.8
Gujarat	29.3	650.0	22.2
Chhattisgarh	29.2	365.8	12.5
Madhya Pradesh	20.4	306.7	15.0
Tamil Nadu	22.0	277.7	12.6
Haryana	17.1	257.3	15.0
Others	64.1	959.1	15.0
Total	534.5	9361.8	17.5

(Source: Indian Horticulture Database 2006)

3.2.1. Insect Pests

3.2.1.1. Fruit borer, *Helicoverpa armigera*

Tomato fruit borer is a polyphagous pest infesting tomato, reducing marketable yield and market value of the crop. This pest has already developed resistance to a number of insecticides in different crops, and is thus difficult to manage. Losses up to 50% in Tamil Nadu (Srinivasan, 1958); 65% in Punjab (Singh and Singh, 1975) and 22 to 38% in Karnataka (Tewari and Krishna Murthy, 1984) have been reported due to this pest attack.

i) Damage: The adult of this pest is stout and medium sized moth and has a dark circular spot in the center on the forewing. They lay small, single, and whitish, round eggs on the trifoliate leaves beneath the top most flower cluster. Eggs hatch in about 3-4 days and the first instar larvae initially feed on the leaves and later migrate to the developing green fruits. Later the larvae bore into the fruits with the posterior end outside the hole (Fig. 3.6). Full-grown caterpillars show characteristic whitish and dark brown longitudinal stripes. Pupation takes place in the soil. Eight or more eggs/30 trifoliate leaves below the top most flower cluster were observed to cause economic loss.

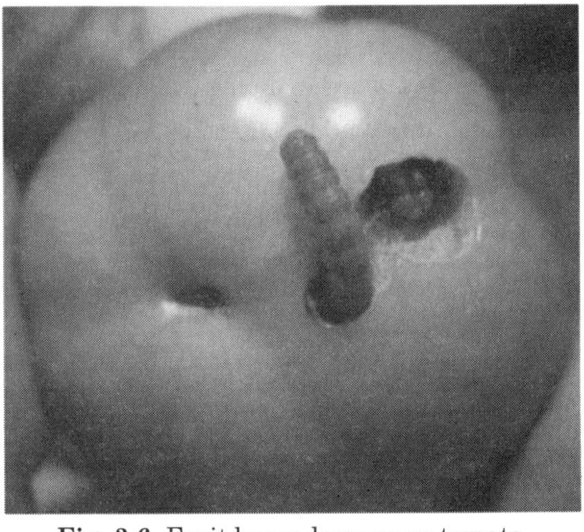

Fig. 3.6. Fruit borer damage on tomato.

ii) Management

a) Cultural Methods: Mechanical collection and destruction of bored fruit at periodic intervals (3-4 times) brings down the borer incidence to less than 2%.

Soil application of neem cake at 250 kg/ha at flowering reduced the borer incidence to 13.2% as compared to 33.2% in control. Foliar sprays of neem soap and pongamia soap both at 1% starting from 25 DAP and repeated four more times at 10 days interval reduced the borer incidence to less than 7% as compared to 33% in unsprayed control plots (Krishna Murthy and Krishna Kumar, 2002) (Table 3.7). The soap sprays were also effective in reducing whitefly, red spider mite and leaf miner incidence.

Table 3.7. Efficacy of neem and pongamia soaps on tomato fruit borer

Treatments	Dose	Mean fruits bored (%)
Neem cake	250 kg/ha	13.21
NSKE	4 %	11.12
Neem oil	1 %	13.24
Pongamia oil	1 %	13.76
Soluneem	600 ppm	7.97
Neem soap	1 %	6.64
Pongamia soap	1 %	6.96
Control	-	33.23

b) Biological Methods: The egg parasitoid, *Trichogramma brasiliensis* gave as high as 90% parasitism when 2.5 lakh/ha were released, while *T. pretiosum* also proved effective at 2.5 lakh/ha and resulted in less than 2% fruit damage with 60-80% parasitism (Krishnamoorthy and Mani, 1996). In Solan, Himachal Pradesh region, *T. chilonis* gave 90-100% parasitism following releases. In Gujarat, fruit borer was checked by releasing *T. chilonis* at 2.5 lakh/ha at weekly interval during Dec-Feb.

Five releases of *T. brasiliensis* at 50,000/ha commencing the first release at the time of flower initiation and repeated at weekly intervals resulted in egg parasitism up to 78.4% as compared to 12.3% in control. Percentage bored fruits in parasitoid released plots was 19% in comparison with 42% in control plots. Releases of *T. brasiliensis* and *T. pretiosum* each at 2.5 lakhs/ha resulted in significant reduction in borer damage to 8.92 and 7.20, respectively as compared to 23.06 and 13.72% in control (Table 3.8).

Table 3.8. Efficacy of *Trichogramma* spp. against *Helicoverpa armigera* on tomato

Trichogramma spp.	Dosage/ha (in lakhs)	Parasitism(%)
T. brasiliensis	5.0	100
	2.5	20-71
	3.0	78.4
T. chilonis	2.5	20-96
	0.63-10.88	76.3
T. exiguum	5.0	100
T. pretiosum	5.0	100
	0.55-10.58	92.4
	2.5	31.3

Larval parasitoids like *Campoletis chloridae*, *Eriborus argentiopilosus* and *Senometopia illota* and the nematode, *Hexamermis* sp. cause 10 to 40% mortality of fruit borer in tomato fields.

Field applications of 5 rounds of *Nomuraea rileyi* at 3.2 x 10^8 conidia/ml along with Triton X -100 (0.01 %) at weekly intervals during evening hours effectively controlled the fruit borer. The treatment reduced the fruit borer damage to 6.76% when compared to control (41.16%) and insecticide (8.02%) (Gopalakrishnan and Mohan, 1996).

The tomato fruit borer can be effectively controlled by sprays of *B. thuringiensis* (Dipel) at 0.5 kg/ha at 10 days interval. The borer infestation was found to be significantly low (7.41%) in 'Dipel' sprayed plots as against 21 to 30% in control (Krishnaiah *et al.*, 1981).

Five rounds of sprays of *Ha* NPV at 250 LE/ha in combination with adjuvants and feeding stimulants at weekly intervals commencing the first spray on flower initiation were effective in increasing the mortality of tomato fruit borer, *H. armigera* to 75 – 85% in the treated plots as against 0 to 5.7% in control plots. The fruit damage was 4% in NPV treated plots as against 15% in control plots. Application of 3 rounds of NPV at 250 LE/ha along with adjuvants *viz.,* 1% soya flour and 0.01% soya oil applied during evening hours at weekly intervals, right from the initial observation of few eggs per plant, recorded significant reduction in tomato fruit borer damage (2.2%) and increase in yield (11.34 kg) as compared to control (21.7% fruit damage and 7.64 kg yield) (Narayanan, 1985).

c) Integrated Methods: Use of African marigold (*Tagetes erecta*) as a trap crop for the management of fruit borer on tomato involves planting one row of 45 day-old marigold seedlings after every 16 rows of 25 day-old tomato seedlings (Fig. 3.7) and spraying of *Ha* NPV at 250 LE/ha or 4% NSKE or 4% pulverized NSPE, 28 and 45 DAP coinciding with peak flowering.

Fig. 3.7. Integrated management of tomato fruit borer using African marigold as a trap crop.

For effective control of *H. armigera*, the egg parasitoid, *T. pretiosum* and *Ha* NPV could be integrated to tackle different stages of the pest. *T. pretiosum* was released at 2.5 lakh/ha and *Ha* NPV was sprayed 2 times at 250 LE/ha for effective control of the pest (4.18% fruit borer damage as compared to 21.79% fruit borer damage in control). The increase in yield was 65.5% over control and the reduction in fruit borer damage was 80.8% over control.

Three releases of *T. pretiosum* + 3 sprays of *B. thuringiensis* at 1 kg/ha was found highly effective against fruit borer in Himachal Pradesh. The release of *T. pretiosum* (2.5 lakh/ha) + 2 sprays of *Ha* NPV (250 LE/ha) was found effective for the management of fruit borer.

Use nylon nets (40 gauge) to avoid insect vectors. *Ha* NPV was sprayed twice at 28 and 35 DAP. Spray of pongamia soap at 1% was given at 40 DAP for the management of fruit borer and leaf miner. The bored fruits were removed mechanically once 40 DAP. The fruit borer and leaf miner incidence were 2% and 0.7-2.7 mines/leaf, respectively.

3.2.1.2. Serpentine leaf miner, *Liriomyza trifolii*

This is a polyphagous pest with a wide host range affecting many vegetable crops like tomato, beans, cowpea, okra, cucurbits *etc*. Damage by *L. trifolii* is severe during summer and rainy months in South India. Damage is more on hybrid tomatoes than open pollinated varieties. A computerized software on serpentine leaf miner called 'SLMINFO' has been developed at the Indian Institute of Horticultural Research, Bangalore, which gives information on host plants, insecticides used, resistance to insecticides, natural enemies, etc.

i) Damage: Serpentine leaf miner is a tiny, metallic fly which punctures the leaf lamina and feed on the oozing sap. Adult is 2-3 mm in size. Damage is severe on cotyledons and young leaves. It lays eggs on the outer margin of leaves. Within 2-3 days, whitish maggots hatch out of these eggs and start mining the leaves and pupate in 6-10 days. Pupation takes place in the soil and occasionally on the leaf surface itself. Pupal period is about 7-8 days. Typical serpentine shaped tunnels are formed in the leaf lamina indicating the path of feeding by the maggots (Fig. 3.8).

Fig 3.8. Serpentine leaf miner on tomato leaves and adult.

ii) Host Range: Economically important crop plants include cucumber, snake gourd, bottle gourd, pumpkin, watermelon, French bean, cowpea, okra, palak, gerbera, marigold, etc. (Virakthmath *et al*., 1993).

iii) Management: As this pest tends to develop pesticide resistance, judicious use of pesticides is recommended.

- Often the incidence starts from nursery itself. Hence, remove infected leaves at the time of planting or within a week of transplanting.

- Apply neem cake to furrows at 250 kg/ha while planting and repeat after 25 DAP (at flowering).
- Spray 4% Neem seed powder extract or 1% neem soap at 15-20 DAP.
- When the incidence is high, remove infected leaves first and mix 0.05% triazophos or 0.015% decamethrin with 7.5g neem/litre and spray.
- Avoid indiscriminate use of insecticides so as to facilitate build up of the natural enemies.
- Avoiding excess application of N-fertilizers.
- Use of yellow sticky traps help in trapping adult flies.
- Tomato variety Sheetal was less attacked followed by Rupali, Rashmi and Naveen.
- *Hemiptarsenus varicornis* is the most predominant indigenous parasitoid with 5-15% parasitism (Krishna Kumar *et al.*, 1998).

Among the hosts, bitter gourd (*Memordica charantia*) is least preferred and this information can potentially be used in the management of *L. trifolii* as a repellant crop around most susceptible tomato plants (Krishna Kumar *et al.*, 1997).

3.2.1.3. Whitefly, *Bemisia tabaci*

White fly is a well known vector which transmits tomato leaf curl virus and causing severe loss in yield. Sastry and Singh (1971) reported loss in yield ranging from 38 to 93% during different months of the year.

i) Damage: They have piercing and sucking mouthparts and both nymphs and adults feed on lower surface of the leaves causing deformation of young leaves (Fig. 3.9). Severe cases leads to chlorotic leaves accompanied by wilting. Whiteflies also excrete honeydew, causing sooty mold.

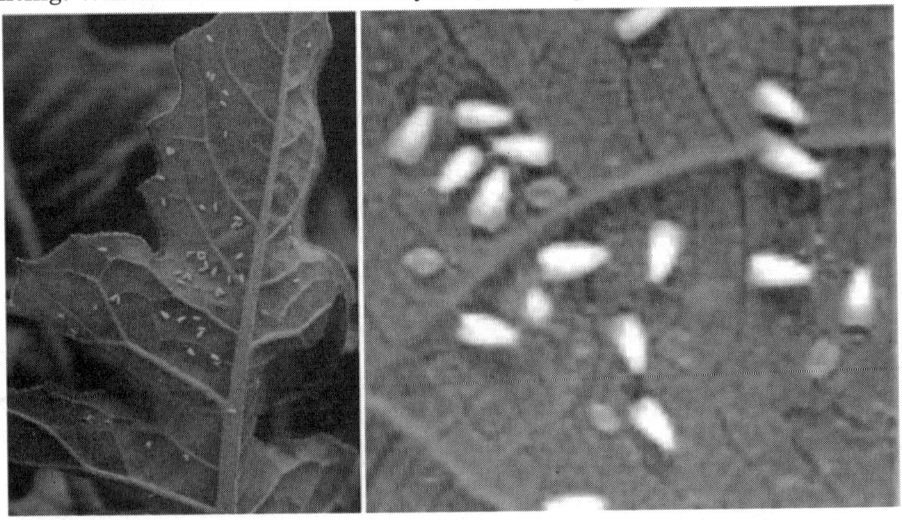

Fig. 3.9. Whiteflies on tomato leaf.

ii) Management

- Use virus resistant hybrids.
- Raise nurseries in nylon nets/polyhouses.
- Dip the roots in a systemic insecticide like imidacloprid (0.3 ml/l) or thiomethoxam (0.3 g/l) for 5 minutes before planting. If portrays are used for raising nursery, drench the portrays with the chemicals one day before transplanting.
- Fifteen days after planting, spray with the systemic insecticides used for root dipping.
- Rogue out the virus affected plants as soon as the symptoms are observed.

3.2.1.4. Thrips, *Thrips tabaci, Frankliniella occidentalis*

i) Damage: The characteristic symptoms of thrips (Fig. 3.10) damage are crinkling, cupping and upward curling of leaves. Flowers drop at later stage. The pest transmits tomato spotted wilt virus.

Fig. 3.10. Thrips on tomato

ii) Management : Spraying endosulfan or monocrotophos each at 2 ml/litre of water gives effective control of thrips.

3.2.2. Mite Pests

3.2.2.1. Red Spider Mites, *Tetranychus urticae*

i) Damage: Tiny red mites (Fig. 3.11) feed mainly on the lower surface of leaves by scraping the epidermis causing yellowing of leaves followed by foliage drying. It has become a major pest of many vegetables like, tomato, brinjal, beans, okra, cucurbits etc. in recent times most probably due to indiscriminate use of pesticides and cultivation of vegetables during off season, particularly during summer.

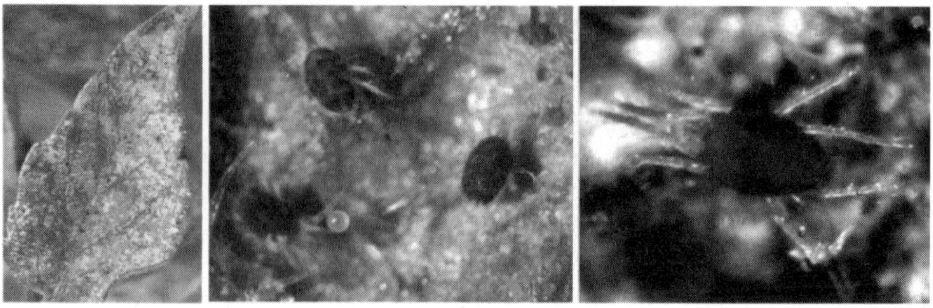

Fig. 3.11. Red spider mite infestation on tomato.

ii) Management

- IPM plots (using marigold as a trap crop and spraying 4% NSKE) were found to have no or very low incidence of this pest. Therefore, following IPM is ideal for preventing the outbreak of mite incidence.
- Generally incidence will be noticed in some patches in a plot. Remove the affected leaves immediately and spray dicofol 18.5 EC at 2.5ml/l or any other acaricide like wettable sulphur 80 WP at 3g/l.
- Spray lower leaves and lower leaf surface thoroughly as mites are generally observed there.
- As an alternative to the chemical acaricides, spray neem oil/neem soap/pongamia soap each at 1%.
- When incidence is severe, first remove and destroy all severely infected leaves and then mix an acaricide with botanical and give a thorough spray.

IPM

The IPM package given below will take care of fruit borer, leaf miner, mite and whitefly (insect vector).

Nursery

- Raise marigold (tall African variety 'Golden Age' bearing yellow and orange flowers) nursery 15-20 days before tomato nursery.
- One week after germination of seeds, spray the seedlings with imidacloprid 200 SL at 0.3 ml/l or thiomethoxam 25 WP at 0.3 g/l.

Before Transplanting

- Apply Neem cake at 250 kg/ha on ridges at the time of preparing land.
- Dip the roots of seedlings (do not dip the foliage as it may cause burning of leaves) with imidacloprid 200 SL at 0.3 ml/l or thiomethoxam 25 WP at 0.3 g/l for 5 minutes.

Main Field

- Transplant 20-25 day old tomato and 45-50 day old marigold seedlings simultaneously in a pattern of one row of marigold for every 16 rows of tomato. However, the first and last row of the plots should be of marigold. Simultaneous flowering of both the crops ensures attraction of fruit borers to marigold flowers.
- Fifteen days after planting, spray imidacloprid 200 SL at 0.4ml/l or thiomethoxam 25 WP at 0.3g/l for leaf curl vector (whitefly) control.
- Apply neem cake at 250 kg/ha to ridges at 20-25 DAP (at flowering) to reduce fruit borer and leaf miner incidence.
- Spray *Ha* NPV at 250 LE/ha with 1% jaggery as sun screen at 28, 35 and 42 DAP in the evening.
- Spray marigold flowers with *Ha* NPV or destroy fruit borer larvae in them.
- As an alternative to *Ha* NPV spray, the egg parasitoids, *Trichogramma chilonis, T. brasiliensis* and *T. pretiosum* at 2.5 lakhs/ha can be released (five releases at 50,000/ha/release). The first release has to be done at the flower initiation of the crop.
- If red spider mite incidence is noticed, spray neem soap 1% or neem oil 1% or any synthetic acaricide like dicofol 18.5 EC (at 1.5 ml/l), or Ethion 50 EC (at 1.5 ml/l) or sulphur 80 WP (at 3 g/l). Spray lower surface of the leaves.
- Mechanical collection and destruction of bored fruits at periodic intervals (3-4 times after fruit set) to minimize fruit borer incidence to the minimum.
- Destroy leaf curl and other virus affected plants as soon as the symptoms appear in a few plants to minimize their spread.

Pesticide Residues

The persistence of terminal pesticide residues and waiting periods for tomato are presented in Table 3.9. (Awasthi, 2001).

Table 3.9. Terminal pesticide residues and their waiting periods for tomato

Pesticide schedule	Dosage/ Concentration (%)	Terminal residues (ppm/days)	Waiting period (days)
Monocrotophos spray at fruiting stage	0.05	0.14/15	10
Malathion spray at fruiting stage	0.10	0.08/7	3
Endosulfan spray at fruiting stage	0.07	0.12/20	7
Cypermethrin spray at fruiting stage	0.01	0.22/10	5
Fenvalerate spray at fruiting stage	0.01	0.08/15	10
Carbofuran soil application at planting	0.5 kg a.i./ha	0.09/60	Nil
Phorate soil application at planting	1.0 kg a.i./ha	0.85/75	Safe at 2nd harvest

3.3. BRINJAL

Brinjal (*Solanum melongena*) is a widely grown vegetable crop in Asian countries. It is being cultivated in 0.553 million hectares producing 9.136 million tonnes of brinjal with an average yield of 16.5 tonnes per hectare (Bijay Kumar, 2007) (Tables 3.10 and 3.11). In India, it is adapted to a wide range of climatic conditions from north to south and east to west. In hilly regions, it is grown only in summer. Brinjal is used in a variety of culinary preparations. Pickles and industrially processed foods are also produced. It is rich in Vitamins A and B and good for diabetic patients.

Table 3.10. Major brinjal producing countries of the world

Country	Area (in '000 ha)	Production (in '000 MT)	Productivity (in MT/ha)	% Share of world production
China	952	17028	17.9	53.5
India	553	9136	16.5	28.7
Egypt	43	1000	23.3	3.1
Turkey	35	930	26.6	2.9
Iraq	23	439	19.1	1.4
Japan	11	396	34.7	1.2
Italy	12	339	27.8	1.1
Sudan	15	272	18.7	0.9
Indonesia	43	252	5.9	0.8
Philippines	21	188	8.8	0.6
Others	120	1856	b5.4	5.8
Total	1829	31836		

Source: FAO (except Indian data) (Indian data source: Indian Horticulture Database 2006)

Table 3.11. Area, production and productivity of leading brinjal growing states in India (2005-06)

State	Area (in'000 ha)	Production (in '000 MT)	Productivity (MT/ha)
West Bengal	152.9	2757.4	18.0
Orissa	127.9	1853.9	14.5
Bihar	53.7	1031.0	19.2
Gujarat	62.7	982.6	15.7
Andhra Pradesh	26.2	549.4	21.0
Maharashtra	32.0	521.6	16.3
Karnataka	14.7	326.0	22.1
Chhattisgarh	19.4	266.4	13.7
Madhya Pradesh	14.8	221.9	15.0
Assam	13.0	171.5	13.2
Haryana	11.5	148.7	12.9

Others	24.5	305.8	12.5
Total	553.3	9136.4	16.5

(Source: Indian Horticulture Database 2006)

3.3.1. Insect Pests

The shoot and fruit borer, *Leucinodes orbonalis;* leafhopper, *Amrasca biguttula biguttula*; ash weevil, *Myllocerus subfacitus;* epilachna beetle, *Epilachna vigintioctopunctata*; red spider mite, *Tetranychus urticae;* gall midge, *Asphondylia* sp.; and stem borer, *Euzophera perticella* are some of the important pests attacking brinjal.

3.3.1.1. Shoot and Fruit Borer, *Leucinodes orbonalis*

This is the most important pest on brinjal not only in India, but also in all South Asian Countries. This pest has developed resistance against all groups of insecticides and management is very difficult. Often the extent of damage due to this pest reaches up to 70-80%. Damage is very severe during rainy season and early winter.

i) Damage: The adult moths lay tiny white eggs singly on flower buds and other plant parts. Eggs hatch into small light brown larvae. Caterpillars feed inside the tender shoots before flowering and cause wilting of the affected shoots (Fig. 3.12). Later, the larvae bore into flowers, flower buds and the grown up larvae migrate and bore fruits contaminating them with excreta (Fig. 3.12). When the incidence is high, unopened flower buds swell and harbour the borer. Just before pupation, the grown up larvae come out of the fruits and flower buds to pupate in silky cocoons on plant parts or debris.

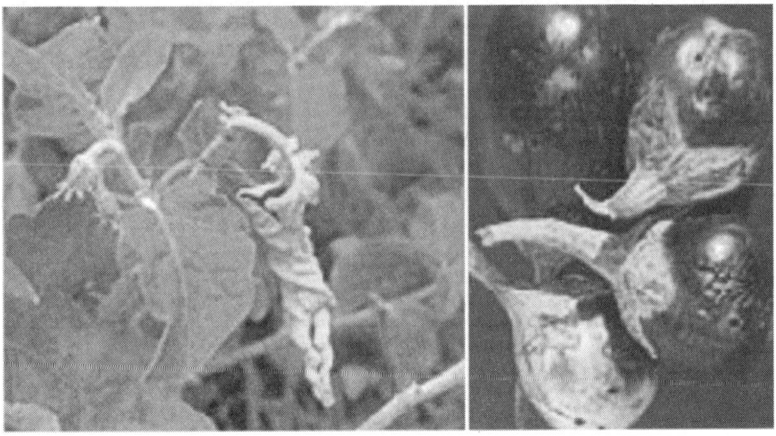

Fig. 3.12. Shoot and fruit borer damage on shoots and fruits of brinjal.

ii) Management

a) Cultural Methods

- Use nylon net barrier for raising nurseries to eliminate pest incidence coming from nursery to main field.

- Weekly monitor wilting shoot tips having borer incidence. Cut and destroy such insect damaged shoot tips during pre-flowering and flowering period.
- Regular destruction of larvae in swollen damaged flower buds and fruits after each harvest is compulsory.
- Grow barrier crops like maize all round the crop coupled with clean cultivation.
- Dusting of ash reduces the incidence of shoot and fruit borer.

Removal and destruction of borer damaged shoots during pre-flowering period has potential for reducing the pest population early in the season. Further, growing a barrier crop such as maize, destruction of borer affected fruits, marginally help in reducing the borer damage.

b) Use of Pheromone Trap: Use of pheromone trap is found to reduce the borer infestation in many parts of India. Use water traps loaded with pheromone for this purpose (Fig. 3.13). About 30 traps per acre is known to bring down the pest incidence. Pheromone lure (Lucinlure) in combination with Portable Water Traps (both of PCI) were effective in mass trapping of *Leucinodes orbonalis* to the tune of 1927/acre in Bangalore, 1022/acre in Belgaum and 590/acre in Jogipet in Andhra Pradesh.

Fig. 3.13. Sex pheromone and funnel pheromone trap to attract shoot and fruit borer of brinjal.

c) Use of Botanicals

Three applications of neem cake during the crop growth decreases the incidence of borer to 8% from 40% in control. Farmers could realize an additional return of Rs. 35,498/ha.

- Apply neem or pongamia cake at 250-500 kg/ha to ridges at flowering. Repeat 2 more times at 30-45 days interval
- Spray NSKE 4% or neem oil 3% at 10 days interval (Table 3.13).

d) Biological Methods: *Eriborus sinicus* gave up to 30% parasitism of shoot and fruit borer at Taiwan (AVRDC, 1996).

Inundative release of *Trichogramma chilonis* at 50,000/ha, 5 times at weekly intervals starting from flower initiation stage is effective.

Application of *B. thuringiensis* at weekly intervals controlled the fruit borer. In plots treated with *B. thuringiensis* (Delfin), there was 30% reduction in larval population and 48.3% yield increase in brinjal. Application of 5 rounds of *B. thuringiensis* var. *kurstaki* (Biobit) at 1 kg/ha at weekly intervals, right from flower initiation was effective in checking the pest. The % fruit damage was low in Biobit (13.44%) as compared to control (19.25%). The yield was also significantly high in Biobit (28.5 kg/plot) as compared to control (14.74 kg/plot).

e) Integrated Methods: A universally applicable IPM for this pest is yet to be developed. However, following IPM package can be adopted to minimize the fruit borer damage. In areas where brinjal is grown in large areas at a stretch at any time, all such growers have to adopt the IPM to get effective control. Adoption of IPM by a few farmers in isolation is not going to bring down the pest population.

For the management of brinjal shoot and fruit borer, combination of clipping of shoots affected by the borer at weekly interval followed by neem cake application (250 kg/ha) at 30 days after planting and 4% pulverized NSPE/1% neem soap/1% pongamia soap sprays at 60, 75, 90 and 105 days after planting was most effective with 22% borer damage as compared to 46% in control (Krishna Murthy and Krishna Kumar, 2002) (Table 3.12).

Table 3.12. Incidence of shoot and fruit borer of brinjal under different treatments

Treatments	Cumulative % of borer infestation	Yield (kgs)
Soil appln. of neem cake at 250 kg/ha at 30, 60 and 90 DAT.	30.05	5.96
Shoot clipping at weekly interval + neem cake at 30 DAT + NSKE (4%) spray at 60, 75, 90 and 105 DAT.	22.04	6.72
Barrier crop of maize sown 10 days before brinjal planting (2 rows at 30 cm from brinjal and maize at 5 cm form plant to plant) + shoot clipping at weekly interval + NSKE (4%) at 45, 60, 75 and 90 DAT.	23.51	6.53
Barrier crop as above + shoot clipping + endosulfan 700 g a.i./ha alternated with cypermethrin at 50 g a.i./ha at 45, 60, 75 and 90 DAT.	29.44	9.01
Cypermethrin at 50 g a.i./ha at 45, 60, 75 and 90 DAT.	36.62	9.59
NSKE (4%) at 45, 60, 75 and 90 DAT.	28.54	14.33
Control.	45.90	5.33
CD (P=0.05).	2.71	3.40

Mixing enodsulfan 35 EC at 2ml/l or cypermethrin 25 EC at 0.75 ml/l with neem soap at 7.5g/l and spraying gives effective control of the pest (Table 3.13).

Table 3.13. Effect of neem products for the control of fruit borer in brinjal.

Treatment	Mean % of infestation (Number basis)	Mean % of infestation (Weight basis)
Neem cake extract 10%	55.74	54.57
Neem seed kernel extract 5%	50.89	49.01
Neem oil 3%	50.85	52.48
Carbosulfan 0.04%	39.77	41.77
Untreated check	61.56	59.35

Spraying *Bacillus thuringiensis* formulation (1%) at weekly interval followed by release of *Trichogramma chilonis* at 2,50,000/ha (50,000/release-5 times at weekly intervals, starting from flowering) was found to reduce the borer incidence.

3.3.1.2. Leaf Hopper, *Amrasca biguttula biguttula*

This is another important pest of brinjal, particularly during summer. This pest also attacks cotton and okra. If adequate control measures are not taken, 50-60% yield loss can be observed in a span of few weeks.

i) Damage: Both adults and nymphs suck the sap from leaves. The eggs are inserted into the midrib or veins. On hatching, the nymphs move along the veins and nymphs suck the leaf sap. On the older leaves, the damage is seen as yellowish-green mosaic symptom followed by brown necrosis and curling along the leaf margin (Fig. 3.14).

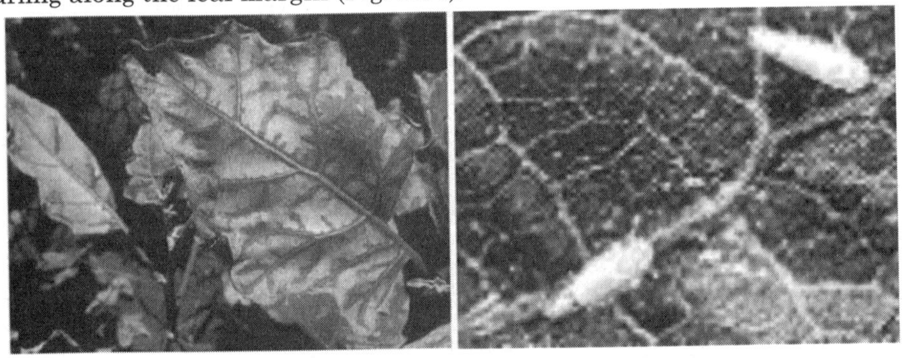

Fig. 3.14. Leaf hopper infestation on brinjal leaves.

ii) Management

- Soil application of neem cake at 250 kg/ha followed by sprays of NSPE 4% or neem soap 1% at 10 days interval brings down the hopper population.

- Sprays of systemic insecticides like dimethoate 30 EC at 2ml/l or imidacloprid 200 SL at 0.3 ml/l or acephate 75 SP at 1 g/l at pre-flowering stage brings down the pest population.

3.3.1.3. Ash Weevil, *Myllocerus subfasciatus*

i) Damage: Ash weevil is another important, endemic pest of brinjal causing plant mortality. Adults feed on the foliage and cause saw teeth like damage along the leaf margin. The adults lay eggs in and around base of the plants in soil and the hatching grubs feed on roots. Affected plants wilt and dry up (Fig. 3.15). Wilting is first noticed in patches in a plot. One can notice very few roots in such wilting plants when uprooted. Both the roots and plant base show scraped damage. Grubs pupate in the soil encrustation.

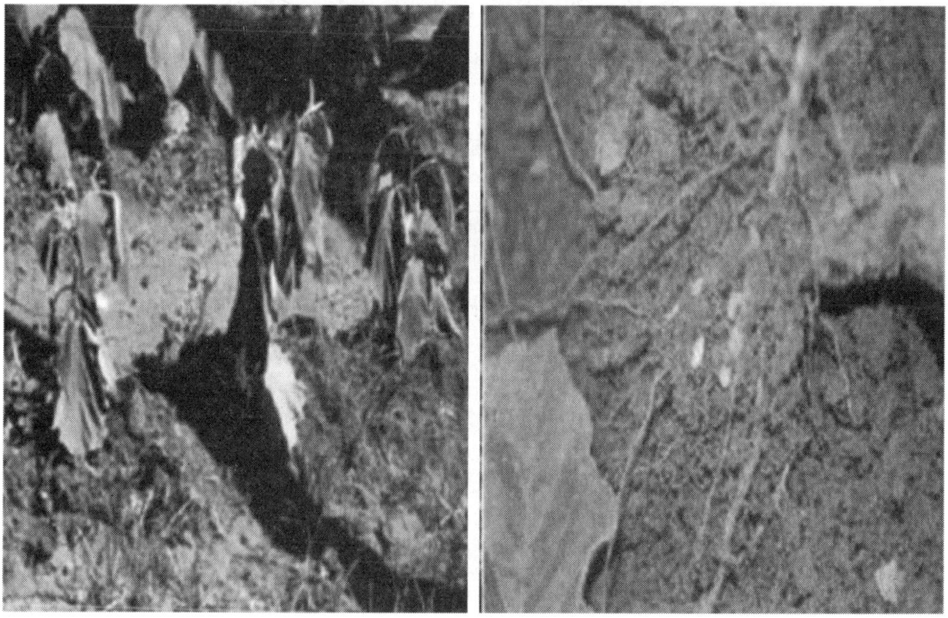

Fig. 3.15. Ash weevil damage on brinjal.

ii) Management

- Soil application of neem cake/pongamia cake at the time of transplanting at 250 kg/ha.
- Repeat the neem cake application for 1 or 2 times at 30 and 60 DAP.

3.3.1.4. Epilachna Beetle, *Epilachna vigintioctopunctata*

i) Damage: Both the grubs and the adult beetles scrape the leaves in semi-circular or half moon shaped fashion (Fig. 3.16). Females lay torpedo-shaped, 15-50 yellowish eggs on the ventral surface of leaves. Heavy infestation results in leaf skeletonization. Pupation takes place on the plant itself.

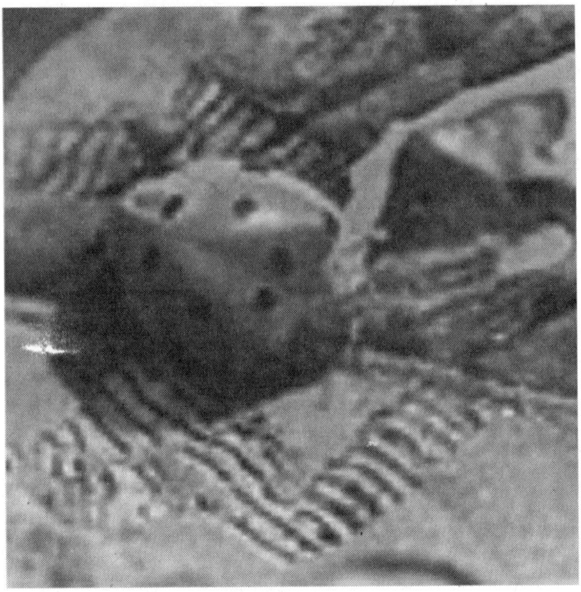

Fig. 3.16. Epilachna beetle on brinjal leaf.

ii) Management

- Collection and destruction of adult beetles.
- Spray any contact insecticide like carbaryl 50 WP at 3g/l or quinalphos 40 EC at 1.5 ml/l, if required.
- It was effectively kept under check by *Pediobus foveolatus* with 60-77% parasitism.
- 'Thuricide' dust (3x10^6 spores/mg) recorded total elimination of the grubs of epilachna beetle on brinjal plants within 10 days (Sundarababu, 1970).

3.3.1.5. Gall Midge, *Asphondylia capparis*

i) Damage: In parts of South India, gall midge, *Asphondylia capparis* is a regular pest on flowers and fruits of brinjal and sweet pepper (Tewari *et al.*, 1987; Krishna Kumar *et al.*, 1997). Maggots of this pest feed on developing ovary/fruits. When young flowers are attacked, stamens and pistils degenerate and fruits fail to form. Following the emergence of the fly, the fruit cracks along the inner side of the curve and is rendered unfit for marketing (Fig. 3.17).

ii) Management

- Spraying of 4% NSKE is effective in controlling this pest.
- The gall midge was found parasitized by *Eurytoma* sp. to the tune of 8-15% (Tewari and Krishnamoorthy, 1986).

Fig. 3.17. Gall midge infestation on brinjal fruits.

3.3.1.6. Stem Borer *Euzophera perticella*, Leaf Roller, *Eublemma olivacea*, Lace Wing Bug, *Urentius echinus* and Thrips, *Sericothrips solanifolii*

i) Damage: In certain parts of India, the stem borer, *Euzophera perticella* is serious. Caterpillars bore downwards killing the affected plants (Fig. 3.18).

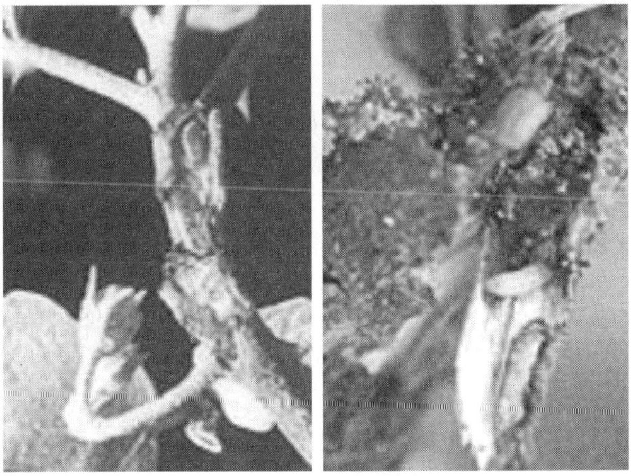

Fig. 3.18. Stem borer on brinjal.

Brinjal thrips, *Sericothrips solanifolii* damage leaves and fruit. Both nymphs and adults suck the sap from leaves and developing buds. Damaged buds later manifest as scarred fruits and lose market value.

Brinjal is attacked by the leaf roller, *Eublemma olivacea* which causes rolling of leaves.

Lace wing bugs, *Urentius echinus* are found in large number on abandoned or less cared plants especially in summer. Both the adults and the nymphs suck the sap and the affected plants turn yellow.

ii) Management

- As far as possible raise the nurseries with the nylon net barrier.
- Most of these pests can be managed by soil application of neem cake at 250 kg/ha.
- Spray 4% NSKE at the time of first flush (Krishna Kumar *et al.*, 1997).
- Repeat the same treatment at 30 days interval for 2-3 times.

3.3.1.7. Mealy Bug, *Coccidohystrix insolita*

i) Damage: The mealy bugs attack isolated old plants which become covered with white mass (Fig. 3.19). Insects suck sap from plants making them pale and weak.

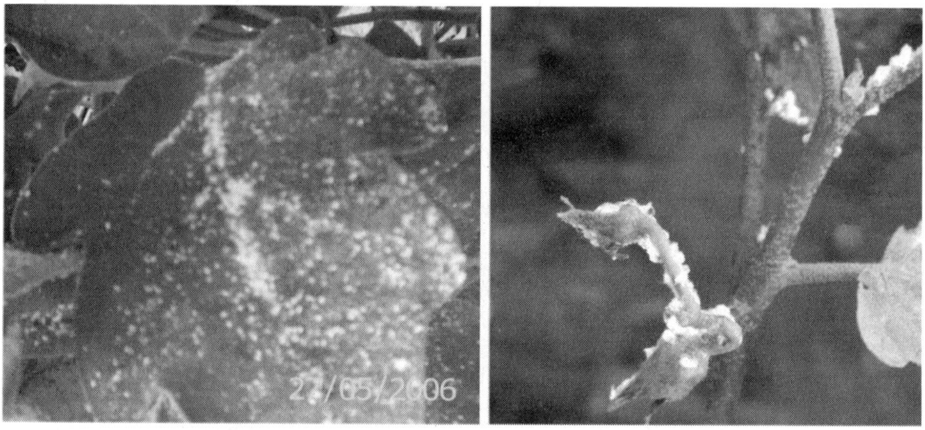

Fig. 3.19. Mealy bug infestation on brinjal.

ii) Management: Spraying 0.15% malathion or 0.4% monocrotophos gives effective control of mealybugs. It was also effectively controlled by release of the predator, *Cryptolaemus motrouzieri* (Krishnamoorthy and Mani, 1996).

3.3.1.8. White Fly, *Bemisia tabaci*

i) Damage: Numerous white winged fly-like insects (Fig. 3.20) feed on the plant. They move at slight disturbance. The characteristic symptoms include yellowing of leaves. Sooty mould develops on the honey dew excreted by this insect. The pest is serious during summer.

ii) Management: Spraying of neem or pongamia soap at 1% gave effective control of white flies on brinjal.

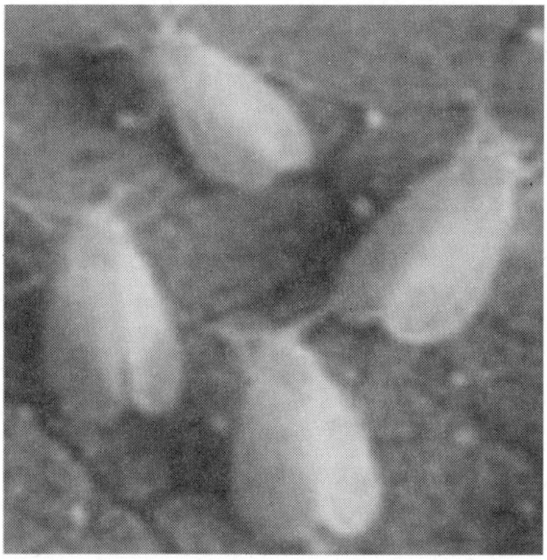

Fig. 3.20. White flies on brinjal leaf.

3.3.1.9. Aphid, *Aphis gossypii* (Fig. 3.21)

Fig. 3.21. Aphids on brinjal.

i) Management: The seedling bare root dip in 3% neem oil followed by 5% NSKE spray significantly reduced the aphid infestation in brinjal (Table 3.14). Release of *Chrysoperla carnea* at 2 grubs/plant was also found effective.

Table 3.14. Effect of seedling bare root dip in neem products on brinjal aphid.

Treatment	Mean population of aphids / 3 leaves	% Reduction over untreated check
Neem cake extract 10%	1.36	24.86
Neem seed kernel extract 5%	1.25	30.93
Neem oil 3%	1.18	34.81
Monocrotophos 0.1%	0.67	62.98
Untreated check	1.81	---

3.3.1.10. White Grub, *Holotrichia consanguinea*

i) Management: Collection of beetles from plants during night hours and killing them in kerosene water reduced the infestation. Fungal pathogens, *Metarrhizium anisopliae* and *Bacillus popillae* var. *holotrichiae* are also effective in reducing the infestation.

3.3.2. Mite Pests

3.3.2.1. Red Spider Mite, *Tetranychus urticae*

i) Damage: Brinjal is often severely affected by mites especially under dry conditions. Larvae are pink in colour while nymphs are greenish red in colour. Spider mites form small web on undersurface of the leaves where they colonize and multiply (Fig. 3.22). Tiny mites (larvae, nymphs and adults) feed mainly on the lower surface of the leaves by scraping the epidermis causing curling, shriveling and yellowing of leaves followed by foliage drying. The affected plants bear few fruits.

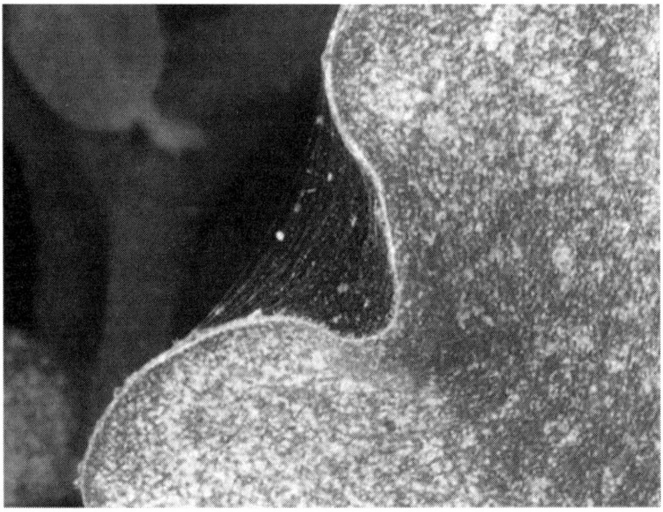

Fig. 3.22. Res spider mite damage on brinjal leaves.

ii) Management

- Sprays of dicofol 18.5 EC at 2.5ml or any other acaricide like wettable sulphur 75 WP at 3g/l.
- As an alternative to the chemical acaricides, spray neem soap/ pongamia soap at 1%.
- Thorough spray on the under surface of the leaves is to be assured.

Pesticide Residues

The persistence of terminal pesticide residues and waiting periods for brinjal are presented in Table 3.15 (Awasthi, 2001).

Table 3.15. Terminal pesticide residues and their waiting periods for brinjal

Pesticide schedule	Dosage/ Concentration (%)	Terminal residues (ppm/days)	Waiting period (days)
Quinalphos spray at fruiting stage	0.05	0.32/15	21
Monocrotophos spray at fruiting stage	0.05	0.90/3	10
Cypermethrin spray at fruiting stage	0.01	0.003/10	2
Fenvalerate spray at fruiting stage	0.01	0.004/10	6
Carbofuran soil application at planting	0.5 kg a.i./ha	0.11/50	Safe at 1st harvest
Phorate soil application at planting	1.0 kg a.i./ha	1.50/85	Safe at 2nd harvest

3.4. CHILLI AND BELL PEPPER

Chilli (*Capsicum annuum*) is valued for its diverse commercial uses. India is a major producer, exporter and consumer of chilli. Indian chillies reach over 90 countries in the world. Bangladesh, Bahrain, Canada, Sri Lanka, Saudi Arabia, USA and UAE are the leading importing countries. It is being cultivated in 0.834 million hectares producing 0.847 million tonnes of chillies with an average yield of 1.0 tonne per hectare. Andhra Pradesh has been the largest chilli-growing state followed by Karnataka and Maharashtra. Productivity of chilli is highest in Andhra Pradesh, followed by Arunachal Pradesh and Punjab. The productivity of chilli in Andhra Pradesh and Punjab is higher since the crop is raised under irrigated conditions compared with Maharashtra and Karnataka. Chilli is rich in vitamins A and C.

Bell pepper (*Capsicum annuum*), also known as capsicum or sweet pepper, is a popular vegetable in India. It is relatively a new entrant into our country. It is mainly cultivated in Himachal Pradesh, Uttar Pradesh, parts of Gujarat, Maharashtra, Karnataka, Ranchi region of Bihar and hilly regions of Tamil Nadu. It grows well in summer season in hills and cooler season in the plains.

3.4.1. Insect Pests

Chilli thrips (*Scirtothrips dorsalis)*, yellow mite (*Polyphagotarsonemus latus)* and fruit borers, *Helicoverpa armigera* and *Spodoptera litura* are the major pests of this crop. The thrips and mite damage often occur on the same leaf and is called churda murda complex.

3.4.1.1. Thrips, *Scirtothrips dorsalis*

i) Damage: Both adults and nymphs suck the sap from young leaves. Affected leaves curl upwards along the margin, and get crinkled and reduced in size (Fig. 3.23). When the incidence is severe, leaves drop and cause heavy reduction in yield. The pest is serious during dry monsoon periods and summer months.

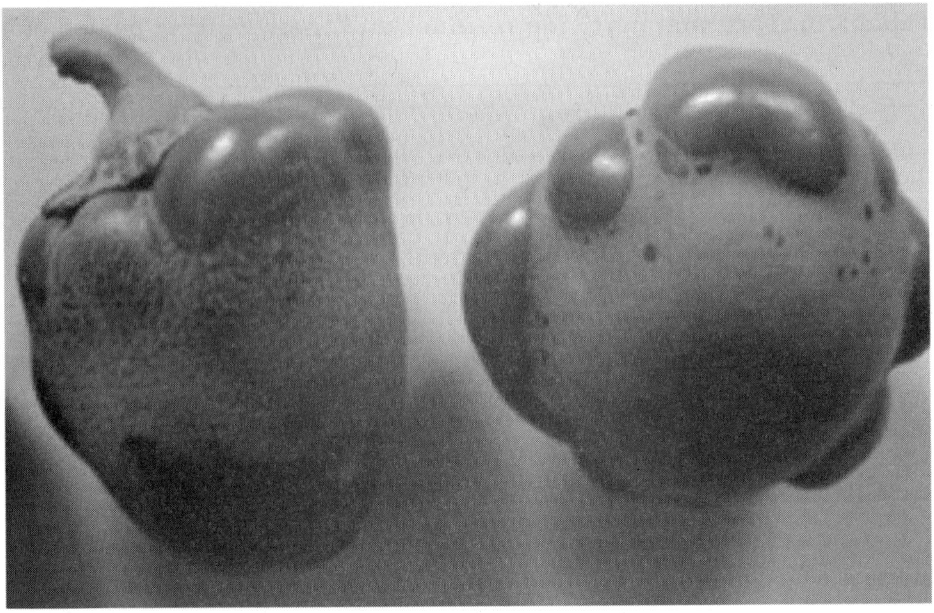

Fig. 3.23. Thrips on fruits of bell pepper.

ii) Management

- Spray acephate 75 SP at 1.5 g/l or fipronil at 1 ml/l or lamda cyhalothrin 5 EC at 0.75 ml/l or imidacloprid 200 SL at 0.3 ml/l alternately at fortnightly interval. Never repeat the same chemical again and again.
- Spray of difenthiuron at 225g a.i./ha and monocrotophos at 150g a.i./ha effectively checked the population of thrips on chilli with higher yield.
- Mix acephate 0.5g, 2 ml of pongamia oil and 1 ml sticker, emulsify (add a little water and shake thoroughly in a bottle) and make the volume to 1 litre and spray.
- Spraying of vertimec at 0.3 to 0.4 ml/l was found effective against thrips and mites for 45 days.
- Soil application of aldicarb at 2.8 kg a.i./ha 15 days after transplanting effectively checked thrips and mites and resulted in minimum leaf curl incidence with higher yield of marketable green chilli fruits (Patel *et al.*, 1983).
- Soil application of carbofuran at 0.5 kg a.i./ha 15 days after transplanting followed by spray of 0.1% triazophos commencing from 40 days after transplanting at 10 days interval effectively controlled thrips and increased marketable green chilli fruits (Patel *et al.*, 1997).
- Chilli cvs. NS 1101, NS 1701 (Namdhari seeds), NP 46A, Pusa Jwala, Arka Lohit and K 2 (Kovilpatti) are tolerant to thrips.
- *Chrysoperla carnea* can be released at 2 grubs/plant.

Integrated Methods: Pongamia oil 2 ml when mixed with 0.5 g acephate + 2 ml sticker gave excellent control of thrips in chillies, even during summer (Krishna Murthy and Krishna Kumar, 2002). The thrips damage rating was reduced to 20.67 in the above treatment as compared to 45.67 in control. Similarly, the yield was also increased in treated plots over control (Table 3.16).

Table 3.16. Thrips damage and yield under different management programmes in chilli (Summer 2002)

Treatments	Doses	Rating of thrips damage/5 leaves	Yield kg/plot
Acepahte + Pongamia oil	0.5 g+2 ml	20.67	1.69
Dimethoate + neem oil	2 ml+2ml	30.00	1.26
Pongamia oil	20 ml	45.67	1.05
Neem oil	20 ml	40.67	3.50
Dimetoate	2 ml	25.00	0.90
Neem seed powder	4%	40.00	0.05
Acephate	0.75 g	10.67	1.35
Control	---	45.67	0.02
CD 5%	---	7.33	0.10

3.4.1.2. Pod Borer, *Helicoverpa armigera*

i) Damage: Tomato fruit borer is also a serious pest in chilli and capsicum. It affects the marketable value of the chilli crop to a great extent, if proper care is not taken (Fig. 3.24).

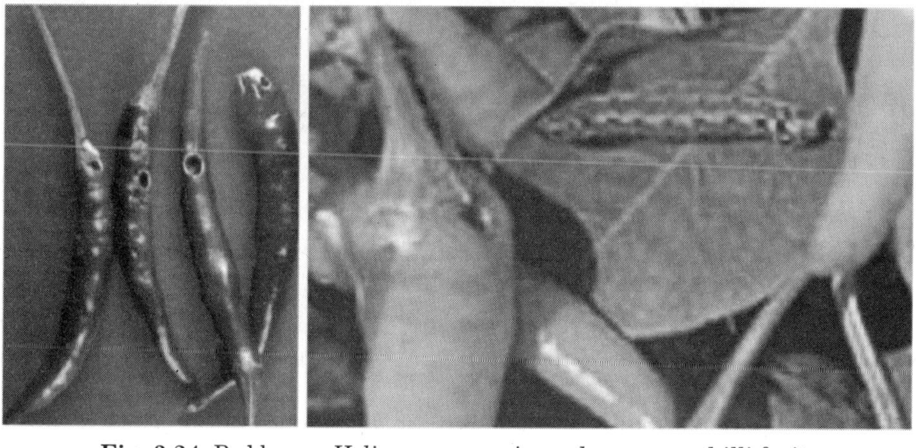

Fig. 3.24. Pod borer, *Helicoverpa armigera* damage on chilli fruit.

ii) Management

- Two rounds of sprays of *Ha* NPV at 250 LE/ha (6 x 10^{12} PIB/ha) at 10 days interval were found effective in reducing the loss caused by borers.

- Inundative release (6 releases) of *Trichogramma brasiliensis, T. pretiosum or T. chelonis* each at 50,000/ha at weekly intervals.
- Use marigold as a trap crop (one row of marigold for every 18 rows of chilli) for managing *H. armigera.*
- Installation of pheromone traps at 5/ha.
- Hand collection of larvae on main and trap crops.
- Spraying with 5% NSKE or *Bacillus thuringiensis* at 500 g/ha.
- Spraying 0.07% endosulfan or monocrotophos or 0.15% carbaryl or 0.05% chlorpyriphos or quinalphos is effective for the management of the pest.

3.4.1.3. Aphid, *Myzus persicae*

i) Damage: The aphid also becomes a serious pest on chilli and sweet pepper occasionally. The infestation may be noticed in a few pockets of a chilli crop. Aphids suck plant sap and cause reduction in plant vigour. Small green insects are clobbered underneath the leaf surface (Fig. 3.25). Sooty mold appears on leaves due to secretion of honeydew by aphids. Feeding of ants on aphids may be observed.

Fig. 3.25. Aphid infestation on chilli leaf.

ii) Management: Sprays of systemic insecticide or 1% neem or pongamia soap gives good control of this pest.

The predators, *Chrysoperla carnea* and *Mallada boninensis*, very effectively controlled aphids on chilli when releases were made at 1: 5 ratio of predator: prey (Mani and Krishnamoorthy, 1997a). Parasitoids, *Aphidius platensis* and *Aphelinus* sp. produced 80-98.8% parasitism in an insecticide-free field (Mani and Krishnamoorthy, 1994).

3.4.1.4. Tobacco Worm, *Spodoptera litura*

i) Damage: Feeding holes on the leaves can be observed due to tobacco worm infestation. Larval entry holes can also be observed near the calyx. The pest infestation causes dropping of young fruits. Mature fruits with seed damage, fecal pellets with seed and watery cavity can be seen.

ii) Management: Collect and destroy egg masses and immature larvae of *S. litura*. Use poison baiting (10 kg rice flour + 1kg of jaggery + 250 g of methomyl 40 SP) for *S. litura*. Repeat the baiting 2-3 times, if necessary. For *S. exigua,* spray indoxacarb 14.5 SC at 0.75 ml/l or spinosad at 0.75 ml/l or thiodicarb 75 WP at 0.75 g/l. Spraying 0.07% endosulfan or monocrotophos or 0.15% carbaryl or 0.05% chlorpyriphos or quinalphos is effective for the management of the pest.

Two rounds of sprays of *Sl* NPV at 250 LE/ha (6 x 10^9 PIB/ha) at 10 days interval were found effective in reducing the loss caused by *S. litura*. It can also be controlled by installation of pheromone traps at 5/ha (Lures should be replaced once in 15 days), hand collection of egg masses and early instar larvae, release of *Telenomus remus* at 50,000/ha, installation of bird perches at 50/ha and spraying with 4% NSKE.

3.4.1.5. White Grub, *Holotrichia consanguinea*

i) Damage: White grubs cause wilting of plants. The plants can be easily pulled out. C-shaped white grubs are found near root region (Fig. 3.26) and cause damage to roots.

Fig. 3.26. White grubs on capsicum.

ii) Management: Only well rotten farm yard manure should be used to prevent root grubs. Collection of beetles from plants during night hours and killing them in kerosene water reduced the infestation. Application of neem cake at 250 kg/ha along with FYM is effective against the pest. In severe cases, application of phorate at 2.5 kg a.i./ha is effective. Fungal and bacterial pathogens, *Metarrhizium anisopliae* and *Bacillus popillae* var. *holotrichiae* are also effective in reducing the infestation.

3.4.1.6. Leaf Curl Complex

i) Symptoms: Chilli leaf curl complex is mainly caused by yellow mite and thrips infestation. It is a major problem on the chilli crop. Sometimes, the Gemini virus also causes enation leaf curl symptom. Reduction in size of leaves, shortening of veins, puckering, sometimes mottling and stunting of plants are common symptoms of the disease. Veins are prominently visible on the lower side of the leaves. The infected plants remain stunted and bear no fruits.

ii) Management: Management of the viruses as well as thrips and mites involves periodical spraying of insecticides and acaricides. Alternate sprays of dicofol at 2.5 ml/litre of water followed by 0.25% wettable sulphur may be taken up to prevent the development of resistance. Two sprays of systemic insecticide are essential for the control of thrips and mite vectors.

3.4.1.7. Gall Midge, *Asphondylia capparis*

i) Damage: Gall midge is an important pest on sweet peppers. Midge damaged sweet pepper lose their market value. Midge infestation may induce shedding of flowers (Fig. 3.27) (Krishna Kumar *et al.*, 1997). Further, pollen germination, seed yield, germination and viability are affected as a result of midge infestation.

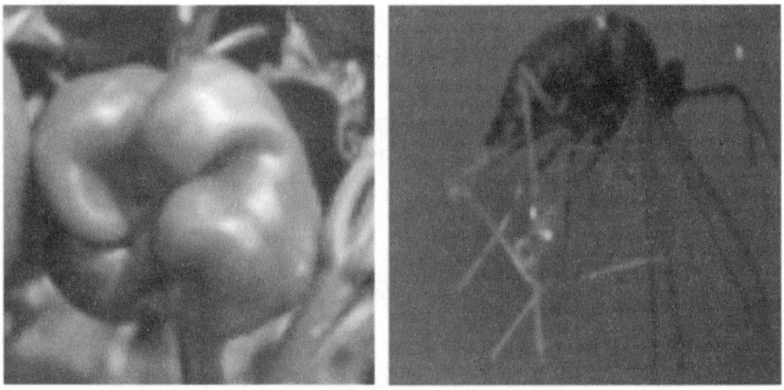

Fig. 3.27. Blossom midge infestation on capsicum.

3.4.2. Mite Pests

3.4.2.1. Yellow Mite, *Polyphagotarsonemus latus*

i) Damage: These infest terminal young leaves like thrips. Mite affected leaves curl downwards and appear twisted (Fig. 3.28). Elongation of petioles is another symptom of white mite infestation. Damage is high in hot and humid seasons.

ii) Management

- Apply wettable sulphur 80 WP at 3g/l or any acaricide (directing the spray on the ventral surface of leaves).
- Mix acephate 0.5g, 2 ml of pongamia oil and 1 ml sticker, emulsify (add a little water and shake thoroughly in a bottle) and make the volume to 1 litre and spray.

- Spraying neem seed powder extract 4% at 10 days interval is effective when the pest incidence is low. As and when the pest incidence increases give synthetic acaricide spray.
- Spraying of vertimec at 0.3 to 0.4 ml/l was found effective against thrips and mites for 45 days.
- *Chrysoperla carnea* can be released at 2 grubs/plant.

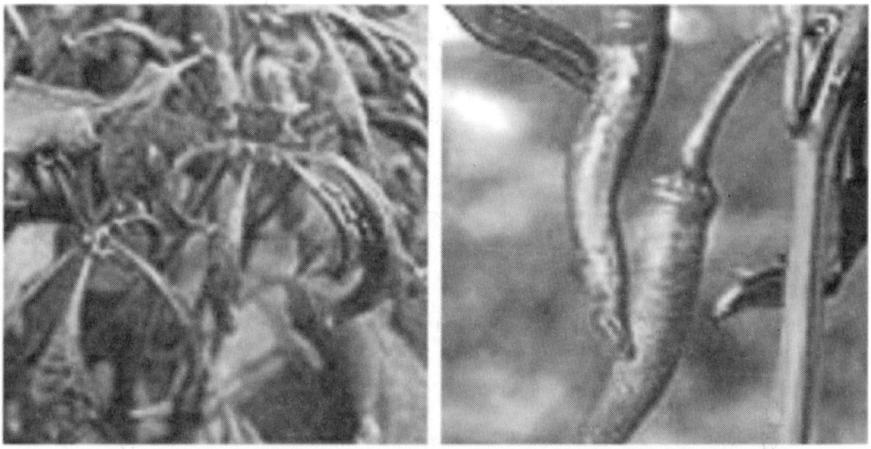

Fig. 3.28. Yellow mites on leaves and fruits of chilli.

IPM

The following IPM can be adopted for managing chilli insect pests:

- In nursery, spray a systemic insecticide like imidacloprid (0.3 ml/l) or acephate (0.75 ml/l).
- Dip the roots in a systemic insecticide like imidacloprid (0.3 ml/l) or acephate (0.75 ml/l) for 5 minutes just before transplanting.
- Apply 250 kg/ha neem cake to soil while preparing land.
- Spray acephate 75 SP at 1.5 g/l or fipronil at 1 ml/l or lamda cyhalothrin 5 EC at 0.75 ml/l or imidacloprid 200 SL at 0.3 ml/l alternately at fortnightly interval. Never repeat the same chemical again and again.
- Mix acephate 0.5 g, 2 ml of pongamia oil and 1 ml sticker, emulsify (add a little water and shake thoroughly in a bottle) and make the volume to 1 litre and spray.
- In endemic areas where *Helicoverpa* is the major problem, use marigold as trap crop and take 3-4 *Ha* NPV sprays at flowering.
- In endemic area of *Spodoptera*, give 3-4 sprays of species specific NPV (either *S. litura* or *S. exigua)*.
- Monitor for *Spodoptera* eggs and leaf damage. Mechanically collect and destroy eggs and gregarious larvae.
- Give poison baits to grown up *Spodoptera* larvae.

Pesticide Residues

The persistence of terminal pesticide residues and waiting periods for chilli are presented in Table 3.17. (Awasthi, 2001).

Table 3.17. Terminal pesticide residues and their waiting periods for chilli

Pesticide schedule	Dosage/ Concentration (%)	Terminal residues (ppm/days)	Waiting period (days)
Carbaryl spray at fruiting stage	0.10	0.14/10	2
Monocrotophos spray at fruiting stage	0.05	0.21/7	7
Quinalphos spray at fruiting stage	0.05	0.20/15	15
Cypermethrin spray at fruiting stage	0.01	0.06/10	3
Fenvalerate spray at fruiting stage	0.01	0.02/15	5
Carbofuran soil application at planting	1.0 kg a.i./ha	0.05/20	60

3.5. ONION AND GARLIC

Onion (*Allium cepa*) is one of the most important commercial vegetables. The bulbs are consumed both in mature and tender stages. Even the leaves and stalks of inflorescence are used as vegetables. It is being cultivated in 0.695 million hectares producing 9.248 million tonnes of onions with an average yield of 13.3 tonnes per hectare (Bijay Kumar, 2007) (Tables 3.18 and 3.19).

Table 3.18. Major onion producing countries of the world

Country	Area (in '000 ha)	Production (in '000 MT)	Productivity (in MT/ha)	% Share of world production
China	901	19040	21.1	30.2
India	695	9248	13.3	14.7
USA	65	3346	51.2	5.3
Turkey	77	2070	26.8	3.3
Pakistan	128	1765	13.8	2.8
Russia	123	1759	14.3	2.8
Iran	50	1685	33.7	2.7
Egypt	43	1302	30.6	2.1
Brazil	58	1138	19.5	1.8
Mexico	40	1115	28.1	1.8
Others	1179	20500	17.4	32.6
Total	3359	62968		

Source: FAO (except Indian data) (Indian data source: Indian Horticulture Database 2006)

Table 3.19. Area, production and productivity of leading onion growing states in India (2005-06)

State	Area (in'000 ha)	Production (in '000 MT)	Productivity (MT/ha)
Maharashtra	154.0	1878.8	12.2
Gujarat	84.9	1876.6	22.1
Bihar	49.8	1011.7	20.3
Karnataka	139.0	1008.5	7.3
Andhra Pradesh	39.4	670.0	17.0
Madhya Pradesh	40.0	639.8	16.0
Rajasthan	44.5	427.2	9.6
Haryana	15.5	317.7	20.5
Uttar Pradesh	21.2	283.1	13.3
Orissa	26.8	243.9	9.1
Tamil Nadu	29.2	233.8	8.0
Others	50.8	657.3	12.9
Total	695.1	9248.4	13.3

(Source: Indian Horticulture Database 2006)

Garlic (*Allium sativum*) is mainly used for flavouring vegetables and meat dishes. It is carminative, stimulant and cures digestive disorders. It has been found to be very effective in curing diabetes and high blood pressure. It is being cultivated in 0.107 million hectares producing 0.436 million tonnes of garlic with an average yield of 4.1 tonnes per hectare. Madhya Pradesh, Gujarat, Orissa, Maharashtra, Uttar Pradesh and Rajasthan are major garlic -growing states. More than 50% production comes from Madhya Pradesh and Gujarat only.

3.5.1. Insect Pests

3.5.1.1. Thrips, *Thrips tabaci*

i) Damage: Thrips are the most important insect pests on onions. Adults are yellowish to yellowish brown with narrow wings. Eggs are laid in clusters in leaf tissues. Numerous nymphs and adults are observed between leaf sheaths and stems lacerating the epidermis of leaves and sucking the exuding coll sap. The affected leaves show silvery blotches which later turn into brownish colour (Fig. 3.29). The leaves get distorted from tips downwards and the plant ultimately wilt and dry away. Seedlings are killed due to heavy infestation and the plant growth is retarded. The shape and size of bulbs as well as yield is affected by thrips.

ii) Management

- Soil application of neem cake at 250 kg/ha at 30 and 60 days after planting.

- Soil application should be followed by 2 to 3 sprays of 4% neem seed powder extract or 1% neem soap at 10 days interval.
- Soil application of phorate or carbofuran granules (1 kg a.i./ha) is also recommended. Alternate spraying of 0.07% endosulfan (2ml/litre of water), or 0.07% nuvacron (2ml/litre of water) or 0.01% cypermethrin (1ml/litre of water) along with a sticker (Triton or Sandovit) at fortnightly intervals is effective.
- Spray fipronil 5 SC at 1ml/l or lamda cyhalothrin 5 EC at 0.5 ml/l or acephate 75 SP at 1g/l alternated with spray of azadirachtin formulation (10,000 ppm) at 2ml/l.
- Thrips are weak fliers and carried to long distances by wind. Blocking adult thrips can reduce the initial and subsequent pest load on onion. Barrier crops can be employed for this purpose. The % blocking was highest in 2 rows of maize planted in zig zag manner (80.45%) followed by one inner row of wheat and one outer row of maize (79.17%). Coccinellid predators (*Scymnus nubilis*, *Chrysoperla* spp. and *Latus* spp.) were higher in plots where maize was used as a barrier crop.
- Onion cvs. Arka Lalima, White Persian, Grano, Sweet Spanish, Crystal Wax, Spanish White, Bombay White, Pusa Red, Udaipur 103, N 780, N 53, and White Poona Red are reported to be resistant/ tolerant to thrips.
- Sprinkling of water through jet nozzles prevents thrips multiplic-ation. Spraying with 5% NSKE during vegetative stage is also effective against thrips.
- Use yellow sticky traps at 10/ha to prevent thrips damage.

Fig. 3.29. Thrips damage on onion leaves.

3.5.1.2. Leaf Eating Caterpillar, *Spodoptera litura, S. exigua*

i) Damage: It is a polyphagous pest which is active in nursery stage. Larvae damage leaves of onion and garlic.

ii) Management: Fields should be properly ploughed to expose and kill the pupae in the soil. Flood irrigation may be done to drown the hibernating caterpillars. Digging trenches around the infested field prevents migration of larvae. Release of *Trichogramma brasiliensis* at 50,000/ha at weekly interval should be done 4-5 times. Anthocorid bugs are potent predators of leaf eating caterpillars.

3.5.1.3. Head Borer, *Helicoverpa armigera*

i) Damage: The head borer is a serious pest of onion seed crop in Northern India. The larva of this insect cuts the pedicel of the flower and feeds on the stalk. Single larva damages many flower stalks. The fully-grown larva is greenish with dark brown/grey lines along the side of body and measures about 35-45 mm in length.

ii) Management: Spraying endosulfan (2-3 ml/litre of water) along with sticker (Triton/ Sandovit) is recommended to control the insect.

3.5.1.4. Onion Maggot, *Hylemia antiqua*

i) Damage: The adults of this insect appear like housefly. The flies lay their eggs on old leaves or on soil and larva enters into the soil and damage disc portion of onion bulb. Infested plants turn yellowish brown and finally dry up. The affected bulbs rot in storage.

ii) Management: Regular crop rotation should be followed and phorate should be applied in the soil before transplanting.

3.5.1.5. Cut Worm, *Agrotis ipsilon*

i) Damage: The larvae are seen in nursery beds and newly transplanted onion fields. The tender plants are found dumped at ground level during the night. Young larvae feed gregariously on foliage but later segregate and enter into soil. Young larvae are yellowish grey and later on become brown, greasy to touch and coil when disturbed. They cut the seedlings at ground level during night and hide during day.

ii) Management: Soil application of carbofuran (1 kg a.i./ha) at the time of planting is recommended. Chlorpyriphos (5ml/litre of water) also gives good control of this pest.

3.5.2. Mite Pests

3.5.2.1. Mites, *Aceria tulipae*

i) Damage: These are small white mites which suck the sap from the leaves making them curl, discolour, and get stunted. They also damage the dormant bulbs in storage. The mites feed upon the epidermal cells of bulb scales, causing the cells to collapse.

ii) Management

- Soil application of neem cake at 30 and 60 days after planting followed by spray of 4% neem seed powder extract or 1% neem soap at 10 days interval.
- Spray acaricides like wettable sulphur 80 WP at 3g/l.

Pesticide Residues

The persistence of terminal pesticide residues and waiting periods for onion are presented in Table 3.20 (Awasthi, 2001).

Table 3.20. Terminal pesticide residues and their waiting periods for onion

Pesticide schedule	Dosage/ Concentration (%)	Terminal residues (ppm/days)	Waiting period (days)
Carbaryl spray at fruiting stage	0.10	1.86/25	20
Malathion spray at fruiting stage	0.05	1.46/5	6
Cypermethrin spray at fruiting stage	0.01	ND*/Bulbs	Nil
Fenvalerate spray at fruiting stage	0.01	ND*/Bulbs	Nil
Carbofuran soil application at planting	1.0 kg a.i./ha	0.42/75	90
Phorate soil application at planting	1.0 kg a.i./ha	0.25/75	60

* ND – Not detectable

3.6. OKRA

Okra (*Abelmoschus esculentus*) is an annual vegetable crop grown in tropical and sub-tropical regions. It is being cultivated in 0.376 million hectares producing 3.684 million tonnes of okra with an average yield of 9.8 tonnes per hectare (Bijay Kumar, 2007) (Table 3.21). Tender green fruits are cooked in curry and soup. The root and stem are used for clearing cane juice in preparation of 'gur'. High iodine content of fruits helps to control goitre while leaves are used in inflammation and dysentery. It is a good source of Vitamins A, B and C. The fruits are also used in cases of renal colic, leucorrhoea and general weakness. The dry seed contains 13-22% good edible oil and 20-24% protein. The oil is used in soap, cosmetic industry and as vanaspati while protein is used for fortified preparations. The crushed seed is fed to cattle for more milk production and the fibre is utilized in jute, textile and paper industry.

Table 3.21. Area, production and productivity of leading okra growing states in India (2005-06)

State	Area (in'000 ha)	Production (in '000 MT)	Productivity (MT/ha)
West Bengal	65.8	718.9	10.9
Bihar	56.2	714.1	12.7
Orissa	71.4	618.8	8.7

Gujarat	41.0	350.3	8.5
Andhra Pradesh	25.6	332.7	13.0
Chattishgarh	18.1	184.7	10.2
Maharashtra	29.0	182.7	6.3
Assam	9.6	119.7	12.5
Uttar Pradesh	8.8	110.2	12.5
Haryana	14.0	93.7	6.7
Karnataka	7.7	63.2	8.2
Others	28.9	194.9	6.7
Total	376.1	3684.0	9.8

(Source: Indian Horticulture Database 2006).

3.6.1. Insect Pests

3.6.1.1. Leaf Hopper, *Amrasca biguttula biguttula*

i) Damage: These are polyphagous and destructive pests of malvaceous crops that lay pear shaped, elongated and yellowish white eggs in the veins on the under surface of leaves. Both the adults and nymphs suck the cell sap (Fig. 3.30) and inject toxic saliva into the plant tissues. Nymphs move diagonally when disturbed. There are 10 to 12 overlapping generations in a year. Plants lose their vitality and affected leaves turn yellow and curl upwards. When the infestation is high, typical hopper burn symptoms are produced. The leaves turn brick red and show large necrotic spots (this is often confused with fungal infection). The pest infestation is more severe during summer. Potentially they can cause 50-100% yield loss if adequate control measures are not taken on time.

Fig. 3.30. Leaf hopper on okra.

ii) Management

- Apply neem cake at 250 kg/ha to soil immediately after germination and repeat it after 30 days.

- In the initial stages of crop before flowering, spray systemic insecticides like acephate 75 SP (1 ml/l) or imidacloprid (0.3 ml/l). Once the fruit harvest starts, avoid systemic insecticides as harvesting is done once in 3-4 days.
- Spray 0.5% neem or pongamia soaps or 4% pulverized neem seed powder extract (NSPE)/NSKE which acts as oviposition deterrent, repellent and growth inhibitory effect on leaf hoppers and result in lower percentage of normal adult emergence. Give the lower surface of the leaves thorough spray, as the pest is generally found there.
- Soil application of neem and pongamia cakes at 250 kg per ha immediately after germination and repeated at 30 days interval twice reduced the hopper incidence to less than 3/plant (Krishna Murthy and Krishna Kumar, 2002).
- Two alternate sprays of deltamethrin at 10 g a.i./ha or fenvalerate, cypermethrin or permethrin all at 50 g a.i./ha proved most effective and recommended for the management of leaf hoppers (Krishna Kumar *et al.*, 1989).
- Lower population of leaf hoppers was noticed with sprays of poultry manure + herbal leaf extract and neem cake + herbal leaf extract.
- Okra cvs. Janardan, EMS-8, Ludhiana Selection-2 (a sister line of Punjab Padmini) and Punjab Padmini had a lower level of jassid population.

3.6.1.2. Shoot and Fruit Borer, *Earias vittella, E. insulana*

i) Damage: Female lays green colored eggs with longitudinal ridges on buds, flowers and fruits. When the crop is young, larvae bore into tender shoots and tunnel downwards, which wither, drop down and growing points are killed (Fig. 3.31). With the formation of fruits, the caterpillars bore inside these and feed on inner tissues which become deformed in shape with no market value (Fig. 3.31). Like brinjal shoot and fruit borer, the infestation is seen on shoots before flowering and after flowering they feed exclusively on fruits.

Fig. 3.31. Shoot and fruit borer damage on shoots and fruits of okra.

ii) Biology: *Earias vittella*, commonly called as spotted bollworm, is very destructive to okra. It is active through out the year reaching peaks during March-May and August-October. Full grown caterpillars are 18 to 24 mm long, stout, spiny and brownish coloured. Adults are 13 to 15 mm long having pale white fore-wings with a broad wedge-shaped horizontal green patch in the middle. Hind wings are silvery-creamy in colour. Wing expanse is about 30 to 34 mm. Eggs are light brown in colour and are laid singly on buds, shoots and fruits. Eggs hatch in about 7 days. Young larvae are spindle shaped with yellow spots on tanned skin colour. The pest has about 8 to 12 overlapping generations in a year. *E. insulana* is found in drier regions. Its fore-wings are completely green. Its moths are smaller in size and larvae almost similar except in size.

iii) Management

- Collect and destroy the affected fruits and stems.
- Apply neem cake at 250 kg/ha after germination and repeat two more times at 30 days interval and spray 1% neem soap or 4% NSPE at 10 days interval. However, please note that botanicals get washed away due to rains. Under such conditions spray effective contact insecticides.
- Spray any contact insecticides like indoxacarb 14.5 SC at 0.75 ml/l. Care must be taken for maintaining a waiting period of at least one week to avoid residues in the produce.
- Avoid staggered planting.
- Two alternate sprays of deltamethrin at 10 g a.i./ha or fenvalerate, cypermethrin or permethrin all at 50 g a.i./ha proved most effective and recommended for the management of fruit borer (Krishna Kumar and Srinivasan, 1984).
- Releases of *T. chilonis* brought about 49.22% reduction in fruit borer (*E. vitella*) in Tamil Nadu. Three applications of *B. thuringiensis* (Dipel) at 0.5 kg/ha at weekly intervals reduced damage by *E. vitella.*
- Okra cvs. Selection-2 (Gujarat Agricultural University, Anand), Parbhani Kranti, EMS-8 and Ludhiana Selection-2 had lower borer infestation on shoots and fruits.

Soil application of neem and pongamia cakes at 250 kg per ha immediately after germination and repeated twice at 30 days interval reduced the borer incidence to less than 7%. Similarly significant reduction in hopper (*Amrasca biguttula biguttula*)(<3/plant) and petiole maggot (*Melanagromyza hibisci*) (<5%) were also noticed. Cakes were better than seed powders in reducing pest incidence and neem cake was the best treatment. This treatment was also found to reduce yellow vein mosaic virus and powdery mildew diseases (Krishna Moorthy and Krishna Kumar, 2002) (Table 3.22).

Table 3.22. Efficacy of neem and pongamia cakes on major pests and YVMV disease in okra

Treatment	Incidence of petiole maggot (%)	Mean no. of hoppers / 3 leaves	Incidence of fruit borer (%)	Incidence of YVMV
Neem seed powder	13.79	3.63	11.21	0.66
Pongamia seed powder	15.73	3.97	16.54	1.33
Neem cake	4.76	2.77	5.22	0.00
Pongamia cake	2.44	3.04	6.22	5.00
Untreated control	16.67	7.24	24.24	80.64

Spraying of 5% NSKE also gives effective control of shoot and fruit borer. Spraying of 5% NSKE and 3% neem oil effectively checked the fruit borer damage in okra (Table 3.23).

Table 3.23. Effect of neem products in the control of fruit borer in okra.

Treatment	Fruit borer damage (%)	
	Weight basis	Number basis
Neem cake extract 10%	7.72	3.26
NSKE 5%	1.87	0.36
Neem oil 3%	1.44	0.23
Carbosulfan 0.044%	2.77	0.53
Untreated check	37.71	33.77

3.6.1.3. Aphids, *Aphis gossypii*

i) Damage: This is a polyphagous pest, feeding in colonies and completely covers the shoot tips, buds and lower surface of leaves (Fig. 3.32). The aphids are very small, soft bodied insects which are whitish-green in colour. Aphids multiply parthenogenetically in large numbers in a very short period. Both nymphs and adults suck the sap due to which plants lose their vitality. Leaves curl downwards and there is retardation of growing shoots. They also excrete honeydew on which sooty mould develops which hampers the photosynthetic activity of the plant.

ii) Management

- Clipping and destruction of infested shoots and plant parts.
- Thorough spraying of 1% neem or pongamia soap or 4% pulverized neem seed powder extract (NSPE).
- During pre-flowering period, spray systemic insecticides like dimethoate 30 EC (2 ml/l) or acephate 75 SP or acetamiprid.

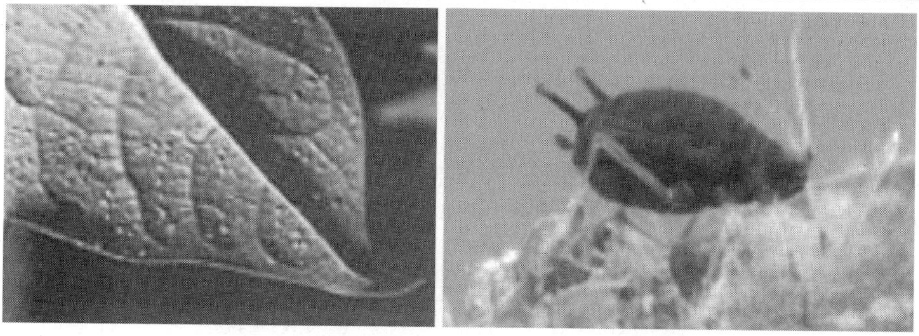

Fig. 3.32. Aphid infestation on okra leaves.

3.6.1.4. Stem Fly/Petiole Maggot, *Melanagromyza hibisci*

i) Damage: This is a dipteran fly and attacks the crop throughout the growth period. Generally drying of leaves is noticed. If petioles of such leaves are opened, yellow maggot and pupae can be noticed (Fig. 3.33). Pest attack is noticed immediately after germination which may either cause mortality or swelling of the particular area of attack. Sometimes cracking of main stem is also observed.

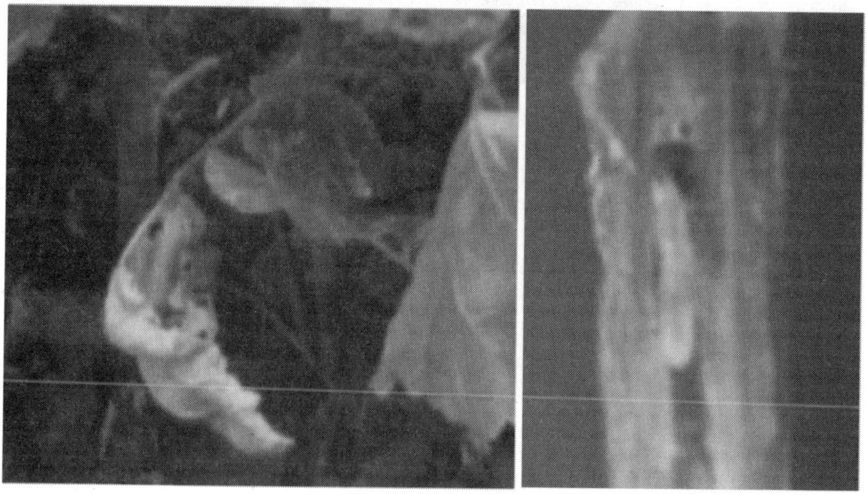

Fig. 3.33. Petiole maggot infestation on okra.

ii) Management

- Apply neem cake at 250 kg/ha immediately after germination and repeat after 30 days. This should be followed by sprays of 4% NSPE or 1% neem soap at 10 days interval after flowering reduced the petiole maggot (*Melanagromyza hibisci*) incidence to less than 5% (Krishna Murthy and Krishna Kumar, 2002).
- Stem fly is heavily parasitized in the field to an extent of 36.7% by *Eurytoma* sp. (Krishna Kumar and Srinivasan, 1988).

3.6.1.5. White Fly, *Bemesia tabaci*

i) Damage: It is a polyphagous vector transmitting yellow vein mosaic virus disease (YVMV). Adults are tiny insects with white wings and pale body. Eggs are attached by thin, small whitish stalks. It lays eggs on the lower surface of leaves. Both the adult and the nymph suck leaf sap and exude honey dew (Fig. 3.34). The plants lose their vitality and become pale yellow. Photosynthetic activity is affected due to sooty mold development.

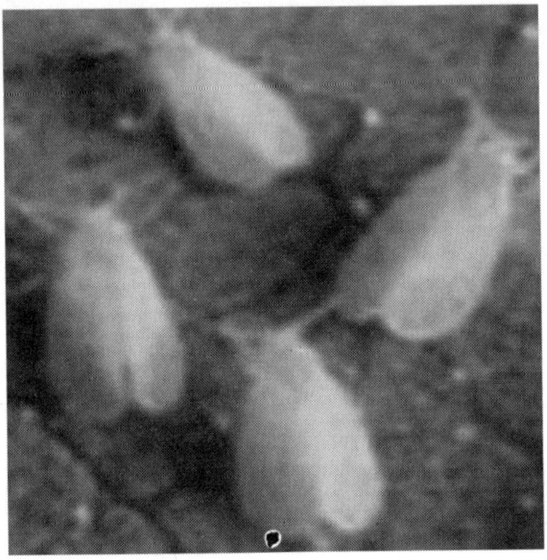

Fig. 3.34. White fly infestation on okra.

ii) Management

- Grow YVMV resistant varieties/hybrids (Arka Anamika, Arka Abhay)
- Apply neem cake at 250 kg/ha at germination and again at 30 DAP followed by sprays of 4 % pulverized neem seed powder extract or 1% neem oil.
- Spray imidacloprid 200 SL at 0.3 ml/l or thiomethoxam at 0.3 gm/l before flowering.

Spraying of 5% NSKE effectively controlled the white fly in okra resulting in over 80% reduction in whiteflies followed by 3% neem oil spray with 78% reduction over untreated control (Table 3.24).

Table 3.24. Effect of foliar application of neem products for the control of white flies in okra.

Treatment	Mean popn. of white flies/ 3 leaves	% Reduction over untreated check
Neem cake extract 10%	9.2	67.9
NSKE 5%	5.7	80.1
Neem oil 3%	6.3	78.0
Carbosulfan 0.044%	23.9	16.5
Untreated check	28.6	---

3.6.1.6. Pod Borer, *Helicoverpa armigera*

i) Damage: This pest is occasionally serious during monsoon season on okra (Fig. 3.35). Damage is restricted to the apical end only. Since okra fruits are harvested once in 2-3 days, damage by *H. armigera* is not very high.

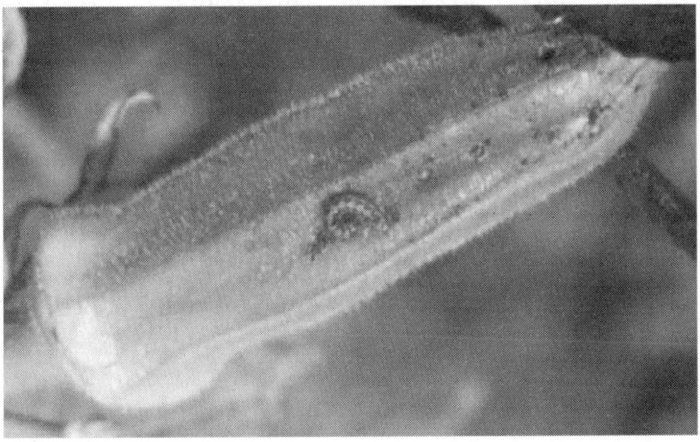

Fig. 3.35. *Helicoverpa armigera* infestation on okra fruit.

ii) Management: Application of 5 rounds of *Ha* NPV at 500 LE/ha at weekly intervals right from flowering gave effective control of the pest.

3.6.2. Mite Pests

3.6.2.1. Mites, *Tetranychus cinnabarinus, Oligonychus coffeae*

i) Damage: The infestation of mites is mostly observed during the warm and dry periods of the summer season. Larvae are pink in colour while nymphs are greenish red in colour. Adults are oval in shape, reddish brown in colour. Colonies of larvae, nymphs and adults feed on the under surface of the leaves (Fig. 3.36). The affected leaves gradually start curling and finally get wrinkled, crumpled and fall down. This in turn affects the growth and fruit-forming capacity of the plants.

Fig. 3.36. Spider mites on leaves of okra.

ii) Management

- Spray dicofol 18.5 EC at 2.5ml/l or any other acaricide like wettable sulphur 50 WP at 3g/l.
- As an alternative to synthetic acaricides, spray 1% neem soap/ pongamia soap.
- Thorough spray on the under surface of the leaves is to be ensured for getting good control.

Pesticide Residues

The persistence of terminal pesticide residues and waiting periods for okra are presented in Table 3.25 (Awasthi, 2001).

Table 3.25. Terminal pesticide residues and their waiting periods for okra

Pesticide schedule	Dosage/ Concentration (%)	Terminal residues (ppm/days)	Waiting period (days)
Monocrotophos spray at fruiting stage	0.05	0.03/7	3
Quinalphos spray at fruiting stage	0.05	0.22/15	12
Cypermethrin spray at fruiting stage	0.01	0.23/5	3
Fenvalerate spray at fruiting stage	0.01	0.19/7	5
Endosulfan spray at fruiting stage	0.07	0.37/3	2
Carbaryl spray at fruiting stage	0.10	0.25/7	Nil
Carbofuran soil application at planting	0.5 kg a.i./ha	0.20/45	50
Phorate soil application at planting	1.0 kg a.i./ha	0.16/52	60

3.7. CABBAGE AND CAULIFLOWER

Cabbage (*Brassica oleracea* var. *capitata*) is an important winter vegetable of cole group. It is a rich source of Vitamins A, B and C and also contains minerals. It is being cultivated in 3.121 million hectares producing 67.315 million tonnes of cabbages with an average yield of 21.568 tonnes per hectare (Bijay Kumar, 2007) (Table 5.1). It covers about 4% of total area under vegetables. India comes next to China in cabbage production. In India, it is being cultivated in 0.267 million hectares producing 5.922 million tonnes of cabbage with an average yield of 22.1 tonnes per hectare (Bijay Kumar, 2007) (Tables 3.26 and 3.27).

Table 3.26. Major cabbage producing countries of the world

Country	Area (in '000 ha)	Production (in '000 MT)	Productivity (in MT/ha)	% share of world production
China	1719	33817	19.5	50.2
India	267	5922	22.1	8.8
Russia	166	4051	24.5	6.0
Korea	42	2603	61.8	3.9
Japan	53	2287	42.9	3.4

Ukraine	71	1475	20.9	2.2
Poland	39	1382	35.6	2.1
Indonesia	64	1293	20.1	1.9
USA	30	1100	36.9	1.6
Romania	55	1012	18.5	1.5
Others	615	12375	20.1	18.4
Total	3121	67315	21.5	

Source: FAO (except Indian data) (Indian data source: Indian Horticulture Database 2006)

Table 3.27. Area, production and productivity of leading cabbage growing states in India (2005-06)

State	Area (in'000 ha)	Production (in '000 MT)	Productivity (MT/ha)
West Bengal	74.7	1982.7	26.5
Bihar	33.7	931.6	27.6
Orissa	36.6	578.3	15.8
Assam	24.3	480.3	19.8
Maharashtra	15.0	381.0	25.4
Gujarat	19.0	329.2	17.3
Uttar Pradesh	7.0	209.9	30.0
Haryana	12.3	172.2	14.0
Karnataka	6.4	134.2	20.8
Chattishgarh	6.8	114.0	16.8
Madhya Pradesh	3.8	75.0	20.0
Others	27.7	533.1	19.2
Total	267.4	5921.6	22.1

(Source: Indian Horticulture Database 2006)

Cauliflower (*Brassica oleracea* var. *botrytis*) is the most popular winter vegetable among cole crops. It is rich in Vitamin C. It is being cultivated in 0.292 million hectares producing 5.260 million tonnes of cauliflowers with an average yield of 18.0 tonnes per hectare (Bijay Kumar, 2007) (Tables 3.28 and 3.29). With the development of new varieties, it is now being grown in non-traditional areas like Andhra Pradesh, Tamil Nadu and Kerala.

Table 3.28. Major cauliflower producing countries of the world

Country	Area (in '000 ha)	Production (in '000 MT)	Productivity (in MT/ha)	% share of world production
China	363	7369	20.3	40.5
India	292	5260	18.0	28.9
USA	69	1288	18.6	7.1
Spain	25	442	17.4	2.4
Italy	18	431	23.7	2.4
France	27	375	13.8	2.1
Mexico	25	359	14.2	2.0

Poland	14	239	17.0	1.3
UK	19	219	11.7	1.2
Pakistan	11	206	18.2	1.1
Others	130	1992	15.3	11.0
Total	995	18180		

Source: FAO (except Indian data) (Indian data source: Indian Horticulture Database 2006)

Table 3.29. Area, production and productivity of leading cauliflower growing states in India (2005-06)

State	Area (in'000 ha)	Production (in '000 MT)	Productivity (MT/ha)
West Bengal	65.6	1666.2	25.4
Bihar	59.8	938.5	15.7
Orissa	45.1	638.9	14.2
Maharashtra	13.0	328.9	25.3
Haryana	21.0	302.8	14.4
Gujarat	16.1	277.8	17.2
Assam	16.7	247.8	14.8
Madhya Pradesh	8.6	137.4	16.0
Uttar Pradesh	6.7	134.4	20.0
Punjab	5.6	131.4	23.5
Chattishgarh	9.6	123.5	12.9
Others	24.0	332.7	13.9
Total	291.8	5260.1	18.0

(Source: Indian Horticulture Database 2006)

3.7.1. Insect Pests

3.7.1.1. Diamondback Moth (DBM), *Plutella xylostella*

i) Damage: Diamondback moth is a major pest of almost all the cruciferous crops and is cosmopolitan in distribution. It is a small grayish brown insect bearing whitish triangular spots on posterior margins of the forewings and while resting, the wings come together and form a diamond pattern. This is a major pest of cruciferous crops, particularly cabbage and cauliflower during January-June months and also during drought periods in monsoon. Eggs are yellowish white, and are laid singly on tender leaves. A single female may lay about 40-60 eggs. Incubation period is 3-6 days. The first instar larvae mine the epidermal surface of the leaves. Second instar onwards the larvae feed externally by making holes in the leaves (Fig. 3.37). Pupation occurs inside silken cocoon on the plants, which lasts for 4-6 days. The complete life cycle from egg to egg takes place in about 20-25 days.

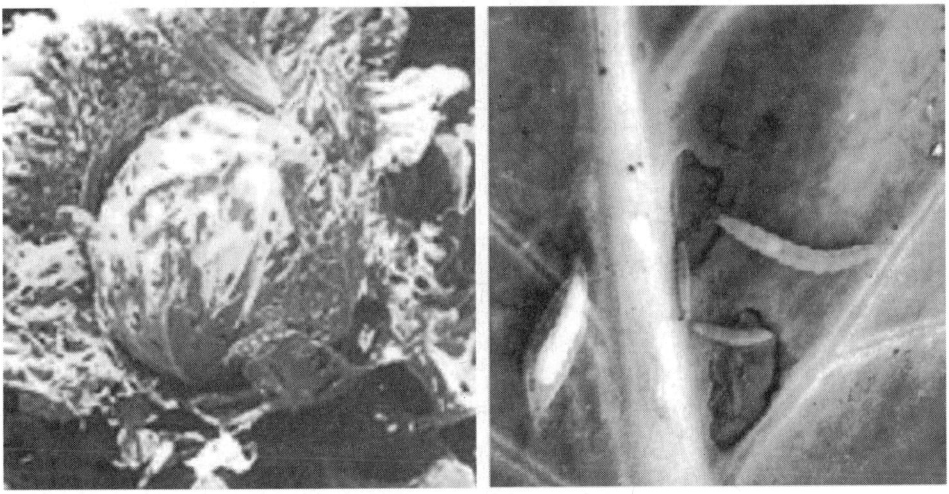

Fig. 3.37. DBM damage on cabbage.

ii) Management

a) Cultural Methods: Sprays of 4% NSKE/4% pulverized NSPE were found to be highly effective against DBM. Field sprayed with NSKE/NSPE had obtained higher yield, incurred 15% less cost and realized 43% higher net returns. The performance of powder formulation of 'Soluneem' was on par with 4% NSKE in controlling DBM. Sprays of neem soap or pongamia soap both at 1% were highly effective in reducing DBM (Table 3.30), aphids (*Brevicoryne brassicae*) and leaf webber (*Crocidolomia binotalis*). Further, the soap sprays significantly increased the yield of cabbage and cauliflower and were relatively safe to the parasitoid, *Cotesia plutellae* (Srinivasan and Krishna Moorthy, 1993; Krishna Moorthy *et al.*, 1998; Krishna Murthy and Krishna Kumar, 2002).

Table 3.30. Efficacy of neem and pongamia soaps on DBM and yield of cabbage

Treatment	DBM incidence/plant	Yield (t/ha)
NSKE	3.00	109.60
Bt	3.00	92.67
Soluneem	3.33	99.60
Neem oil	9.00	66.53
Pongamia oil	7.67	73.33
Neem soap	2.33	111.33
Pongamia soap	4.67	116.67
Control	36.33	43.87

Use of Neem and Pongamia Soaps: Under field conditions, neem and pongamia soap sprays were highly effective in reducing DBM, aphids

(*Brevicoryne brassicae*) and leaf webber (*Crocidolomia binotalis*) and were at par with NSKE. They were superior to many commercial synthetic insecticides like dichlorvos, cypermethrin, combination formulation of cypermethrin and profenfos in controlling DBM. Further, the soap sprays significantly increased the yield of cabbage. On cauliflower too, the neem and pongamia soap sprays were effective and increased the yield significantly. The data on DBM, its parasitoid *Cotesia plutellae* and yield of cabbage and cauliflower are given in Tables 3.31 and 3.32. The soap sprays were also found to be relatively safe to *C. plutellae* (Krishna Murthy and Krishna Kumar, 2002).

Table 3.31. Effect of different botanicals and chemicals on the incidence of DBM, yield and the incidence of larval parasitoid *Cotesia plutellae* in Cabbage (2001-2002)

Treatments	Dose	No. of DBM larvae /10 plants		Yield t/ha		*Cotesia plutellae* cocoons/10 plants
		Winter	Summer	Summer	Winter	
Neem soap	0.5%	10.66	5.33	60.92	37.73	1.66
Neem soap	0.75%	1.33	10.67	85.00	38.73	1.33
Neem soap	1%	0.33	15.00	83.89	51.00	2.33
Pongamia soap	0.5%	1.33	2.67	70.00	37.73	2.00
Pongamia soap	0.75%	0.33	7.00	74.11	47.00	0.33
Pongamia soap	1%	0.00	14.33	87.78	52.33	1.33
NSPE	4%	0.00	2.33	88.00	51.00	3.66
Diclorvos	0.1%	0.00	---	48.89	-	0.00
Bt	0.15	0.00	12.00	79.33	41.46	0.00
Cypermethrin	0.0125%	4.00	65.33	51.89	27.20	0.00
Control	---	5.67	92.33	33.78	13.53	2.67
C.D. (P=0.05)	---	6.15	12.15	16.34	6.13	2.14

Table 3.32. Effect of different botanicals on Diamond back moth population and yield of cauliflower (2001-2002)

Treatments	Dose (%)	DBM/10 plants	Yield (t/ha)
Neem soap	0.5	4.33	6.18
Neem soap	0.75	5.00	5.27
Neem soap	1.0	4.33	7.83
Pongamia soap	0.5	6.67	3.23
Pongamia soap	0.75	3.00	5.85
Pongamia soap	1.0	1.33	7.90
NSPE	4	0.00	7.62
Bt	0.1	5.33	7.53
Cypermethrin	0.0125	13.33	4.82
Control	---	33.00	3.24
CD (P=0.05)		3.69	2.73

The use of both neem and pongamia soaps in the cabbage IPM was successfully validated in more than 50 farmer's field during 2001 and 2002. Among them, 15 farmers used only soap sprays while others used 1-2 insecticide applications (which was not really necessary) and managed the pests very well and reduced the cost towards pesticides. These farmers were previously spraying insecticides more than 12 times for controlling insect pests of cabbage.

The soaps lack systemic effect and the residual toxicity is limited. Therefore, the coverage on the crop surface is very important while spraying to get effective control. During summer months if adults of DBM are found in large numbers, then they have to be controlled first using light traps during night. Otherwise, larval population may not get reduced as the eggs laid by the high adult population hatch and develop without any problem due to low residual toxicity.

b) Biological Methods: Under natural conditions in an undisturbed ecosystem (insecticide-free environment), the parasitoid, *Cotesia plutellae* is capable of exercising about 70-80% control of DBM in India (Nagarkatti and Jayanth, 1982; Chandramohan, 1994) (Fig. 3.38).

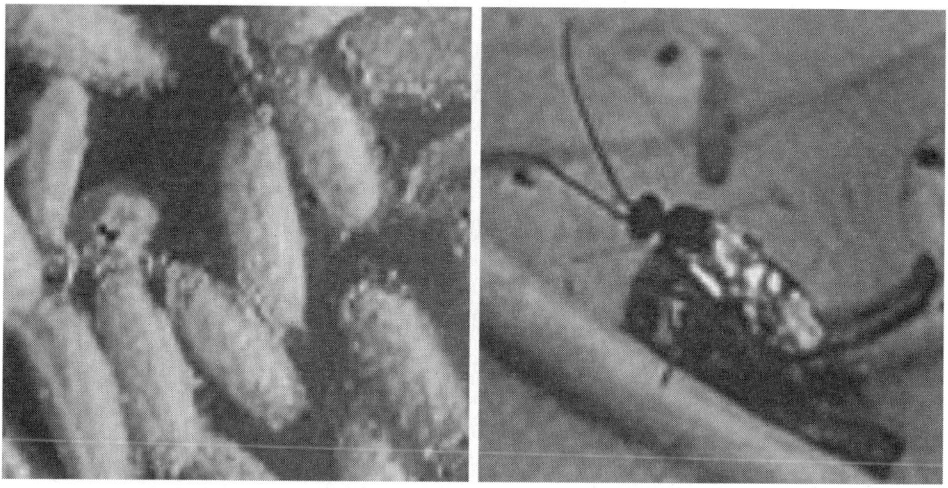

Fig. 3.38. Pupae and adult of DBM parasitoid, *Cotesia plutellae.*

The parasitoid, *Didegma semiclausum* producing excellent parasitism up to 68% in Nilgiris, showed its full potential in places of cooler temperatures ranging from 15-20°C and high hills (Chandramohan, 1994).

Bacillus thuringiensis formulations have been found effective for the management of DBM. 'Biotrol' WP ($25x10^6$ spores/mg) was effective in controlling the DBM on cabbage (Rajamohan and Jayaraj, 1978). 'Thuricide' HPSC and 'Dipel' WP (16 $x10^6$ IU/mg) at 1 and 1.5 g/l could effectively check *P. xylostella* on cabbage (Varma and Gill, 1976). Weekly sprays of 'Dipel' at 0.5 kg/ha were recommended for the control of *P. xylostella* and *Crocidolomia binotalis* (Krishnaiah *et al*., 1981). Application of 5 rounds of *B. thuringiensis* var. *kurstaki* (Halt) at 1 kg/ha at weekly interval gave effective control of larval population (0.61-0.81 larva/plant) (79.9% reduction in larval population over check) as compared to control (3.02 larvae/plant). The yield was also high in Halt (70.2 kg/plot) as compared to control (53.8 kg/plot).

The highest mortality of DBM (52.5% in WP and 70.0% in Emulsified suspension formulation) was recorded in case of *Beauvaria bassiana* isolated from DBM. Application of 5 rounds of *Paecilomyces farinosus* at weekly intervals significantly brought down the larval population of DBM and increased the marketable yield. Natural epizootic of *Zoopthora radicans* was recorded on cabbage causing 68.6% mortality to the insect larvae (Gopalakrishnan, 1998). Application of 5 rounds of *P. farinosus* at 1.7 x 10^8 spores/ml + Triton X-100 (0.01%) at weekly interval significantly brought down the larval population to 1.07/10 plants (as compared to 5.15/10 plants in control) and increased the fruit yield to 43.69 kg/plot (as compared to 30.88 kg/plot in control) (Gopalakrishnan, 1998).

c) Integrated Methods

- Sow Indian mustard as a trap crop.
- Spray 4% neem seed powder extract or 1% neem or pongamia soap by giving thorough coverage to the crop canopy.
- Install light traps to control adults if required. For one acre plot, use 3-4 light traps (60 or 100 W bulbs). They have to be hung above a bucket half filled with water. Otherwise hung the bulb above a gunny bag (slanting below). Gunny bag has to be smeared with grease or oil. Illuminate the bulbs for full night. Adults of DBM will get attracted to light and trapped in the water/oil. Use the light traps for 3-4 days for effective control of DBM adults (Fig. 3.39).

Fig. 3.39. Light trap to attract DBM larvae.

IPM using Indian Mustard as a Trap Crop: IPM using Indian mustard as a trap crop involves planting of paired rows of mustard after every 25 rows of cabbage/cauliflower and spraying of 4% NSKE at primordial formation. Two more sprays of 4% NSKE may be given at 10-15 day interval after the first spray. The IPM gave 152% more returns than pure cabbage crop. IPM controls diamondback moth, leaf webber (*Crocidolomia binotalis*); stem borer (*Hellula undalis*); aphids (*Brevicorne brassicae, Hyadaphis erysimi*) and bug (*Bagrada cruciferarum*) (Srinivasan and Krishna Moorthy, 1991) (Fig. 3.40).

Fig. 3.40. IPM in cabbage and cauliflower using Indian mustard as a trap crop.

3.7.1.2. Leaf Webber, *Crocidolomia binotalis*

i) Damage: This pest is widely distributed in the Indian Subcontinent, South Asia and Australia. It is a serious pest of all cruciferous crops. Full grown caterpillars are pale brown in colour with a paler dorsal stripe and two dorso-lateral, one lateral and one sub-lateral series of black spots and specks. The moths lays eggs in clusters of 40-100 on leaf surface which hatch in 5-7 days. Young larvae feed gregariously and later web the leaves together and feed within (Fig. 3.41). These leaves contain excreta of the leaf webber rendering either total loss or giving overall dirty appearance to the heads and hence the less market value for the produce. This results in rotting of cabbage heads and cauliflower curds. The larval period lasts for 25-50 days while pupation occurs in soil, which varies from 15-40 days.

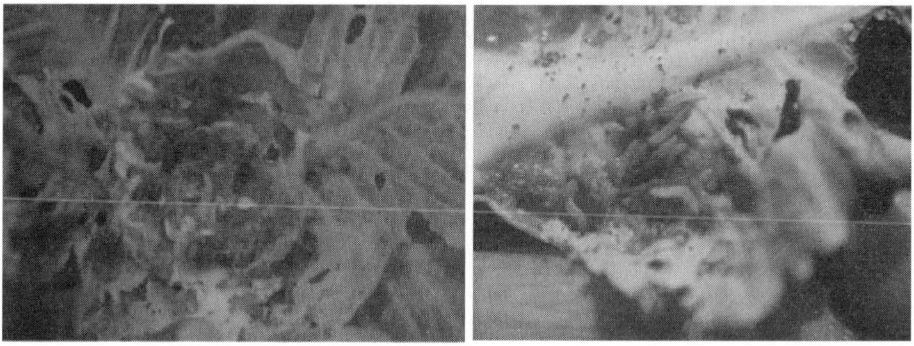

Fig. 3.41. Leaf webber damage on cabbage leaves.

ii) Management

- Collection and destruction of egg masses and gregarious larvae.
- Spray any contact insecticide on the foliage when the larvae are observed on the crop.
- Spray 4% neem seed kernel extract (NSKE) or 4% pulverized neem seed powder extract (NSPE).

3.7.1.3. Stem Borer, *Hellula undalis*

i) Damage: Though this pest is sporadic in nature, its infestation prevents the head formation (Fig. 3.42) and hence is serious. Many times more than 50% of the plants show multiple head formation. Caterpillars are pale whitish brown in colour and have 4 to 5 purplish brown longitudinal stripes. The adult moths lays eggs singly on the leaves, the hatched larvae mine the leaves along side veins. The grown up larvae bore into stems preventing head initiation which results in multiple shoots and multiple head formation with low market value. The damage is more severe during March-July months.

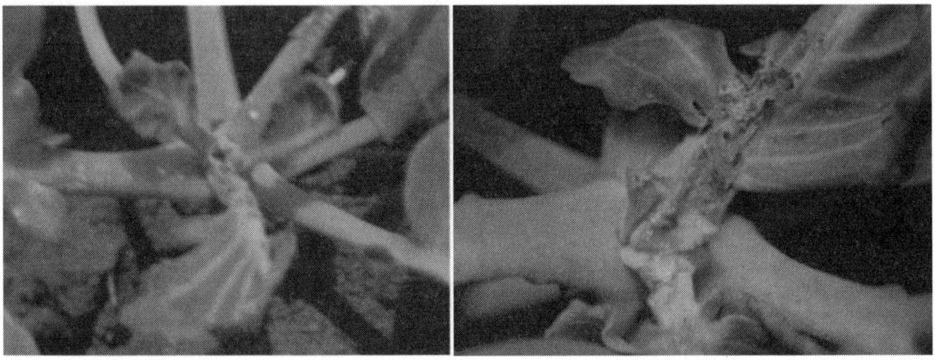

Fig. 3.42. Stem borer damage on cabbage.

ii) Management

- Spray the seedlings with an effective contact insecticide before planting followed by one more spray at 10 DAP, particularly for summer and kharif crops (between March-July).
- Spray 4% pulverized neem seed powder extract (NSPE).

3.7.1.4. Aphids, *Brevicoryne brassicae, Hyadaphis erysimi*

i) Damage: Aphid is very serious during dry and hot periods of the year. *Brevicoryne brassicae* is more dominant than *Hyadaphis erysimi* and its incidence is high during February-June. *B. brassicae* have white mealy covering while, it is lighter in colour in *H. erysimi*. These are parthenogenetic. Both nymphs and adult aphids suck the sap and devitalize the plants affecting the quality of head or curds formed (Fig. 3.43). Aphids also produce copious amount of honey dew which makes the plants sticky favouring the growth of sooty mold as a result of which a black coating is formed on affected plants, hindering the photosynthesis. These aphids become severe when contact insecticides are sprayed regularly.

ii) Management

- Spray any systemic insecticide like dimethoate 30 EC at 2ml/l when the aphids are observed on the crop.

- Spray 4% pulverized neem seed powder extract (NSPE).
- Follow mustard trap crop IPM as mustard will attract all aphids after flowering and main crop will be spared of the pest incidence.
- Cauliflower intercropped with non-crucifer host plants like sunflower, tomato and marigold were highly effective in reducing the aphid incidence and enhancing the number of natural enemies resulting in higher yields.
- Natural epizootic of fungus *Erynia neoaphidis* causing 80.6% mortality was reported on cabbage (Gopalakrishnan and Mohan, 1993).

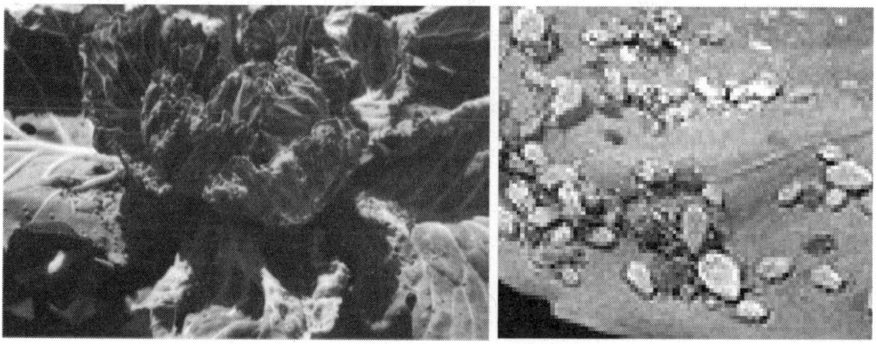

Fig. 3.43. Aphid infestation on cabbage leaf.

3.7.1.5. Tobacco Caterpillar, *Spodoptera litura*

i) Damage: It is a polyphagous pest. The moths lay dirty white eggs in groups and young larvae up to third instar feed gregariously and skeletonize the leaves (Fig. 3.44). Caterpillars are pale brown in colour with a greenish to violet tinge. Grown up larvae completely devour the leaves and other plant parts in case of severe incidence. When the incidence is high they attain cut worm status, hide during daytime and come out during night and devastate the crop. During severe infestation, the entire crop may be defoliated overnight.

Fig. 3.44. Cabbage head damaged by *Spodoptera litura*.

ii) Management

- Collection and destruction of egg masses and gregarious feeding larvae is the most effective and cheapest method. Regular scouting has to be followed to locate egg masses and location of skelitonized leaves.
- Poison baiting (10 kg rice/wheat flour + 1 kg of jaggery + 500 g of methomyl formulation per acre). Mix the flour with jaggery in the morning. In the evening add insecticide and mix carefully by wearing gloves and using a stick. Sprinkle the mixture in furrows. The fermenting jaggery attracts the caterpillars and after feeding the bait they get killed. Baiting has to be done 2-3 days until complete control is achieved.
- NSKE spray at 4% controls early stage larvae. Large larvae have to be controlled by piercing with a sharp thick iron needle in head and killing.
- The exotic egg parasitoid, *Telenomus remus* alone was found to give effective control (Krishnamoorthy and Mani, 1985).

3.7.1.6. Gram Caterpillar, *Helicoverpa armigera*

i) Damage: This is a minor pest of cruciferous vegetables which feed on the foliage deteriorating the market quality of heads and curds. Solitary larvae with hairs and big holes plugged with excreta are found on heads.

ii) Management

- Mechanical control will be very effective for gram caterpillar than spraying insecticides as the grown up caterpillars are likely to hide inside the hole and insecticide will not reach it by spraying.
- Young larvae may be reduced by spraying 0.07% endosulfon or 0.1% dichlorvos.

3.7.1.7. Mustard Saw Fly *Athalia lugens proxima*

i) Damage: This pest is occasionally serious on cabbage and cauliflower. One can see black larvae gregariously feeding on leaves.

ii) Management

- As and when observed, spray quinalphos or chlorpyriphos both at 0.05% for the management of mustard saw fly.

3.7.1.8. Cabbage Butterfly, *Pieris brassicae*

i) Damage: This is widely distributed in North America, Europe, China, Burma and India. In India, it is found along Himalayan region and parts of

North India. It is not found in South India. Eggs are laid in clusters on undersurface of leaves. A female may lay more than one cluster of eggs. Young larvae are gregarious. They disperse later and skeletonize the leaves or even bore into heads. Pupation is on leaves and stems. Incubation, larval and pupal period are 3, 5 and 7 days, respectively.

ii) Management

Mechanical removal of egg masses and gregarious larvae is effective. Spraying of *Bacillus thuringiensis* also gives good control of this pest. 'Thuricide' (30x10^6 spores/mg) at 0.4% was found effective for the control of *Pieris brassicae* and *Plutella xylostella* on cauliflower.

Cabbage butterfly is found heavily parasitized (80-90%) by a gregarious parasitoid, *Cotesia glomerata* and an ichneumonid, *Hypersota abeninus* in Solan region of Himachal Pradesh (Sood *et al.*, 1995).

Pesticide Residues

The persistence of terminal pesticide residues and waiting periods for cabbage and cauliflower are presented in Tables 3.33 and 3.34 (Awasthi, 2001).

Table 3.33. Terminal pesticide residues and their waiting periods for cabbage

Pesticide schedule	Dosage/ Concentration (%)	Terminal residues (ppm/days)	Waiting period (days)
Endosulfan spray at head formation stage	0.07	1.40/7	5
Monocrotophos spray at head formation stage	0.05	0.08/3	2
Phosalone spray at head formation stage	0.05	1.00/10	10
Metalaxyl spray at head formation stage	0.20	0.57/20	21
Cypermethrin spray at head formation stage	0.01	0.17/10	7
Fenvalerate spray at head formation stage	0.01	0.06/15	11
Carbofuran soil application at planting	1.0 kg a.i./ha	0.44/50	60
Phorate soil application at planting	1.0 kg a.i./ha	0.60/50	60

Table 3.34. Terminal pesticide residues and their waiting periods for cauliflower

Pesticide schedule	Dosage/ Concentration (%)	Terminal residues (ppm/days)	Waiting period (days)
Endosulfan spray at curd formation stage	0.07	0.44/10	3
Quinalphos spray at curd formation stage	0.05	0.65/15	20
Phosalone spray at curd formation stage	0.05	0.65/10	6
Cypermethrin spray at head formation stage	0.01	0.11/10	8

Fenvalerate spray at head formation stage	0.01	0.05/15	9
Phorate soil application at planting	1.5 kg a.i./ha	0.42/60	70

3.8. PEA

Pea (*Pisum sativum*) is an important vegetable crop widely grown throughout the world. As a cool season crop, it is extensively grown in temperate zone; but restricted to cooler altitudes in the tropics and winter season in the sub-tropics. A rich source of proteins (25%), amino acids and sugars (12%), green peas are an all time favourite vegetables. It is rich in Vitamins A,B,C and minerals. It is being cultivated in 0.314 million hectares producing 2.299 million tonnes of peas with an average yield of 7.3 tonnes per hectare (Bijay Kumar, 2007) (Tables 3.35 and 3.36).

Table 3.35. Major peas producing countries of the world

Country	Area (in '000 ha)	Production (in '000 MT)	Productivity (in MT/ha)	% Share of world production
India	314	2299	7.3	29.5
China	241	2210	9.2	28.3
USA	86	859	10.0	11.0
France	31	360	11.7	4.6
Egypt	28	290	10.4	3.7
UK	32	133	4.2	1.7
Turkey	17	122	7.0	1.6
Algeria	24	110	4.5	1.4
Hungary	14	85	6.2	1.1
Netherlands	5	84	16.5	1.1
Others	303	1248	4.1	16.0
Total	1094	7799		

Source: FAO (except Indian data) (Indian data source: Indian Horticulture Database 2006)

Table 3.36. Area, production and productivity of leading peas growing states in India (2005-06)

State	Area (in'000 ha)	Production (in '000 MT)	Productivity (MT/ha)
Uttar Pradesh	139.0	1146.8	8.3
Madhya Pradesh	19.4	212.9	11.0
Himachal Pradesh	16.3	177.0	10.9
West Bengal	21.7	129.4	6.0
Jammu & Kashmir	7.3	112.2	15.4
Punjab	17.2	103.3	6.0
Haryana	11.1	84.9	7.6

Uttarakhand	9.6	60.9	6.3
Chattishgarh	7.2	50.8	7.1
Bihar	8.1	50.5	6.2
Orissa	4.8	42.0	8.7
Others	52.4	128.0	2.4
Total	314.0	2298.7	7.3

(Source: Indian Horticulture Database 2006)

3.8.1. Insect Pests

3.8.1.1. Bean Fly, *Ophiomyia centrocematis*

i) Damage: This species is specific to peas and lays eggs in the base of the stem. This causes drying of leaves, starting from the base. Hence, the first symptom is drying of the first primitive leaf at the base. This occurs after about 20 days of sowing. A database on bean flies has been developed by the Indian Institute of Horticultural Research, Bangalore (Anon, 2000).

ii) Management

- Apply neem cake at 250 kg/ha to the ridges immediately after germination. This should be followed by sprays of 4% pulverized neem seed powder extract or 1% neem soap or neem formulation with more than 10000 ppm azadirachtin at 2-3 ml/l 20 and 30 DAS.
- Spray acephate at 0.75 ml/l or dimethoate 30 EC at 2 ml/l or 0.05% phosphamidon at 20 and 30 DAS.
- Sow the crop in second fortnight of October to escape the damage from stemfly in Punjab.
- Application of 0.05% phosphamidon at 20 DAS gives effective control of the pest. Second spraying may be repeated after 10 days.

3.8.1.2. Lycaenid Pod Borer, *Lampides boeticus*

i) Damage: This is a butterfly pest and commonly known as pea blue. The eggs are laid singly in flowers and flower buds. In peas, the larvae bore into young pods. The entry hole gets blocked after some time. When the larvae attain fourth instar, they migrate and cause extensive damage to flower buds and pods (Krishna Moorthy and Srinivasan, 1989). Hence, control measures have to be taken in the early stages of the pest incidence.

ii) Management

- In peas, the first spray has to be given at flowering as soon as eggs are noticed with cypermethrin 25 EC at 0.5 ml/l or chloripyriphos 20 EC at 2.5 ml/l or indoxacarb 14.5 SC at 0.5 ml/l.
- Weekly sprays of 5% NSKE after flowering reduces the pest incidence to a great extent.

3.8.1.3. Gram Pod Borer, *Helicoverpa armigera*

i) Damage: This polyphagous pest also attacks peas occasionally along with *Lampides boeticus*. The habit of egg laying and larval feeding are the same as pea blue butterfly.

ii) Management

- In peas, the first spray has to be given at flowering as soon as eggs are noticed with indoxacarb 14.5 SC at 0.5 ml/l or novaluron at 0.75 g/l.

3.8.1.4. Aphids, *Aphis craccivora, Acrythosiphon pisum*

i) Damage: Both nymphs and adults suck the sap from young plant parts like shoot tips and inflorescence. *A. craccivora* is greenish black and A. *pisum* is greenish in colour. *A. pisum* generally occurs on peas.

ii) Management

- Clip infected shoots and spray 1% neem or pongamia soap.
- Spray acephate 75 WP at 0.75 ml/l or dimethoate 30 EC 2ml/l or 0.05% phosphamidon or 0.05% monocrotophos.

Pesticide Residues

The persistence of terminal pesticide residues and waiting periods for peas and beans are presented in Table 3.37. (Awasthi, 2001).

Table 3.37. Terminal pesticide residues and their waiting periods for peas and beans

Pesticide schedule	Dosage/ Concentration (%)	Terminal residues (ppm/days)	Waiting period (days)
Endosulfan spray at pod formation stage	0.07	0.14/10	1
Monocrotophos spray at pod formation stage	0.05	0.13/7	6
Quinalphos spray at pod formation stage	0.05	0.06/15	8
Cypermethrin spray at pod formation stage	0.01	0.03/15	3
Fenvalerate spray at pod formation stage	0.01	0.07/15	3
Deltamethrin spray at pod formation stage	0.002	0.07/10	8
Carbofuran soil application at sowing	0.5 kg a.i./ha	0.15/55	70
Phorate soil application at sowing	1.0 kg a.i./ha	0.26/60	70

3.9. FRENCH BEAN

French bean (*Phaseolus vulgaris*), also known as common, haricot, kidney, string, salad, runner or snap bean, is an important leguminous vegetable. It is consumed as tender pods, shelled green beans and dry beans. It is a nutritious vegetable which contains proteins (1.7g), calcium (50 mg), phosphorus (28 mg), iron (1.7 mg), carotene (132 mg), Thiamine (0.08 mg), Riboflavin (0.06 mg) and Vitamin C (24 mg/100 g of edible pods). It is largely grown in hilly areas of Himachal Pradesh, Jammu and Kashmir and north-eastern states during summer, winter and autumn crop in parts of Uttar Pradesh, Maharashtra, Karnataka and Andhra Pradesh.

3.9.1. Insect Pests

3.9.1.1. Bean Fly, *Ophiomyia phaseoli*

i) Damage: This species is serious on French beans. Some times this species is also noticed on dolichos beans. This pest is particularly serious during dry periods of *Kharif* (June-August) and after January-February months when temperature increases above 30°C. At times more than 70% mortality is caused by this pest attack. The adults are small flies. These lay eggs in the unifoliate leaves that come immediately after germination (Fig. 3.45). The oviposition is done by puncturing the leaf and eggs are laid under the leaf epidermis and these later turn into white spots, often confused with disease. These dried ovipositional cavities are often observed within 2-3 days of germination of seeds. The larvae mine the leaf lamina, veins, midrib, petiole and enter the stem. Larval feeding in stem results in mortality or reduction of the plant growth (Fig. 3.45). Generally the incidence is high during prolonged dry spells between rains, particularly in June-July months. When the incidence is early and high, plants invariably die. This mortality is noticed by farmers only after about 25-30 days, when the plants are about to flower. Any control measure taken at that time does not control the mortality, as it is too late for the plant to recover.

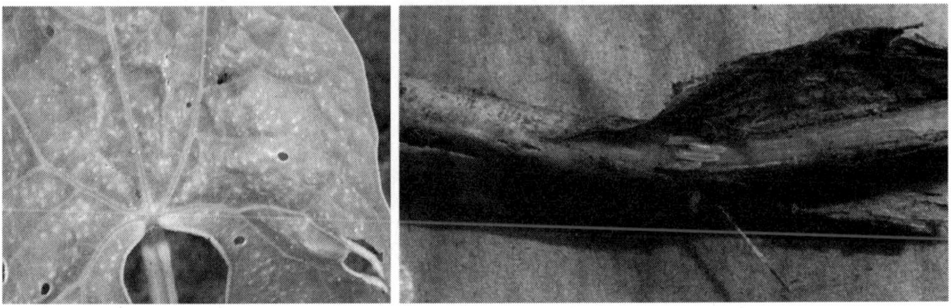

Fig. 3.45. Stem fly damage on leaves and stem of French bean.

ii) Management

- Apply neem cake at 250 kg/ha immediately after germination. Do not delay, particularly during *kharif* period.
- Monitor the plants for adult activities, puncture marks and petiole mining soon after germination.
- As soon as a few adults are noticed hovering over the crop, spray endosulfan 35 EC at 2 ml/l or acephate 75 WP at 0.75 g/l or 4% pulverized NSPE or 1% neem soap or neem formulation with 10000 ppm azadirachtin at 2 ml/l.

- Second spraying is given when on an average 5 leaves have petiole mining symptoms per 10 leaves (about 12-20 days after sowing).

As soon as few puncture marks are noticed on unifoliate leaves caused by oviposition by the adults inside leaf lamina, spraying of 0.07% endosulfan or 4% NSKE/4% pulverized NSPE/1% neem soap/1% pongamia soap is to be given. Second spray can be given after about 15 to 20 DAS. Soil application of neem cake at 200 kg/ha is also effective (Krishna Moorthy and Srinivasan, 1989, 1990, 1997).

3.9.1.2. Serpentine Leaf Miner, *Liriomyza trifolii*

i) Damage: The accidentally introduced pest serpentine leaf miner has become a pest on French beans.

ii) Management : NSKE 4% applied at 10 and 15 DAS French bean was able to reduce serpentine leaf miner incidence up to 21 DAS (15.20 mines as compared to 62.93 mines in control) and also increased pod yield (5.05 t/ha as compared to 1.05 t/ha in control) (Krishna Kumar, 1998).

3.9.1.3. Aphid, *Aphis craccivora*

i) Damage: Small soft bodied insects suck the sap from leaves, tender shoots and flower buds which curl, deform and dry up (Fig. 3.46).

Fig. 3.46. Aphid infestation on petiole and leaves of French bean.

ii) Management: Single spray of water suspension of *Fusarium pallidoroseum* (7 x 10^6 spores/ml) gave effective control of aphids under field conditions. Reinfestation of plant by aphids was not observed while the plants sprayed with quinalphos at 0.05% showed reinfestation necessitating repeated application of the insecticide (3 sprays).

3.9.2. Mite Pests

3.9.2.1. Spider Mite

i) Damage: Webbed colonies of spider mites feed on underside of leaves causing whitish grey patches. The leaves dry up when the infestation is high.

ii) Management : The predatory mite, *Phytoseilus persimilis* gave excellent control of spider mites when released at 10 adults/plant under open field conditions (Krishnamoorthy and Mani, 1989b).

3.10. COWPEA

Cowpea (*Vigna unguiculata*) also known as southern pea and black-eye pea, is one of the most important vegetables. It is cultivated for its long, green or purplish pods to be cooked as vegetable or for dry seeds used as pulse. Its foliage is used as fodder or green manure. In India, cowpea is grown almost throughout the country.

3.10.1. Insect Pests

3.10.1.1. Spotted Pod Borer, *Maruca testulalis*

i) Damage: This is a serious pest on cowpeas (Fig. 3.47). The pyralid moth lays eggs on small flower buds as soon as flowering starts. Flower buds can be examined for holes caused by young larvae. Therefore, control measures have to be taken at flower bud stage itself to prevent the larvae from developing and feeding on pods.

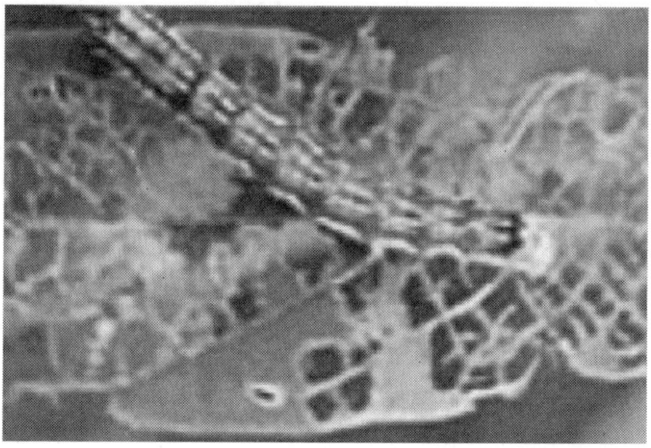

Fig. 3.47. Pod borer damage on cowpea.

ii) Management

- Apply neem cake at 250 kg/ha at flowering and the first spray has to be given at flower bud formation with 4% pulverized neem seed powder extract (NSPE) or 1% neem soap. Repeat the spray after one week.
- Spray cypermethrin 25 EC at 0.5 ml/l or chloripyriphos 20 EC at 2.5 ml/l or indoxacarb 14.5 SC at 0.5 ml/l.
- *Crotalaria* sp. as a trap crop has been proposed in cowpea fields against pod borer in Nigeria (Jackai and Singh, 1981).

3.10.1.2. Bugs

i) Damage: Many bugs are known to attack leguminous crops. Of these, *Nezara viridula* and *Coptosoma cribraria* are often serious, mostly on cowpeas and lablab. Eggs are laid on tender plant parts. Nymphs and adults suck the sap from foliage, flower parts and pods and emit characteristic foul smell.

ii) Management

- Apply neem cake at 250 kg per ha immediately after germination and repeat at flowering. Spray 1% neem soap or 4% pulverized neem seed powder extract at 10 days interval.
- Dust 4% endosulfan or spray with indoxacarb at 0.5 ml/l if the incidence is high.

3.10.2. Vertebrate Pests

i) Damage: Small green barbet, Red vented bulbul and Rose ringed parakeet, *Bandicota bengalensis* and *Funambulus palmarum* fed on cowpea flower buds and seeds.

ii) Management: Intercropping of cowpea with ridge gourd and erecting pandal to 0.5 m height reduced damage, resulting in higher pod yields (Anon, 1992).

3.11. PIGEON PEA

3.11.1. Insect Pests

3.11.1.1. Pod Borer, *Helicoverpa armigera*

i) Damage: Pod borer is one of the major pests in tropical and subtropical regions. The adult moths have yellowish to orange fore wings in females and greenish grey in males, with a light dark transverse band in the distal third. The kidney-shaped marking is slightly distinct and smoky. Hind wings are pale grey with a broad, darker marginal band and small brown marking near the base. Females lay several hundred eggs on all parts of plant. On hatching, the larvae feed for short time on the tender leaflets, flower buds and tender roots. Slowly it enters and feed on the seeds inside the pods (Fig. 3.48). The half portion of larva remains inside while feeding on the developing seeds.

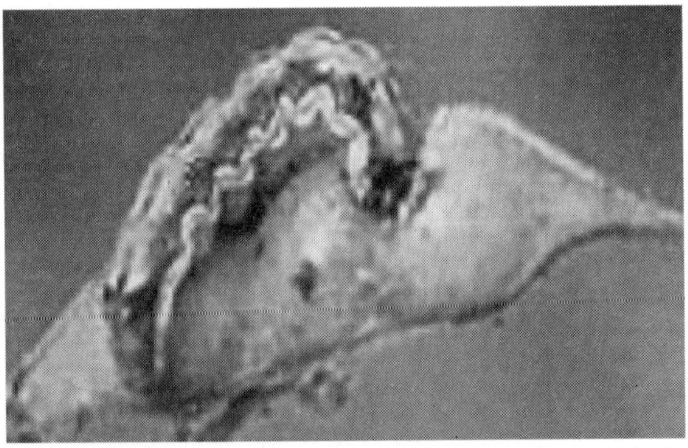

Fig. 3.48. Pod borer on pigeon pea.

ii) Management: Bhatnagar *et al.* (1982) reported that intercropping pigeon pea with sorghum resulted in reduction of *Helicoverpa armigera* damage and increased the combined yield.

Soil incorporation of neem cake at 250 kg/ha at the time of flowering gave significant reduction in borer incidence in pigeon pea to 6% as compared to more than 20% in other non-IPM plots (Krishna Moorthy and Krishna Kumar, 2002).

Integrated Methods: Integrated management of pigeon pea pod borer using *Ha* NPV at $2x10^6$ PIB's/ml + nematode DD-136 (*Steinernema feltiae*) at $3x10^3$ infective juveniles/ml was attempted. The least population of *H. armigera* was observed in virus + nematode treated plots (0.2) when compared to control (12.6). The percentage of pod damage was significantly lower in virus + nematode (1.07) when compared to control (40.31). Significant increase in pod yield was obtained in virus + nematode treated plots (10.7 kg) when compared to control (3.2 kg) (Narayanan and Gopalakrishnan, 1987).

IPM

Validated IPM Technology (Karnataka)

- Fall ploughing to expose pupae to hot sun and natural enemies.
- Mixing of sorghum and mesta seeds at 250 g/ha with pigeon pea seeds, which act as live bird perches.
- Installation of pheromone traps at 5/ha to monitor the pests.
- Erecting the branched twigs at 20/ha which act as bird perches.
- First spraying with ovicide indoxacarb or methomyl at 300 g/ha or profenophos at 2 litres/ha.
- Second spraying with 5% NSKE or commercial neem formulation (1500 ppm Azadirachtin) at 2 litres/ha.
- Third spraying with *Ha* NPV at 250 LE/ha.
- Fourth spraying with indoxacarb at 300 ml/ha or chlorpyriphos at 2.5 litres/ha or quinalphos or endosulfan at 2 litres/ha.
- If necessary, spraying of synthetic pyrethroids at 500 ml/ha.
- In case of scarcity of water, dust 4% endosulfan followed by 1.5% quinalphos or 0.4% fenvalerate at 25 kg/ha.

Validated IPM Technology (Maharashtra)

Pre-sowing

- Deep ploughing and exposure of soil to hot summer to kill pupating larvae.
- In areas where termite and cut worm is a problem, seed treatment with chlorpyriphos at 8 ml/kg or pre-sowing mixing of soil with

chlorpyriphos dust is recommended as they are relatively low cost and initially protect the seed and seedlings from cut worms.

- Synchronized sowing of multiple pest resistant varieties such as Sharad, Asha, Maruthi, Bahar, Abhaya, BSMR-736.

Sowing Time

- Intercrop sorghum to conserve beneficial insects and serve as bird perches.
- Early planting around mid–June in North West Plain Zone to avoid *Helicoverpa armigera.*
- Use marigold as a trap crop on border or interspersed with crop for pod borer control.

Post-Sowing

- Monitoring through regular field scouting along with pheromone traps (5-10/ha) to assess population build up of borers and their management. A threshold level of 5-6 moths/trap/day is indicative of its peak activity period and warrants initiation of management practices.
- Mechanical shaking of plants and collection of larvae for preparation of *Ha* NPV locally to cut down the cost.
- Spray 5% NSKE or neem based formulation against *H. armigera* and other Lepidopterous pests.
- Spray with *Ha* NPV at 450 LE/ha along with UV retardant.
- Spray eco-friendly pesticide like endosulfan if pest complex exceeds ETL.

Validated IPM Technology (Uttar Pradesh)

- Deep summer ploughing to destroy immature stages.
- High yielding varieties with pest/disease tolerance - Asha and Maruthi against PSM for Andhra Pradesh and Karnataka, and Bahar for Uttar Pradesh (Varanasi).
- Installation of pheromone traps at 10/ha in the month of September.
- Erection of bird perches at 25/ha for facilitating predation of *Helicoverpa* larvae.
- One spray with 2% neem oil.
- Two applications of 5% NSKE in September and October.
- Spray *Ha* NPV at 500 LE/ha (1.5 x 10^{12} POB)/ha in September and October – when larvae are small.
- Shaking of plants 5 times a day starting from October for short duration pigeon pea.

- Need based spray with endosulfan at 2 litres/ha.

3.12. CLUSTER BEAN

Cluster bean (*Cyamopsis tetragonoloba*) is also popularly called as 'guar' in India. Its tender pods being edible and are used in preparation of curry. The green pods are rich in Vitamins A and C. Since it is draught tolerant, it can be grown in summer. It is also grown as a forage and green manure crop. Its seeds are reported to be good cattle feed.

3.12.1. Insect Pests

3.12.1.1. Pod Gall Midge, *Asphondylia cyamopsii*

i) Damage: Cluster bean is found severely infested with pod gall midge causing severe reduction in yield. The pest remained active during July to November. However, maximum damage was found in September-October. Summer crop grown in March-April remained completely free from gall midge infestation. The total life cycle of *Asphondylia cyamopsii* is completed in 26 days.

ii) Management: The parasitism by parasitoids (*Dioxybracon* sp.) varied from 47 to 61%. Carbaryl at 0.2% and endosulfan at 0.1% spray gave effective control of the pest (Chari, 1970).

3.13. FIELD BEAN

3.13.1. Insect Pests

3.13.1.1. Pod Borer, *Adisura atkinsoni*

i) Damage: This is the major pod borer in field bean. The eggs are laid on tender pods. The young larvae bore into tender pods. They develop inside the bored pod and come out after attaining fourth instar, which is a migratory stage. It causes heavy damage by way of feeding the developing pods and reducing the marketable yield.

ii) Management

- Apply neem cake at 250 kg/ha at flowering and the first spray with monocrotophos has to be given at flowering and repeat after 20 days. This should be followed by sprays of 4% pulverized neem seed powder extract (PNSPE) or 1% neem soap at tender pod formation. Repeat sprays at 10 days interval. The borer incidence was 14% due to neem cake application as compared to more than 36% in control (Krishna Murthy and Krishna Kumar, 2002).
- Spraying monocrotophos at tender pod stage based on peak egg laying and again when on an average 1.5 flat pods are damaged per inflorescence was found effective (Krishna Moorthy *et al.*, 1988).
- Spray cypermethrin 25 EC at 0.5 ml/l or chloripyriphos 20 EC at 2.5 ml/l or indoxacarb 14.5 SC at 0.5 ml/l at peak egg laying stage or tender pod stage and repeat after 15 days.

- 'Biotrol' (*Bacillus thuruingiensis*) at 1 g/l reduced the pod borer infestation on field beans (Surulivelu *et al.*, 1977).
- Crop rotation with cereals (sorghum, maize) reduced the incidence of pod borer.
- Spraying of NPV at 250 LE/ha in combination with endosulfan at 0.035%, twice at fort-nightly intervals significantly reduced both larval population and pod damage by *Adisura atkinsoni* and *Sphaenanches anisodactylus* (Narayanan, 1987).

3.13.1.2. Lycaenid Pod Borer, *Lampides boeticus*

i) Damage: This is a butterfly pest and commonly known as pea blue. The eggs are laid singly in flowers and flower buds. In lablab, the larvae bore into flower buds. The pest can be monitored by the presence of the bored holes. When the larvae attain fourth instar, they migrate and cause extensive damage to flower buds and pods.

ii) Management

- In lablab, spraying has to be done at flower bud stage, when on an average about 1 flower bud is damaged per inflorescence. Spraying with any synthetic pyrethroid or 0.05% chlorpyriphos or 5% NSKE after flowering is effective.

3.13.1.3. Bugs

i) Damage: Many bugs are known to attack leguminous crops. Of these, *Nezara viridula* and *Coptosoma cribraria* are often serious, mostly on cow-peas and lablab. Eggs are laid on tender plant parts. Nymphs and adults suck the sap from foliage, flower parts and pods and emit characteristic bad smell.

ii) Management

- Apply neem cake at 250 kg per ha immediately after germination and repeat at flowering. Spray 1% neem soap or 4% pulverized neem seed powder extract at 10 days interval.
- Dust 4% endosulfan or spray with indoxacarb at 0.5 ml/l if the incidence is high.

3.14. AMARANTHUS

3.14.1. Insect Pests

Amaranth is susceptible to damage by foliar insects such as leaf miners, leaf roller caterpillars, cutworms, aphids, flea beetles, and mites. An effective method of controlling insect pests is to cover the bed with a fine screen or nylon mesh netting (32-mesh or finer).

3.14.1.1. Leaf Webber/Caterpillar, *Hymenia recurvalis*

i) Damage: Eggs are laid in and around the leaf veins and the larvae join leaves together and after enclosing itself inside, scrape the green part thus making leaves unfit for consumption (Fig. 3.49). Green slender caterpillar feed on leaves.

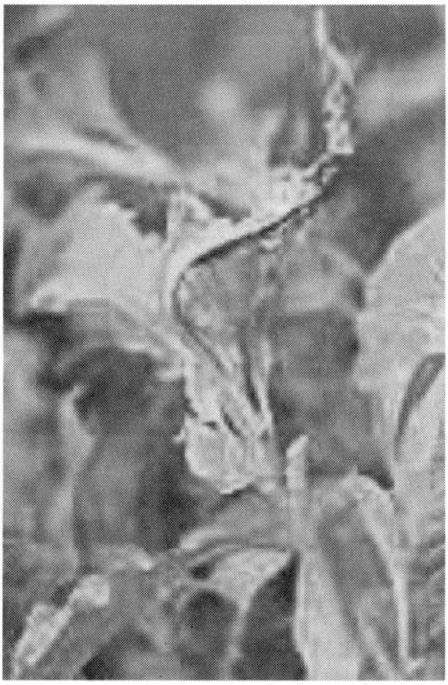

Fig. 3.49. Leaf caterpillar on amaranthus leaf.

ii) Management: The population of leaf webber caterpillars and the extent of leaf damage in yellow oleander (*Thevetia nerrifolia*) plant extract (2% seed and fresh leaf extract) treated plots were significantly low compared to control plots. Seed extract sprayed plots gave an yield of 4.10 kg/plot as compared to 2.75 kg/plot in control.

'Thuricide' dust ($3x10^6$ spores/g) at 20 to 25 kg/ha, 'Thuricide' WP ($3x10^6$ spores/g) at 0.1 and 0.2% and 'Thuricide' EC ($15x10^9$ spores/g) at 0.1 and 0.2% were found to be highly effective in controlling the leaf caterpillar (Sundarababu, 1968).

The leaf webber was found to be parasitized to the tune of 62% by the parasitoid, *Apanteles* sp.

3.14.1.2. Stem Weevils, *Hypolixus truncatulus, H. haerens*

i) Damage: Several species of weevils feed on amaranth. Adult weevils feed on leaves, but the larvae (grubs) are more damaging because they bore into roots and stems, causing rotting and potentially lodging and predisposition to diseases. Stem-boring weevils such as the pigweed weevil (*Hypolixus haerens*) are the most damaging causing plants to wither and lodge. The adult weevil lays its eggs in leaf veins, stem or branches, and the larvae bore through stems to the root collar hollowing the stems (Fig. 3.50).

Feeding by larvae results in stems that are more susceptible to wind breakage resulting in increased crop losses. The larvae pupate in the stem. Attack by this weevil has been associated with extensive tissue discolouration, decay and cankers in branches, stems, and root collars of *Amaranthus hybridus*. This weevil has been found to be associated with fungi (mainly *Fusarium* spp.) that cause tissue decay and a canker disease.

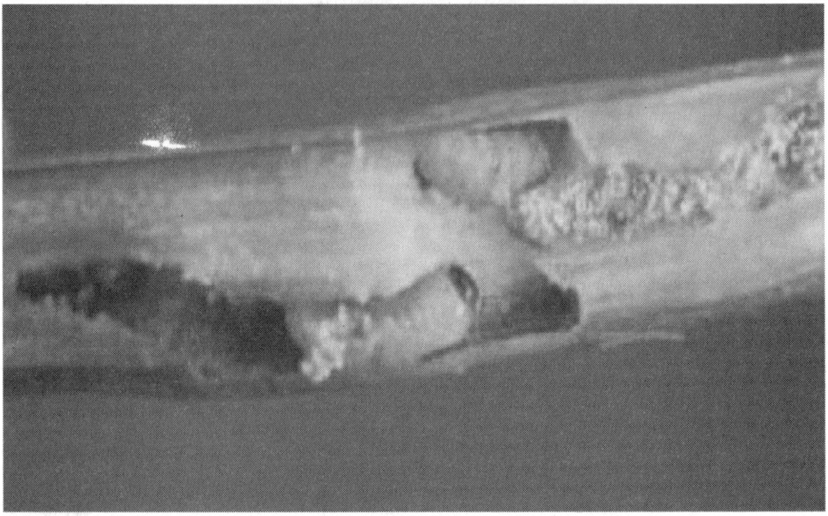

Fig. 3.50. Weevil larvae and adult feeding in amaranth stem (Courtesy: A. M. Varela).

ii) Management : Uproot and destroy attacked plants to reduce number of weevils and prevent damage to healthy plants.

3.14.1.3. Aphid, *Lipaphis erysimi*

i) Damage: Aphids suck plant sap (Fig. 3.51) resulting in stunted growth of the plant.

Fig. 3.51. Aphid infestation on amaranthus leaf.

ii) Management: Spraying 4% NSKE gives effective control of aphids.

3.14.1.4. Bugs

i) Damage: Bugs can cause severe damage to flowering head and seeds, and may be particularly damaging to grain amaranth when present in large numbers during the critical seed fill stage (Fig. 3.52). They are usually of minor importance in vegetable amaranth, and no control measures are needed.

Fig. 3.52. Bugs feeding on amaranth flowering head (Courtesy: A. M. Varela).

3.15. CURRY LEAF

3.15.1. Insect Pests

3.15.1.1. Lemon Butterfly, *Papilio demoleus*

i) Damage: Caterpillars of lemon butterfly feed on leaves (Fig. 3.53).

Fig. 3.53. Larva and adult of lemon butterfly.

ii) Management: Hand pick and destroy caterpillars.

3.15.1.2. Scale Insect

i) Damage: Scale insects affect branches of curry leaf plant (Fig. 3.54).

ii) Management: Spraying 4% NSKE gives effective control of the pest.

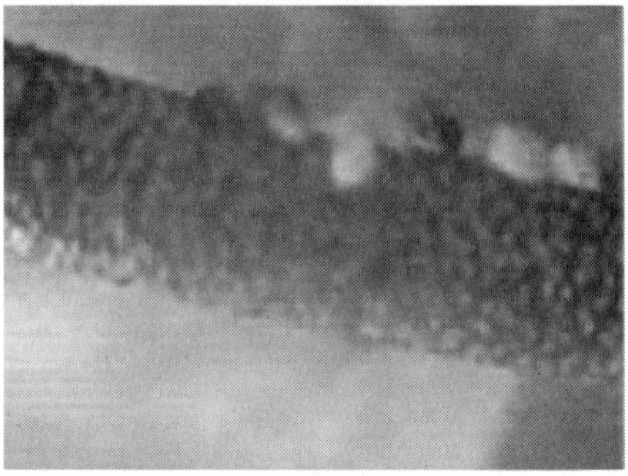

Fig. 3.54. Scale infestation on stem of curry leaf plant.

3.16. DRUMSTICK

3.16.1. Insect Pests

3.16.1.1. Bud Worm, *Noorda moringae*

i) Damage: Larvae bore into flower buds (Fig. 3.55) and cause shedding of flowers.

Fig. 3.55. Bud worm infestation on drumstick.

ii) Management: Dusting 10% carbaryl at 25 kg/ha gives effective control of the pest.

3.16.1.2. Leaf Caterpillar, *Noorda blitiealis*

i) Damage: The leaf caterpillar causes defoliation of trees (Fig. 3.56).

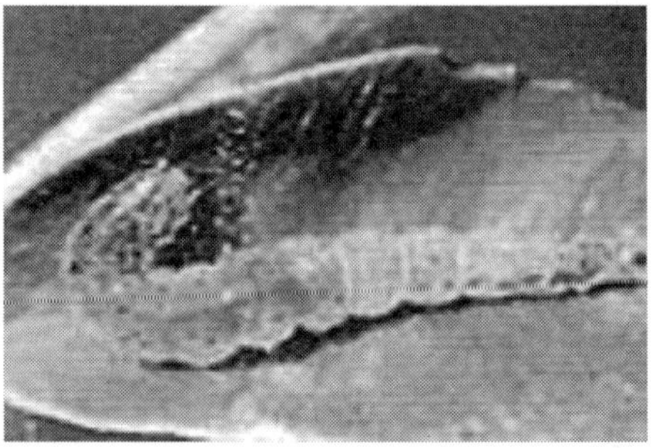

Fig. 3.56. Leaf caterpillar infestation on drumstick.

ii) Management: Dusting 10% carbaryl at 25 kg/ha gives effective control of the pest.

3.16.1.3. Hairy Caterpillar, *Eupterote mollifera*

i) Damage: The hairy caterpillar causes defoliation of trees (Fig. 3.57). Larvae collect on branches in a gregarious mass.

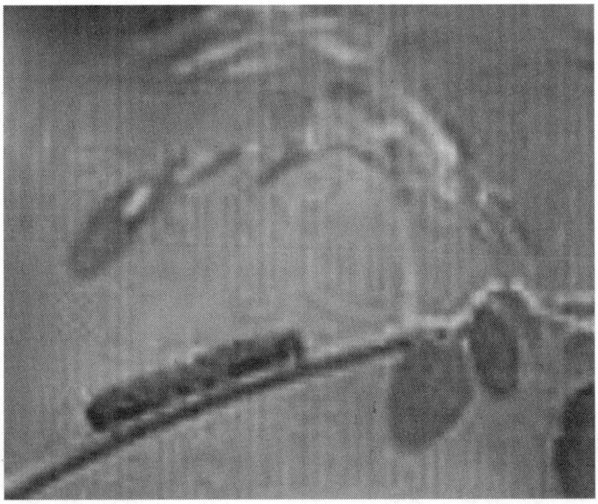

Fig. 3.57. Hairy caterpillar infestation on drumstick.

ii) Management: Destroy larvae by torching.

3.16.1.4. Fruitfly, *Gitonia* sp.

i) Damage: Fruitfly maggots cause fruits to dry from the tip.

ii) Management: Spraying dichlorvos at 1 ml/litre of water gives effective control of fruitfly.

Pesticide Residues

The persistence of terminal pesticide residues and waiting periods for leafy vegetables are presented in Table 3.38. (Awasthi, 2001).

Table 3.38. Terminal pesticide residues and their waiting periods for leafy vegetables

Pesticide schedule	Dosage/ Concentration (%)	Terminal residues (ppm/days)	Waiting period (days)
DDVP spray on foliage	0.05	0.25/3	3
Malathion spray on foliage	0.06	0.03/3	Nil
Phosphamidon spray on foliage	0.05	0.24/15	11
Dimethoate spray on foliage	0.05	0.19/21	10
Endosulfan spray on foliage	0.07	0.14/22	8
Fenvalerate spray on foliage	0.01	0.28/15	20

3.17. PUMPKIN

Pumpkin (*Cucurbita moschata*) occupies a prominent place among vegetables owing to its high productivity, nutritive value, good storability, long period of availability, better transport qualities and extensive cultivation in sub-tropical and tropical parts of the world. Fruits are rich in Vitamin A. It is also consumed as processed and stock feed. The fruits can be used in preparing sweets, candy, fermented into beverages. Yellow or orange-fleshed pumpkins are rich in carotene. In India, it is grown mainly in Assam, West Bengal, Tamil Nadu, Karnataka, Madhya Pradesh, Uttar Pradesh, Orissa and Bihar.

3.17.1. Insect Pests

3.17.1.1. Red Pumpkin Beetle, *Aulacophora foveicallis, A. lewisii*

i) Damage: The adults are small, elongated yellow beetles. The yellow brown adults flying near plants defoliate the leaves, immediately after germination (Fig. 3.58). The larvae feed on roots and plant parts touching the ground.

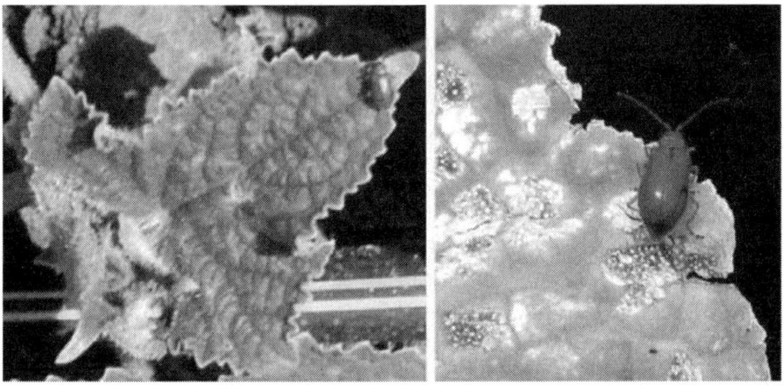

Fig. 3.58. Red pumpkin beetle on pumpkin leaves.

ii) Management: Mechanically collect and destroy the pest if incidence is low. If the incidence is high, spray indoxacarb 14.5 SC at 0.5 ml/l or 0.2% carbaryl or 0.05% quinalphos or 0.05% chlorpyriphos.

The red pumpkin beetle was found to be highly susceptible to *Bacillus thurigiensis* (Sundarababu, 1969). Spraying with *Beauvaria bassiana* (2×10^6 cfu/g) is also effective against red pumpkin beetles.

3.17.1.2. Serpentine Leaf Miner, *Liriomyza trifolii*

i) Damage: This is an introduced pest occurring on many cucurbit vegetables. Heavy incidence is noticed in watermelon, pumpkin, cucumber *etc*. However, bitter gourd seems to be resistant. Mines having serpent shape with small larvae are noticed all over the leaves (Fig. 3.59). The incidence starts after about 15 days of germination of seeds. Left alone, it gets suppressed by natural enemies.

Fig. 3.59. Serpentine leaf miner damage on pumpkin leaf.

ii) Management

- Left alone, the leaf miner gets suppressed by natural enemies. A native larval parasitoid, *Hemiptarsenus vericornis* has been observed to be the major parasitoid on this pest.
- Soil application of neem cake at 250 kg/ha immediately after germination.
- Destroy cotyledon leaves with leaf mining at 7 days after germination.
- Spray 4% Pulverized NSPE/NSKE or 1% neem soap or neem formulation with 10,000 ppm or more (2 ml /l) 15 days after sowing and repeat after 15 days if necessary.
- If the incidence is high, first remove all severely infected leaves and destroy. Then mix neem soap 5 gm and hostothion 1 ml/l and spray. After one week, spray 1% neem soap or 4% pulverized NSPE or neem formulation with 10,000 ppm or more (2 ml /l).
- Never repeatedly spray any one insecticide again and again.

3.17.1.3. Fruit Fly, *Bactrocera cucurbitae*

i) Damage: This is the major pest of cucurbits and is widely distributed. The damage by maggots results in rotting of young and ripened fruits or drying and shriveling of fruits before maturity. Sometimes even the base of the plant gets attacked and plants start wilting. The incidence is more in wet climate.

Eggs are white and are laid in groups and are 1.0-1.5 mm long and taper at both ends. Full grown larvae are 5-10 mm in length. Pupae are brown and 5-8 mm long. The life cycle may take about 10-18 days, but gets prolonged during winter in North India (Butani and Jotwani, 1984).

ii) Management

- Use of resistant varieties e.g. Arka Suryamukhi (pumpkin).
- Soil application of neem cake at 250 kg/ha immediately after germination and repeat at flowering followed by sprays of 1% neem soap or 4% pulverized NSPE/NSKE at 10 days interval after flowering were found effective.
- Crush pumpkin 1 kg and add 100 gm jaggery and 10 ml malathion and keep in the plot (10-15 places per ha). This will attract the adults to the fermenting pumpkin and they lay eggs and get killed. Repeat the process 2-3 times in a cropping season.
- Erect cue lure (para pheromone trap) 8 per ha to attract and trap male fruit flies.
- Spray carbaryl 50 WP at 3 g/l or indoxacarbat 0.5 ml/l. Foliar spray of 0.05% fenthion with 5% jaggery at fruit formation/ripening stage is very effective.

The fruits of cucurbits can be protected from fruit flies by planting maize as a trap crop. The maize is then treated with insecticides to kill the adult flies. This way, susceptible cucurbit crops escape damage and do not have any insecticidal residues. Since the trap area encourages the population of natural enemies, fruit flies will be further reduced by these natural enemies (Matthews, 1984).

3.17.1.4. Thrips, *Thrips palmi*

i) Damage: Both nymphs and adults feed on the young shoot tips and floral parts and are suspected vectors of tospo virus. They are serious during summer months and are suspected to transmit bud necrosis disease in watermelon.

ii) Management

- Soil application of neem cake at 200 kg/ha (once immediately after germination and again at flowering) followed by 4% NSPE and 1% neem soap alternately at 10-15 days interval.
- Spray any systemic insecticides like 0.05% acephate or dimethoate 30 EC at 2ml/l.

3.17.1.5. Leaf Eating Caterpillar, *Diaphania* (=*Margaronia*) *indica*

i) Damage: Long shining caterpillars feed on leaves and fruits.

ii) Management

- Soil application of neem cake (once immediately after germination and again at flowering) followed by 4% NSPE and 1% neem or pongamia soap alternately at 10-15 days interval.
- Spray carbaryl 50 WP at 3 gm/l or indoxacarb at 0.5 ml/l.

- Dusting the plants with ash temporarily repels the beetles. Sow the crop in November to avoid the damage from the pest. Pumpkin leaf beetles and the leaf feeding caterpillars on gourds were found to be highly susceptible to *B. thurungiensis*.

3.17.2. Mite Pests

3.17.2.1. Red Spider Mite, *Tetranychus neocaledonicus*

i) Damage: This pest is serious on cucurbits during warmer climate. Colonies of mites within silk web can be observed on ventral leaf surface when the incidence is high (Fig. 3.60).

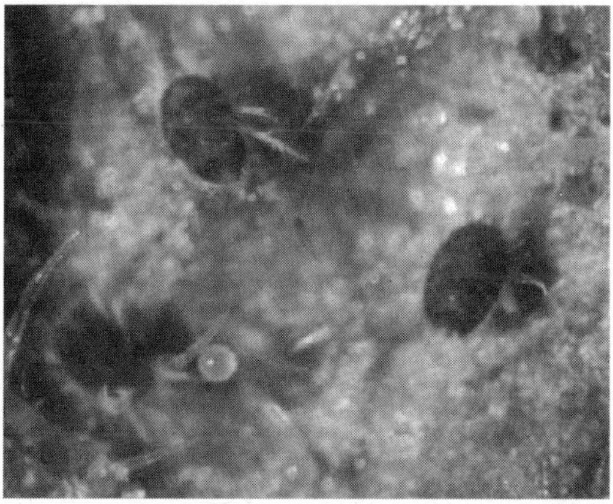

Fig. 3.60. Red spider mite on cucurbits.

ii) Management

- Spray of 1% neem or pongamia soap on lower leaf surface thoroughly.
- Alternately, spray dimethoate 30 EC at 2ml/l or ethion 50 EC at 1ml/l or wettable sulphur 80 WP at 3g/l.

3.17.3. Vertebrate Pests

i) Damage: Damage to cucurbits was noticed at sowing, flowering and fruiting stages. At sowing, Blue rock pigeon (*Columba livia*), squirrel (*Funambulus palmarum*) and rabbit (*Orictolagus cuniculus*) damaged the seedlings. An estimation revealed that on an average 17.2 to 30.5% seedlings are lost due to these vertebrates. Birds even consumed petals, detach androecium and damage the reproductive parts. The birds preferred flowers of cucumbers to that of snake gourd, ridge gourd and bitter gourd that are less tasty with high cucurbitacins (Srihari and Chakravarthy, 1998).

IPM

- Soil application of neem cake at 250 kg/ha after germination and repeat once at flowering.

- Removal of cotyledon leaves infected with leaf miner one week after germination, followed by spraying of 4% neem seed powder extract or 1% neem soap reduces the incidence of leaf miner substantially.
- Erect pheromone traps cue lure at 8 traps/ha and change after 60 days

3.18. CUCUMBER

India is considered to be the home of cucumber (*Cucumis sativus*). Cucumber is cultivated for fresh consumption or as pickling cucumber for preservation, marinated with vinegar, salt, dill or other spices. It is an important salad crop cultivated both in north and south and lower as well as higher hills in India.

3.18.1. Insect Pests

3.18.1.1. Fruit Fly, *Dacus cucurbitae*

i) Damage: Fruit fly infestation causes oozing of resinous fluid from fruits which become distorted and malformed (Fig. 3.61). Maggots feed on fruit pulp causing premature dropping of fruits.

Fig. 3.61. Fruit fly infestation on cucumber.

ii) Management: Application of neem cake or sprays of 4% NSKE were very effective in controlling fruit fly in cucumber (Table 3.39). Soil application of neem cake reduced the incidence of fruit fly to 6%, whereas insecticide sprayed plots recorded 15% incidence.

Table 3.39. Efficacy of neem cake and NSKE on fruit fly in cucumber

Treatment	Dose (%)	Incidence of fruit fly (%)
NSKE	4.00	11.08
Carbaryl	0.15	15.17
Metasystox	0.05	22.77
Monocrotophos	0.05	20.84
Phosphomidon	0.05	22.36
Neem cake	100 g/pit	6.26
Pongamia cake	100 g/pit	21.45
Control	-	48.96

3.18.1.2. Serpentine Leaf Miner, *Liriomyza trifolii*

i) Damage: Mines having serpent shape with small larvae are noticed all over the leaves. The incidence starts after about 15 days of germination of seeds. Leaf miners make small tunnels in the leaves of plants. Other symptoms include discolouration of the leaves. Left alone, it gets suppressed by natural enemies.

ii) Management: NSKE 4% was effective in reducing the serpentine leaf miner damage and increasing the fruit yield of cucumber (9.1 kg/12 sq. m.) compared to control (2.4 kg/12 sq. m.) (Krishna Kumar, 1998).

3.18.2. Vertebrate Pests

i) Damage: Small green barbet caused a loss of about 38% fruit in cucumber, a local delicacy when grown as a sole crop on wooden platforms (called pandals) in South Canara district, Karnataka.

ii) Management: The damage was reduced to 8.5% when cucumber was grown on raised basins (2 m diameter, 0.3 m height) mixed with pumpkin, bottle gourd and ridge gourd (Srihari and Chakravarthy, 1998).

3.19. BITTER GOURD

3.19.1. Insect Pests

3.19.1.1. Fruit Fly, *Bactrocera cucurbitae*

i) Damage: Fruit fly infestation causes oozing of resinous fluid from fruits which become distorted and malformed. Maggots feed on fruit pulp (Fig. 3.62) causing premature dropping of fruits.

ii) Management: The incidence of fruit fly was reduced to less than 9% by sprays of neem/pongamia soap as compared to 32% incidence in unsprayed control. Sprays of 4% pulverized neem seed powder recorded the lowest (0 to 3.60%) fruit fly incidence (Table 3.40). Soil application of neem cake followed by foliar sprays of neem seed powder extract is effective for managing fruit fly in bitter gourd (Krishna Murthy and Krishna Kumar, 2002).

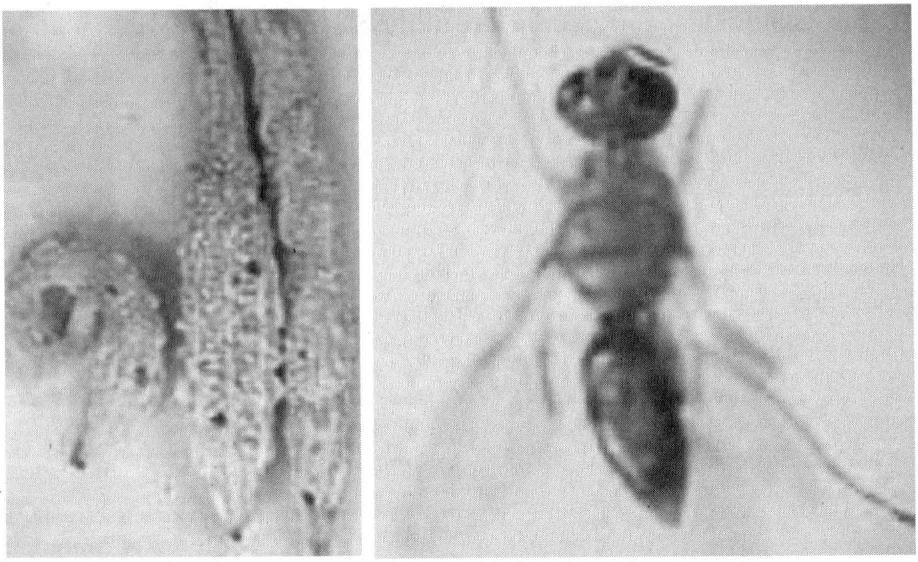

Fig. 3.62. Fruit fly damage on bitter gourd fruits.

Table 3.40. Management of fruit fly in bitter gourd

Treatments	Dose	Fruit fly incidence %
Neem seed powder extract	7.5%	0.00
Carbaryl	0.15%	21.31
Methyl demeton	0.05%	24.09
Monocrotophos	0.05%	26.33
Phosphamidon	0.05%	39.97
Neem cake	100 g/pit	8.33
Pongamia cake	100 g/pit	18.66
Control	---	49.51
CD (P=0.05)		14.98

Bitter gourd cvs. Hissar II and Ghoti possessed field resistance to fruitfly attack.

3.20. WATERMELON

Watermelon (*Citrullus lunatus*) is an important cucurbitaceous summer vegetable grown for ripe fruits. Though it can be grown in garden land, it is a major river-bed crop of Uttar Pradesh, Rajasthan, Gujarat, Maharashtra and Andhra Pradesh. An excellent desert fruit, it is relished by rich as well as poor. The fruit juice makes an excellent refreshing and cooling beverage after adding a pinch of salt and black pepper. The fruits contain 92% water, 0.2% protein, 0.3% minerals and 7.0% carbohydrates in a 100g edible flesh.

3.20.1. Insect Pests

3.20.1.1. Thrips, *Thrips palmi*

i) Damage: The nymphs and adults are found moving on young shoot tips. If the tip of the plant is tapped, many young ones and adults can be collected on a white sheet of paper. Thrips are very serious during summer months and are suspected to transmit bud necrosis disease in watermelon.

ii) Management: Spraying of systemic insecticide like 0.05% acephate reduces the pest incidence. Spray of 1% neem or pongamia soap was found to reduce thrips damage in watermelon. Sprays of 4% NSKE/4% pulverized NSPE were also found very effective. Soil application of neem cake once immediately after germination and again at flowering also reduces the pest incidence.

3.21. MUSK MELON

3.21.1. Insect Pests

3.21.1.1. Red Pumpkin Beetle, *Raphidopalpa foveicallis; Fruit Fly, Bactrocera cucurbitae*

i) Damage: Adult beetles damages the cotyledonary leaves. Sometimes severe damage results in death of plants. Grubs feed on roots. Damage by beetle to grown up leaves does not cause economic damage.

ii) Management: Soil application of carbofuran at 200 to 300 g a.i./ha at sowing, followed by 4 sprays with 250 to 500 g a.i./ha of carbaryl at 10 days interval (beginning 10 days after sowing) gave effective control of both red pumpkin beetle and fruit fly even under high insect pressure. Spraying of endosulfan at 15 ml or carbaryl at 2 g/litre of water at cotyledonary leaf stage gives effective management of the pest.

Musk melon cv. Kharda is reported to be resistant, while Lakhazda is resistant to fruit fly.

3.22. GHERKEN

3.22.1. Insect Pests

3.22.1.1. Fruit Fly

i) Damage: Fruit fly infestation causes oozing of resinous fluid from fruits which become distorted and malformed. Maggots feed on fruit pulp causing premature dropping of fruits.

ii) Management: The incidence of fruit fly was reduced to less than 9% by sprays of both neem and pongamia soaps as compared to 32% incidence in unsprayed control (Table 3.41). Spraying of monocrotophos and phosphamidon recorded more than 10% fruit fly incidence (Krishna Murthy and Krishna Kumar, 2002). The soap sprays were also effective on leaf miner.

Table 3.41. Management of fruit fly in Gherkins (2001)

Treatment	Dose	Fruit fly incidence
NSPE	7.5%	2.25
Carbaryl	0.15%	10.77
Metasystox	0.05%	12.28
Monocrotophos	0.05%	11.95
Phosphamidon	0.05%	28.68
Neem cake	100 g/pit	5.31
Pongamia cake	100 g/pit	9.50
Neem soap	1%	9.44
Pongamia soap	1%	9.40
Control	---	32.30
CD (P=0.05)		4.07

3.23. RADISH

Radish (*Raphanus sativus*) is a root vegetable which can be grown as an intercrop with other crops. Its leaves are rich in Vitamins A and C. It can be eaten raw and also cooked. It has some medicinal value. Radish is cultivated throughout India. West Bengal, Bihar, Uttar Pradesh, Punjab, Assam, Haryana, Gujarat and Himachal Pradesh are major radish-growing states.

3.23.1. Insect Pests

3.23.1.1. Mustard Sawfly, *Athalia lugens proxima*

i) Damage: The activity of mustard sawfly is found from last week of July to end of March. The egg, larval, prepupal and pupal period last for 3, 10, 3 and 10 days, respectively. The pest hibernates in the pre-pupal stage in an earthen cocoon (Patel, 1967). Besides attacking radish (Fig. 3.63), the pest also attacks mustard, cabbage and cauliflower.

Fig. 3.63. Mustard sawfly damage on radish leaf.

ii) Management: Spraying of 5% NSKE showed higher antifeedant and larvicidal effect. Repellin at 1%, margocide at 0.1% and neemark at 1% showed their higher oviposition deterrent, antifeedent and larvicidal action, respectively on sawfly. Synthetic insecticides like triazophos + decamethrin at 0.036% and quinalphos at 0.05% gave higher larval mortality (Patel, 1995).

3.23.1.2. Aphids, *Brevicoryne brassicae, Lipaphis erysimi, Myzus persicae*

i) Damage: Colonies of aphids are seen clustered on tender stems (Fig. 3.64) and ventral side of leaves. They suck the plant sap. The affected plants curl and show yellowing symptoms. The seed crop of radish suffers comparatively more as compared to crop grown for edible root.

Fig. 3.64. Aphids on radish stem.

ii) Management: Three sprays of monocrotophos or endosulfan at 2 ml/litre of water on attainment of 50 aphids (*Lipaphis erysimi*)/plant provided the highest cost: benefit ratio (Bath and Singh, 1989).

Soil application of aldicarb, phorate or carbofuran all at 1.5 kg a.i./ha as and when aphid infestation is noticed gave effective control of the pest.

3.24. CARROT

Carrot (*Daucus carota*) is another root vegetable rich in Vitamin A and some minerals. It is grown all over India. Main carrot growing states are Uttar Pradesh, Assam, Karnataka, Andhra Pradesh, Punjab and Haryana. It is taken raw as well as in cooked form. It is made into pickles and sweet meat. Carrot juice is a rich source of carotene (a precursor of Vitamin A) and contains appreciable quantities of thiamine and riboflavin. It is sometimes used for colouring butter and other foods. Green carrot leaves are highly nutritive, rich in protein, minerals and vitamins and used as fodder and also for preparation of poultry feed.

3.24.1. Insect Pests

3.24.1.1. Thrips, *Thrips tabaci*

i) Damage: Adults and larvae suck the cell sap from the leaves of the plants (Fig. 3.65). White blotches appear on leaves. Tips of leaves wither and turn brown.

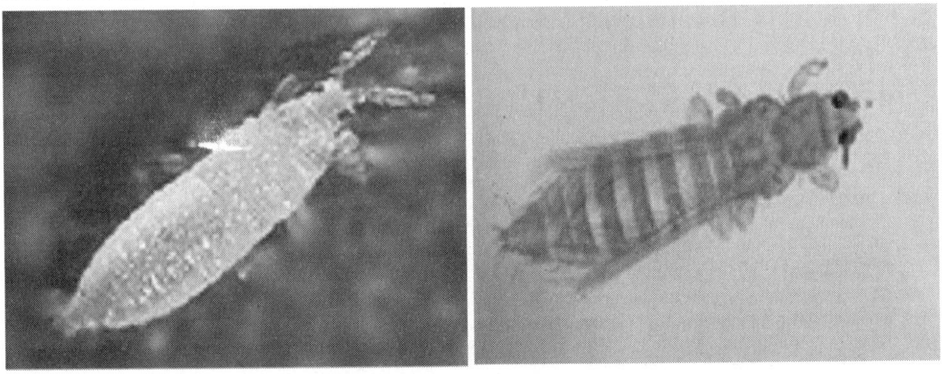

Fig. 3.65. Nymph and adult of thrips on carrot.

ii) Management: Spraying monocrotophos or phosphamidon or malathion at 0.5 ml/litre of water gives effective control of the pest.

3.24.1.2. Rust Fly, *Psila rosae*

i) Damage: Larva mines into the root and causes holes, which rots and leads to secondary infection (Fig. 3.66). Drooping and discoloration of foliage from green to rusty brown are the other symptoms.

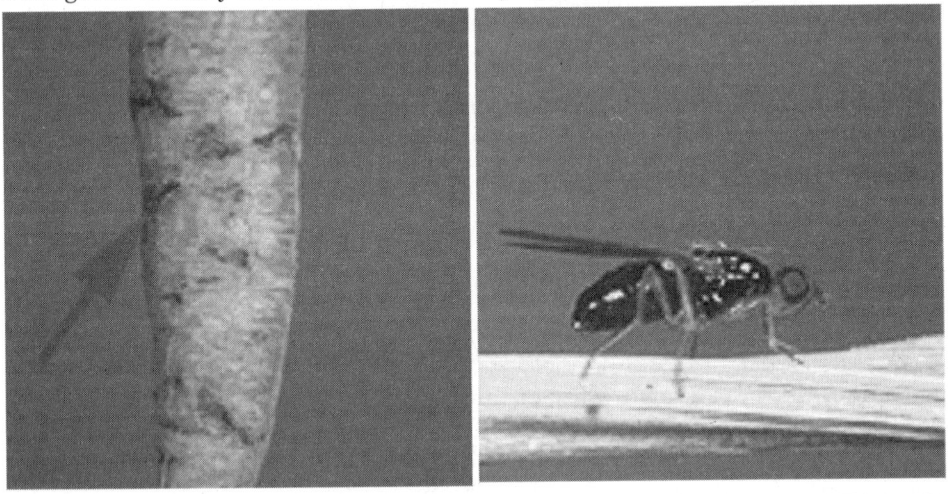

Fig. 3.66. Rust fly damage on carrot and adult of rust fly.

ii) Management: Deep summer ploughing, rotation of crops, destruction of wild hosts are the cultural methods that can be used to reduce the pest population. Application of carbofuran at 1 kg a.i./ha at a depth of 10-15 cm during land preparation is effective (Table 3.42).

Table 3.42. Terminal pesticide residues and their waiting periods for root vegetables

Pesticide schedule	Dosage	Terminal residues (ppm/days)	Waiting period (days)
BHC dust soil application at sowing	5.0 kg a.i./ha	ND*/120	105
Aldrin dust soil application at sowing	3.0 kg a.i./ha	0.005/90	90
Carbaryl dust soil application at sowing	1.5 kg a.i./ha	ND*/60	60
Methyl parathion dust soil application at sowing	1.0 kg a.i./ha	ND*/60	60
Carbofuran soil application at planting	1.0 kg a.i./ha	ND*/55	60

* ND – Not detectable

3.25. MUSHROOMS

Mushrooms have achieved significant importance due to their nutritive and medicinal values. Mushrooms, the edible fleshy fungi, represent one of the few valuable sources of nutritive and palatable food. They are rich in proteins (35-48%), vitamins and minerals. They are weight reducers and useful for diabetic patients. Though, these fungi have been naturally occurring and being utilized in India since time immemorial, the commercial cultivation is only of recent times. It was only in 1964 that the first successful crop of mushroom was grown at Solan in Himachal Pradesh. Since then, research on various aspects of mushroom cultivation has been strengthened and today it is one of the important cottage industries, not only in the state of Himachal Pradesh but its commercial cultivation has also been initiated in most of the other states and union territories of India.

At present, it is estimated that 40,000 tonnes of mushrooms are produced annually. The major share (85%) is contributed by button mushroom (*Agaricus bisporus* and *A. bitorquis*). Oyster mushroom (*Pleurotus* spp.) is popular in southern states and Maharashtra and milky mushroom (*Calocybe indica*) in Tamil Nadu, Andhra Pradesh and Karnataka (Table 3.43).

Table 3.43. State-wise production of cultivated mushrooms

State	Production in MT	Type of mushroom
Tamil Nadu	10000	*Agaricus/Volvariella*
Maharashtra	7000	*Agaricus/Pleurotus*
Haryana	4000	*Agaricus*
Punjab	4000	*Agaricus*
Uttar Pradesh	4000	*Agaricus/Pleurotus*
Andhra Pradesh	2500	*Agaricus/Pleurotus*
Madhya Pradesh	2000	*Agaricus/Pleurotus*
Goa	1000	*Agaricus*
North Eastern States	1000	*Agaricus/Lentinus/*

Orissa	700	*Pleurotus/Volvariella*
Karnataka	600	*Agaricus/Volvariella*
Himachal Pradesh	500	*Agaricus*
Jammu & Kashmir	400	*Agaricus*
West Bengal	250	*Agaricus/Pleurotus*
Kerala	100	*Pleurotus/Volvariella*
Rajasthan	50	*Pleurotus*
Others	1900	*Agaricus/Pleurotus*
Total	40000	

Source : NCMR&T, Solan (1996-97)

3.25.1. Insect Pests

Pests are a constant threat to the successful commercial production of mushrooms. Uncontrolled populations of any of the main mushroom pests can result in substantial losses in yield due to both direct larval action and associated disease spread by adult flies.

Flies found in cropping houses cause reduction in yield by:

- Damaging the compost.
- Feeding on the mycelium.
- Act as vectors of diseases.
- Nuisance to pickers.
- They can cause crop rejection at market in pre-packs or chips.

3.25.1.1. Sciarids - *Lycoriella castanescens, L. ingenua*

i) Damage: Sciarids are compost feeders and really prefer unspawned compost to that colonised by *Agaricus* mycelium. The adult flies are attracted by volatiles given off by the compost so are a danger at any time after pasteurisation when the compost is cool. The adults are usually 3-4 mm long (although some species recently found associated with mushrooms have been slightly bigger) and with the naked eye can be distinguished from the other fly pests of mushrooms by their long antennae (Fig. 3.67). The females deposit between 50 to 200 eggs in the compost and the developing larvae pass through four moults varying in size from 1 to 8 mm long before pupating. They are easily recognised by their black shiny heads. They feed on rotting vegetation, so mushroom compost is an ideal substrate for them but the accumulation of their waste (called frass) renders it unsuitable for mushroom growth. Although exudates produced by the growing *Agaricus* mycelium inhibit the larvae, they have been observed browsing on the growing tips and when there is a large infestation the larvae will burrow up into the mushroom stipes.

Besides causing direct damage to the crop, the flies can also spread mites and diseases, especially *Verticillium* as the spores of this disease are sticky and easily adhere to the bodies of the flies.

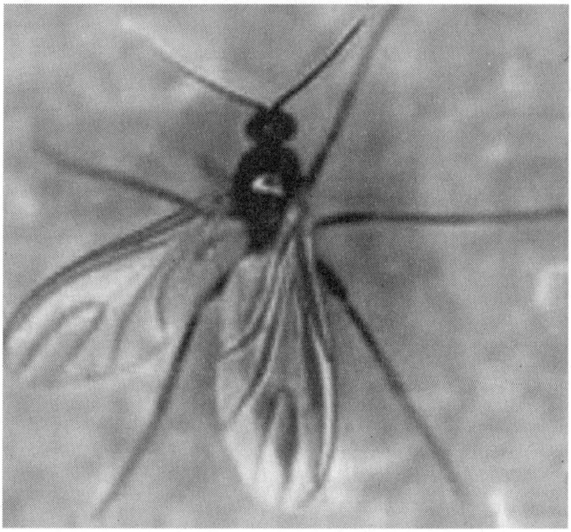

Fig. 3.67. Adult sciarid fly

ii) Management

- Efficient composting and peak heating.
- Exclusion of adults by filters and tight fitting doors. It is important that doors are not left open as an aid for ventilation.
- Disposal of crops debris at a distant location from the farm.

Sciarid larvae can be treated with the growth regulator Dimilin and with Nemasys M-an insect parasitic nematode. The adult flies of both sciarid and phorid can be sprayed with pyrethrums although the flies have often laid at least some of their eggs before leaving the beds.

3.25.1.2. Phorids, *Megaselia halterata*

i) Damage: Phorid larvae are obligate mycelial feeders and therefore the adult flies are not attracted to oviposit in the crop until after spawning. The larvae are white, 1 to 6 mm long, are stubby at one end and have a pointed head at the other. They feed on the growing mushroom mycelium but rarely feed on the fruiting body itself, although different species of phorid are known to do so. They can be distinguished from sciarid larvae by the absence of the black head and they develop more rapidly into a pupa. The adult fly is 2 to 3 mm long, has very short antennae and a characteristic hump-back. It can be distinguished from the sciarid fly by the short antennae and by its rapid, jerky, running movement. They are unable to fly when the temperature falls below 12°C and are therefore unlikely to re-infest mushroom houses between late autumn and early summer.

The female can lay 50 eggs close to growing mycelium, and at a temperature of 24°C these eggs will reach the adult stage inside two weeks. The larvae are white legless maggots with no black head and tend to have a blunt tail region. These larvae feed entirely on growing mycelium although some species can tunnel into mushrooms. Adults are usually annoying to pickers. In addition, these flies can spread disease such as *Verticillium*.

ii) Management

- Efficient composting and peak heating.
- Exclusion of adults by filters and tight fitting doors especially during the first three weeks after spawning.
- Disposal of stems immediately from the growing house, these should be dumped at a distant location. When the bags are emptied they should also be discarded well away from the farm.

Since the withdrawal of diazinon, there is no pesticide currently available to treat an infestation of phorid larvae.

3.25.1.3. Cecids, *Heteropeza pygmaea*

i) Damage: The three most common species of cecids which infect the mushroom crop are *Heteropeze pygmaea, Mycophila speveri* and *M. barnesi.* Cecid flies are rarely seen as the larvae reproduce paedogenetically - i.e. new generations are produced within the body of a 'mother' larva without sexual reproduction - and the flies are very small. It is usually the larvae which are about 1 mm long that are first noticed through the plastic bag. The commonest species is *Heteropeza pygmaea* which has white larvae whereas the *Mycophila* species have orange larvae. Each larva becomes a 'mother larva' which can give birth to 12-20 daughter larvae within a week of its own birth, without any adult midge being present. This ability to multiply so rapidly results in enormous numbers of larvae in the compost and when the food supply becomes limiting within the compost the larvae swarm and go onto the mushrooms, especially after the crop has been watered. Large numbers of larvae especially of the orange type make the mushrooms unsaleable. The larvae also carry a bacterium which causes increasing losses due to the degradation of mushrooms.

When there is a bad infestation, the larvae clump together and flow over the sides of the mushroom beds onto the floor where they can be inadvertently transported to healthy crops on feet and equipment.

ii) Management

a) Efficient Composting and Peak Heating: Cecids and their larvae are only killed when temperatures throughout the compost reach 55°C for 5 hours or 60°C for 3 hours.

b) Strict Hygiene: If all larvae are killed during pasteurization, infestations at cropping usually indicate that pests have entered the bags after filling. If there is a cecid problem on the farm it is vital to isolate infested houses by disinfecting tools, knives, buckets etc. Clean overalls should be worn in each house. Filters should be replaced at the end of each crop and doors should be tight fitting and not left open as an increased aid for ventilation. The area where casing is to be unloaded should be disinfected. Any bags that are visibly infected with cecids during cropping should be sprayed with a disinfectant, sealed and dumped at a distant location; the same applies for crop debris.

c) Efficient Cook Out and Disinfection of Premises: Before the bags are emptied they should be sprayed with disinfectant. After emptying, the interior of the house should be washed out.

3.25.2. Mite Pests

3.25.2.1. Tarsonemid Mite, *Tarsonemus myceliophagus*

i) Damage: These mites are pale brown and are so minute that they are only visible with the aid of a microscope. They cause damage by feeding entirely on hyphae of mushrooms and the grower will know if he has these mites present, as the base of the stem of the mushroom will show a reddish brown discolouration. Where severe infestations occur the whole base of the mushroom may be detached from the growing surface.

ii) Management

- As with eelworms little can be done when mites are present in the growing house, therefore efficient composting and peak heating must take place to ensure that they are killed during the pasteurization process.
- Good hygiene should be practiced around the farm, especially in the clearance of crop debris.

3.25.2.2. Tyroglyphid Mites, *Tyrophagus* spp.

i) Damage: These mites can be identified as they are slow moving, translucent, with long hairs on their bodies. If these mites are present in abundance they eat small pits in the caps and stalks. These pits then suffer from bacterial decomposition, which breaks down tissues just below the surface. This results in the skin collapsing which leaves an open pit. Tyroglyphids may also feed on mushroom mycelium, where they are present in large numbers, crop reductions can be caused. Mites usually gain entry into the compost by clinging onto Sciarid flies when the mites are in migratory stage. These migratory stages are normally produced when mites become over crowded.

ii) Management

- The mites should not be a problem where efficient composting and peak heating takes place.

- Organic debris should not be allowed to accumulate around the farm as it provides a breeding ground for mites.

3.25.2.3. Red Pepper Mites, *Pygmephorous* spp.

i) Damage: These mites are not regarded as primary pests; their presence is usually an indicator that *Trichoderma* (green mould) is present in the compost. These mites feed on various weed moulds but not mushrooms, thus their presence indicates that the compost is unsatisfactory. The mites are yellowish-brown in colour, 0.25 mm in length and have a flattened appearance. They also are capable of rapid rates of reproduction. As already stated these mites are secondary pests and they often swarm on the casing and mushroom surfaces. Where this happens their presence makes the mushrooms unsaleable. These mites can also spread spores of *Trichoderma* from bag to bag.

ii) Management

- Efficient composting and peak heating.
- Good husbandry around the farm.
- Where mites appear on individual bags during cropping, these should be sealed up and disposed of.

Chapter 4

ORNAMENTAL CROPS

India is known for growing of traditional flowers such as jasmine, marigold, chrysanthemum, tuberose, crossandra and aster. Commercial cultivation of cut flowers-rose, orchids, gladiolus, carnation, gerbera, anthurium and lilies-has also become popular.

With the liberalization of Indian economy, the domestic floricultural trade, which hitherto remained an unorganized sector, became an organized sector with the active participation of corporate giants. The establishment of state-of-art hi-tech floricultural units in early 90's for intensive cultivation of cut flowers contributes about Rs. 300 crores of valuable foreign exchange today. Various government agencies are actively involved in the promotion of this sector and are highly committed to resolve the teething problems. The world demand for floricultural products is approximately US $ 40 billions and cut flowers contribute nearly 60% (US$ 25 billions) to the global trade. India's share in the world flower market is about 0.7% which is contributed mainly from open cultivation.

Floriculture is estimated to cover an area of 126,000 ha with production of 694,000 MT of loose flowers during 2006. The area under cut flowers has increased significantly in recent years and the production has reached 2762 million numbers (Table 4.1). Another 500 ha of climatic control greenhouses are available for growing quality flowers for export. Many corporate houses (more than 100) have set up 100% export oriented units with an investment of more than Rs. 1000 crores. India is bestowed with natural advantages like favourable weather conditions, soil, cheap labour which are likely to make a real mark in international trade. With this, India is emerging as one of the powerful forces in world flower trade.

Table 4.1. State-wise area and production of flowers (1999-2000)

State / UT's	Area (in ha)	Production (in MT)	
		Loose	Cut
Andhra Pradesh	18087	126609	---
Delhi	3450	21007	---

Haryana	2250	40500	760
Himachal Pradesh	133.3	976	181
Jammu & Kashmir	42	7	17.4
Karnataka	20801	124440	---
Madhya Pradesh	1387	832	---
Maharashtra	6600	28336	---
Orissa	175	868.5	9.3
Tamil Nadu	18120	142250	---
U.P. Plains	160	---	160
West Bengal	13227	14644	5678
Others	4174.7	8723.5	1.0
Total	**88607**	**509193**	**6806**

Floriculture is an intensive type of agriculture wherein the income per unit area is much higher than any other agricultural product if it is done in a scientific way. The area used for production of ornamental plants is low, but the cash returns are high because of the intensive cropping. The flowers grown in large areas include jasmine, rose, marigold, chrysanthemum, crossandra, champak, tuberose, gladiolus and aster. New crops like lilies, tulips, anthurium, carnation, gerbera, etc., are also finding potential place in the foreign market. With the rapid development in urban planning and hotel industry, there has been an increasing demand for cut flowers and ornamentals. The cut flowers followed by live plants, cut foliage, seeds and bulbs have got an immense potential for export to Western countries where the cost of production is comparatively high. India exports roses, gladioli, carnations, chrysanthemums, lilies and jasmines to Gulf, Germany, France, England, Netherlands and Singapore. Our share in the export market has registered a sharp increase from Rs.14.45 crores in 1991-92 to Rs.299.41 crores in 2005-06.

Large scale cultivation of ornamental crops face many pest problems. Attack by insects, mites and other pests is one of the key limitations in production of these crops. Pest scenario in ornamental crops varies with the growing conditions. Closed structures and uniform growing conditions favour multiplication of pests round the year on protected ornamentals like rose, carnation, gerbera, chrysanthemum, orchids and anthurium. Red spider mites, thrips, whiteflies, aphids and leaf miners are serious pests on these crops. Open cultivated ornamentals are mainly damaged by bud borers, foliage caterpillars, beetles and weevils, besides thrips and aphids. Damage by insects causes severe yield loss in terms of quantity by making holes and eating on leaves, buds, flowers and fruits. Curling, distortion, discolouration, browning and drying of plant parts as a result of damage cause qualitative loss.

Strategies adopted to control pest problems in ornamental crops mainly include use of insecticides specially imported and costly ones in case of protected ornamentals. Cultural methods like pruning, removal and destruction of infested parts and raking of soil are followed in some crops. Plant products such as neem and pongamia oils and kernel extracts are also used to some extent.

Understanding of the pest incidence, biology, distribution and symptoms of damage caused and the present control methods followed is important to develop ideal integrated management methods in these crops.

4.1. ROSE

Rose (*Rosa* spp.) is the most ancient and popular flower grown the world over. In India, it is commercially grown for cut flowers and rose oil. Rose flowers without stems and loose flower petals are used in traditional markets for making garlands, for offering in temples, while the florist shops sell cut roses with stems mainly for bouquets and floral arrangements. In recent times, about 60 units have been established under joint ventures around Bangalore, Pune, Nasik, Mumbai, Hyderabad, Gurgaon (Haryana), Chandigarh and Saharanpur (Uttar Pradesh) for growing roses in greenhouses for export of cut flowers to Japan, The Netherlands, Germany and other European countries. Besides, the Damask rose (*Rosa damascena*) and Edward rose (*R. bourboniana*) are cultivated for rose *attar* and other products – *gulkand, gulabjal,* and *pankkurj.* It is cultivated in about 6,000 ha area in the states of West Bengal, Uttar Pradesh, Gujarat, Haryana, Jammu and Kashmir, Tamil Nadu, Karnataka, Rajasthan, Madhya Pradesh, Andhra Pradesh and Punjab.

4.1.1. Insect Pests

4.1.1.1. California Red Scale, *Aonidiella aurantii*

i) Damage: Scales appear in severe form before and after the monsoon season. Both adults and young scales suck the sap from mature shoots and deplete vigour of the plant. Infested shoots bear reddish brown encrustations (Fig. 4.1), bear few small flowers and dry in case of severe damage.

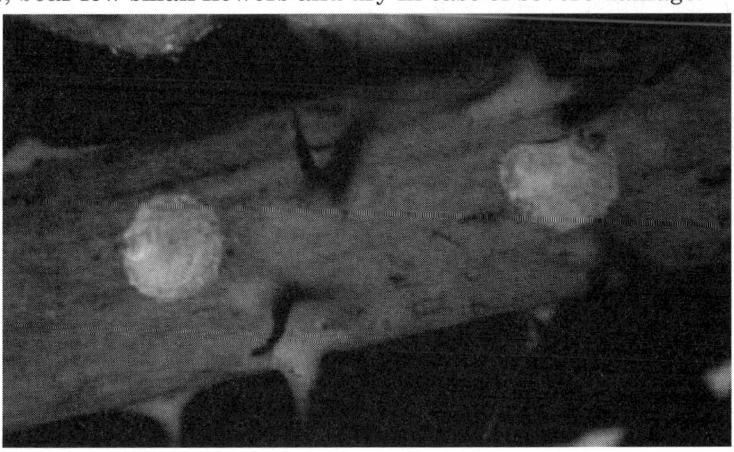

Fig. 4.1. Scale infestation on stems of rose.

Besides infesting roses, these scale insects are serious pests of citrus fruit and they infest a wide range of ornamentals. They inject a toxic saliva as they feed and cause leaf-yellowing, leaf-fall, and die-back of twigs and stems.

Female red scales are circular, about 2.5 mm in diameter and have a slightly raised centre. The scale is reddish brown, and the insect under the scale is yellow. Nymphs are light-yellow, oval, and less than 0.5 mm long. They emerge from under the female scale and crawl about for a few hours before settling permanently. Scale numbers build up quickly in dry-dusty areas or dry seasons.

ii) Management

- Selection of planting material free from scales is essential.
- Cutting and burning or burying of infested parts is also effective.
- Spraying of malathion 50 EC at 1ml/l or chloropyriphos 20 EC at 2.5 ml/l + dimethoate 30 EC at 2ml/l or ethion 50 EC at 1ml/l causes significant reduction of the scale.
- Soil application of carbofuran at 1.0 kg. a.i./ha (33 kg/ha of Furadan 3G) followed by application of 10% pongamia oil to infested shoots, immediately after pruning provides complete control of red scale within 3 weeks (Jhansi Rani and Jagan Mohan, 1997).
- In case of severe infestation, remove shallow layer of bark of stem above soil surface. Tie a cotton swab dipped in monocrotophos (10 ml/l of water) in the exposed portion. The scales will get killed. Afterwards, the exposed portion is to be swabbed with Bordeaux paste.

4.1.1.2. Thrips, *Rhipiphorothrips cruentatus, Scirtothrips dorsalis*

i) Damage: Thrips are slim-winged insects 1 mm in length, resembling fine black slivers of wood. Preferring light-coloured blooms and often appearing in large numbers, flowers are often left looking bruised and lusterless.

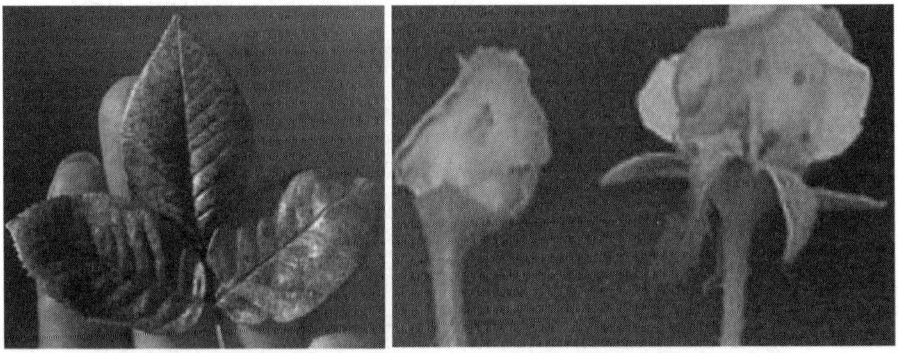

Fig. 4.2. Thrips damage on leaves and flowers of rose.

Thrips are sap-sucking insects that can cause deformities in flowers, leaves (Fig. 4.2), stems, and shoots. The insects are difficult to see because not only are they small but also they hide deep in the flower or on the undersides of the leaves, which become silvery on the upper surface. Thrips are most active during spring and autumn.

Female thrips lay eggs within leaf tissue which hatch out and cause damage to foliage by rasping the tissue and sucking the cell sap that oozes out. Matured nymphs pupate in the soil and remain inactive during winter.

ii) Management

- Spraying of 0.1% acephate (75 SP at 1.5 g/l) or 0.05% dimethoate (30 EC at 2ml/l) or 0.05% oxydemeton methyl along with 0.5% pongamia oil 2-3 times at 15 days interval with onset of new flush control thrips on open cultivated roses (Jhansi Rani and Jagan Mohan, 1997).
- In case of severe infestations, applying fipronil 5 SC at 1.5 ml/l is effective.
- Raking up of the soil during winter and drenching the soil with chlorpyrifos 20 EC at 5ml/l kills pupae in the soil.
- Spraying pongamia oil 0.5% when early symptoms are observed control thrips in open cultivated roses.
- Spraying Vertimac (biopesticide of *Streptomyces avermitilis*), 4% NSKE, 2% neem oil, 1% jatropha oil or annona oil was found effective.
- Nair *et al.* (1991) reported effective control of thrips by soil application of carbofuran at 0.18 g a.i./plant and spraying of quinalphos or phosphamidon both at 0.05%.

4.1.1.3. Aphid, *Macrosiphum rosae*

i) Damage: Likely to be found on new shoots and buds, aphids are soft bodied insects 1-2 mm long. Often green but occasionally light-brown, and sometimes with wings, they may cover (in a colony) the complete growing tip of the plant (Fig. 4.3). Aphids are most active in spring and summer and multiply at a prodigious rate feeding on the sap of the plant by piercing the plant cells via a proboscis. In large numbers they may seriously retard the growth of the plant and ruin buds. They are particularly damaging to the new shoots with subsequent damage to the emerging leaves which become malformed with much the same appearance as leaf-curl in peaches.

Aphids feed on roses by puncturing and sucking plant fluids from the veins. Aphids cause damage by secreting large amounts of honey dew which collects on leaves and buds. Honey dew encourages the growth of sooty mould which is unsightly and inhibits photosynthesis by reducing the amount of light reaching the leaves.

Only pregnant females survive after winter. In spring, females reproduce without mating and produce nymphs which form social broods and feed near their mothers. Winged females develop and disperse and aphid populations build up rapidly with spring growth.

Fig. 4.3. Aphid infestation on buds and stems of rose.

ii) Management : Small populations can be removed by hosing at high pressure, but this needs to be repeated every two or three days. For large infestations, the following recommendations may be adopted:

- Spray *Verticillium lecanii* at 3g/l.
- When infestation begin, spray neem or pongamia oil at 10 ml/l or dimethoate 30 EC at 2ml/l to control aphids in open cultivated roses (Jhansi Rani and Jagan Mohan, 1997).
- If the incidence is severe, spray imidacloprid 200 SL at 0.4ml/l or cartap hydrochloride 50 SP at 1g/l.
- Rajagopal and Kadire Gowda (1994) reported effective control of aphids with 0.1% liminool or 0.25% methyl parathion.

4.1.1.4. Bud Borer, *Helicoverpa armigera*

i) Damage: Infestation of the bud borer is seen on open cultivated roses in severe form from January to March. Female moths lay cream coloured eggs singly on young buds. Hatched larvae bore into buds by making holes and feed on growing petals (Fig. 4.4). Caterpillars also damage flowers by eating flowers and leaving excreta.

ii) Management

- Collection and killing of grown up larvae reduces further damage and population.
- NSKE 4% or neem oil 1% sprays offer protection to plants from borer damage (Jhansi Rani and Jagan Mohan, 1997).
- Spraying methyl parathion or methomyl each at 0.05% or endosulfan at 0.07% or ethofenprox at 0.01% controls the pest effectively.

- Application of *Ha* NPV at 250 LE/ha causes considerable mortality of early instar larvae.

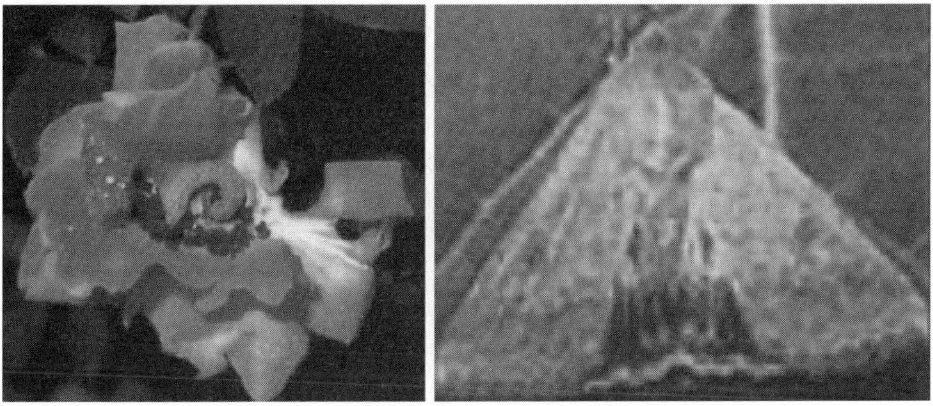

Fig. 4.4. Bud borer damage on flowers of rose.

4.1.1.5. Leaf Caterpillars, *Euproctis fraternae, Spodoptera litura, Achaea janata*

i) Damage: Pest incidence is severe from September to October. Female moths of *E. fraternae* and *S. litura* lay eggs in clusters on lower side of leaves. Early instars are gregarious in habit and cause skeletonization and drying of leaves. *A. janata* adults lay eggs in groups on tender parts of plants. There are four larval instars with a total developmental period of 33-41 days (Vyas, 1994). Matured larvae of *A. janata, E. fraternae* and *S. litura* caterpillars feed voraciously on leaves, buds and flowers resulting in severe defoliation and flower loss.

ii) Management

- Deep ploughing in summer exposes pupae to predators.
- Collection and destruction of leaves with egg masses and early instars reduces further infestation.
- Setting up of light traps attract adult moths.
- Spraying of neem oil 1% or NSKE 4% checks damage by caterpillars (Jhansi Rani and Jagan Mohan, 1997).
- Complete control of leaf caterpillars, *E. fraternae* and *A. janata* was achieved with 'Thuricide' (*Bacillus thuringiensis*) dust and spray ($3x10^6$ spores/g) (Sundarababu, 1969).
- Application of *Sl* NPV at 250 LE/ha gives effective control of *S. litura*.

4.1.1.6. Chafer Beetles, *Adoretus* spp., *Apogamia* spp.

i) Damage: Black, brown or metallic green coloured adult beetles emerge from soil with the onset of monsoon showers (May-June) after dusk. They feed on leaves, buds, flowers and tender shoots by making irregular holes and cuts (Rai, 1977) resulting in severe defoliation. Eggs are laid in the soil and grubs feed on roots.

ii) Management

- Deep ploughing after pruning exposes eggs, grubs and pupae to natural enemies.
- NSKE 2.5% spray inhibits feeding by the beetles (Doharey and Singh, 1981).
- Fortnightly spraying of quinalphos or chloropyriphos or carbaryl each at 0.15% during evening protects foliage from beetle damage (Jhansi Rani and Jagan Mohan, 1997).
- Application of phorate or carbofuran each at 1.0 kg a.i./ha kills grubs in the soil.

4.1.1.7. Weevils, *Myllocerus* spp.

i) Damage: Weevils attack on rose starts in April and continue till October. Ash coloured adult weevils damage leaves by feeding from margins in a characteristic fashion. Eggs are deposited in the soil and grubs feed on developing roots and pupate there. Heavy infestation of the weevil results in severe foliage loss.

ii) Management

- Deep digging after pruning exposes grubs and pupae to natural enemies.
- Application of 0.1% carbaryl or malathion dust or phorate at 1.0 kg a.i./ha control soil inhabiting stages.
- Spraying of quinalphos or chloropyriphos or malathion each at 0.05% protects foliage from beetle damage (Jhansi Rani and Jagan Mohan, 1997).

4.1.1.8. White Fly

i) Damage: White flies resemble a tiny, white 1.5-4 mm long moth. Most commonly, they live in the upper regions of the plant. Whiteflies pierce leaf tissues (Fig. 4.5) causing noticeable wilting, chlorosis, loss of leaves and/or stunted growth. In addition, they produce "honey dew" which attracts other pests and produces a medium for "sooty mold" fungus.

ii) Biology: The female white fly will deposit up to 400 eggs in circular clusters on the undersides of leaves. Within 5-10 days, the eggs hatch into scale-like larvae which pass through several stages during the next four weeks, after which they pupate for about a week and emerge as a new adult to renew the cycle. Depending on temperature and other environmental conditions, however, the entire life cycle can span as little as 18 days. Optimum temperatures for infestation are 18°C to 24°C.

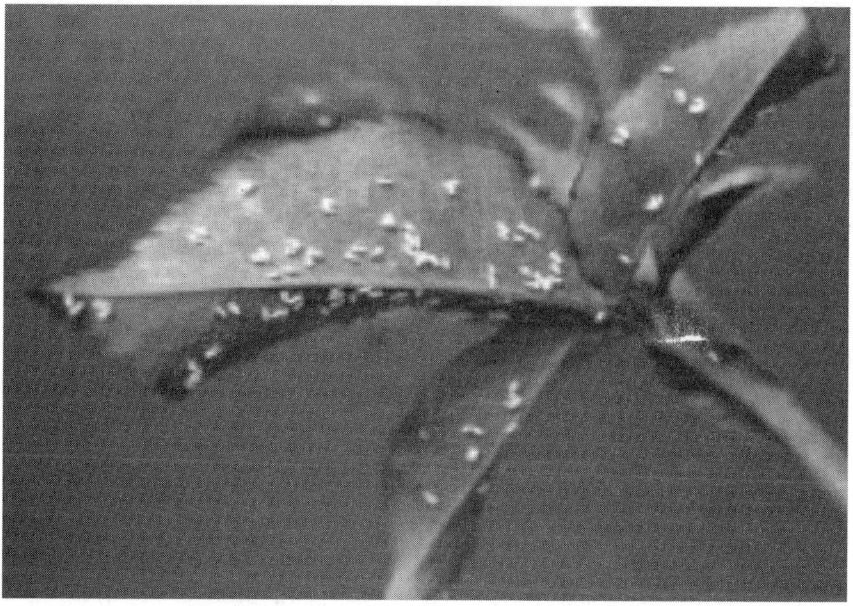

Fig. 4.5. White fly infestation on rose leaves.

iii) Management

- Install sticky yellow traps, non-baited traps, light traps or pheromone traps.
- Biological control agents like *Beauveria bassiana, Paecilomyces fumosoroseus*, or *Verticillium lecanii* fungi (the latter for greenhouse use); parasitoids like *Encarsia formosa* and *Eretmocerus californicus*; and a host of other predators like lacewings, lady beetles, and predatory mites can be used for controlling the pest.
- Extracts of Nicotiana plants and a variety of other organic and synthetic pesticides including neem oil, horticultural oils, pyrethrums and garlic water can be used against white flies. White flies can quickly mutate to pesticide-resistant strains. If using synthetic pesticides, rotate between two or three different chemical types.

4.1.1.9. Mealy Bugs

i) Description: Mealy bugs are soft bodied critters covered with a waxy material that resembles powder or cotton. The adult males have wings and the females look like a barnacle but with multiple legs and a tail. Size varies, but is generally 1.5-3.0 mm long. Because they do not have a hard outer covering, they are very sensitive to temperature and humidity.

ii) Damage: Mealy bugs pierce the plant, sucking its juices to the point of rapid defoliation, deformation and eventual death if not treated in time. In greenhouse conditions where natural predators are scarce, control must be achieved quickly. Fortunately, their cotton-like exudates makes detection reasonably easy. Unfortunately, they tend to congregate on plant surfaces hard to reach with insecticidal sprays.

iii) Biology: Mealy bug is very prolific breeder. One species can produce up to 1000 eggs at a time, but half of that is more common. The mealy bug can have as many as eight life cycles per year in southern climates which means one single female mealy bug can be the progenitor of as many as 4000 offspring. It is no wonder many commercial growers throw out their infested plants rather than try to rescue them with treatment.

The female mealy bug will lay eggs within a cottony mass that surrounds its body. In about a week, the eggs hatch into nymphs and begin crawling to a suitable location on the plant nearby. After four weeks of sucking on your prized specimen, those nymphs that will become males form a mass of cotton-like filaments where they change into winged adults about two weeks later. The females change into their scale-like bodies about two weeks after the males; they mate; and the process renews.

iv) Management

- Hand removal or cotton swabs may be adopted if populations are small.
- Biological control agents such as *Leptomastix* sp. mini-wasp; and natural predators like lady bird beetles (specifically, *Cryptolaemus montrouzieri*, the "Mealy bug Destroyer") and lacewings are effective against mealy bugs.
- Various synthetic and organic pesticides like diatomaceous earth (DE), horticultural oils, insecticidal soaps, and neem oil are also effective.

4.1.2. Mite Pests

4.1.2.1. Red Spider Mite, *Tetranychus urticae*

i) Damage: Red spider mite, also known as two-spotted mite, is the most serious pest of rose. These arachnids prefer the underside of leaves and are difficult to see with an unaided eye. Evidence of their presence is silvering of leaves where the mites have destroyed individual leaf cells. Fine webbing and eggs on the undersides of leaves is further evidence of the presence of *Tetranychus urticae*.

Mites are sap-sucking arachnids that can cause yellowing or bronzing of leaves (Fig. 4.6). If the population is large, leaf drop occurs and the health of the plant is affected. Roses are damaged when nymphs and adults pierce the leaves and suck the contents of cells causing them to collapse and die. Heavily-infested leaves become lifeless and may become bronzed and shriveled. Plants may lose their leaves.

Females lay up to six eggs a day, totaling 70 or more. Eggs are minute, globular and almost transparent pale yellow-white. Males develop from unfertilized eggs and females develop from fertilized eggs. Egg hatch takes from three to 10 days depending on the surrounding temperature. Newly-hatched larvae have six legs and are pale yellow-white, minute and oval. When they molt, larvae become eight-legged nymphs which mature after another two molts. Young mites take four to 12 days to mature and live for about three weeks. In hot weather, the complete life cycle may be as short as two weeks and overlapping generations may occur. Adult females are about 0.5 mm long and adult males are about 0.3 mm long. When feeding in summer, adult females are a yellowish-green with two pronounced dark spots, one on each side of the body. In males, these spots are less obvious and the body is smaller and tapered towards the rear. Both sexes also have two reddish eye-spots on the body.

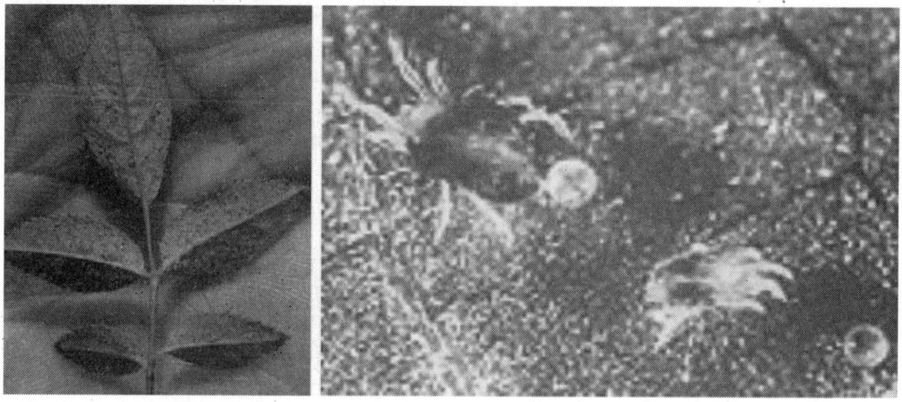

Fig. 4.6. Red spider mite infestation on leaves of rose.

Red spider mite is a profuse web-spinner in its active stages. Under artificial long-day length or in warm situations such as glasshouses, red spider mite may continue to feed and reproduce through winter. Outdoors, when temperatures and day length decrease in autumn, males die and adult females stop feeding and change colour to orange-red. They usually migrate to sheltered sites where they assemble to over winter. When temperatures increase in spring, females leave their shelters, begin to feed, and lose their red colour.

ii) Management

- Cutting and burning of severely infested plant parts should be undertaken.
- Proper ventilation, clean cultivation and frequent irrigation to bring down temperature and to keep the pest under check.
- Spraying dicofol 18.5 EC at 2.5 ml/l or wettable sulphur 80 WP at 3 g/l (Pirone, 1978) or 0.05% oxydemeton methyl or 0.012% fluvalinate (Pokharkar *et al.*, 1986) or profenphos 50 EC at 1 ml/l or ethion 50 EC at 1 ml/l followed by pongamia or neem or jatropha oil at 5 ml/l control mites.
- New miticides like vertimec at 0.0025%, flufenoxuron at 0.01%, diafenthiuron at 0.025%, profenphos at 0.05% and amitraz at 0.05%

offer very good solution for the mite problem (Jhansi Rani and Jagan Mohan, 1998).

- Neem or pongamia oil 0.5 to 1% sprays should be given in evening hours and before onset of summer to avoid mite build up and better control under polyhouse conditions.
- Release of predatory mite, *Phytoseilus persimilis* at 26/m^2 successfully contains mite population.

4.2. CARNATION

Carnation (*Dianthus caryophyllus*) flower is valued for its excellent keeping quality, wide array of colours and forms, and ability to rehydrate after continuous transportation. Moderate climatic control measures that are economical can deliver quality carnations at the internationally competitive prices year round. In India, carnation is commercially grown mostly under cover around big cites like Solan, Shimla, Mandi, Kullu, Chandigarh, Ludhiana, Guragaon, Kalimpong, Pune, Bangalore and Delhi. Carnation has tremendous potential for export market.

4.2.1. Insect Pests

4.2.1.1. Thrips, *Frankliniella schultzei*

i) Damage: The young and adult stages mainly attack growing buds, flowers and feed on sepals and petals causing discolouration (silver streaks) and distortion of affected floral parts. Damage to young buds will not allow buds to open leading to considerable flower loss.

Adult female thrips are narrow-bodied, light-brown or grey and 1-2 mm long. Males are smaller and yellow. Both sexes have two pairs of narrow delicate wings fringed with long hairs. The wings lie along the back when not in use. The colourless eggs are inserted into the tissue of petals, sepals, blossom stems, stamens, pistils, calyx cups and young leaves. When full grown they crawl to the ground and pupate about 50 mm under the surface. Generation times vary from 10 days to more than a month, depending on the temperature.

Thrips have mouth parts adapted for rasping and sucking. Feeding by thrips is usually confined to soft recent growth or flowers where they rasp the surface cells and suck the sap as it oozes out. Petals and anthers turn brown and shrivel. Normal opening of flowers may be affected if thrips enter before the flowers open. Feeding and egg laying on the pistil may cause it to wither. Egg laying in the tender tissue of young leaves may cause the tissue around the eggs to die and fall out leaving small, irregularly shaped 'shot-holes'. In addition to damage done by feeding, crops may also be downgraded by fecal deposits.

ii) Management: Spraying of 0.1% acephate or 0.015% ethofenprox or 0.007% imidachloprid or 0.05% cartap hydrochloride at the initial stage of infestation controls thrips damage. Soil drenching with 0.01% chloropyriphos reduces further incidence by killing thrips present in the soil.

4.2.1.2. Bud Borer, *Helicoverpa armigera*

i) Damage: Infestation of bud borer is observed from May to September on carnation. Bud worms are the larvae of noctuid moths. Female moth lays cream coloured eggs singly on young buds. Larvae bore into growing points by making holes in buds and damage flowers by feeding on developing petals and reproductive parts resulting in petal drop. They also feed on the young shoots and foliage. They make characteristic round holes in buds and flower heads. The caterpillars are about 40 mm long when fully grown, have various shades of yellow, green, pink or brown, with dark flecking and longitudinal paler and darker stripes.

ii) Management: Collection and killing of grown up larvae by putting them in 0.1% chloropyriphos solution reduces borer population. Spraying of 0.07% endosulfan or 0.05% methyl parathion or 0.01% ethofenprox when eggs are observed on buds controls damage by the pest effectively.

4.2.2. Mite Pests

4.2.2.1. Red Spider Mite, *Tetranychus urticae*

i) Damage: This is a devastating pest on carnation. Infestation is severe during summer. Mite colonies are found on upper surface of leaves beneath leaf folds and are covered by webs and they spread to entire plant in severe cases. More population is seen on lower part of the plant and economic threshold is 0.5 mite/leaf (Jesitor, 1976). Damaged leaves and buds (Fig. 4.7) become discoloured and dry. Stunted growth and drying of plant are the symptoms in case of heavy damage.

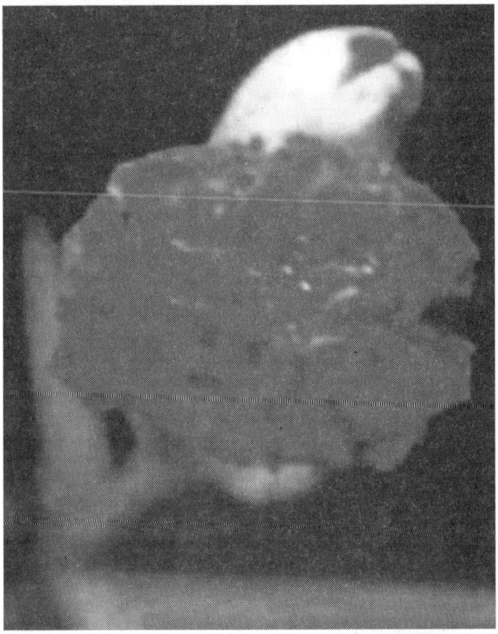

Fig. 4.7. Mite infestation on carnation flower.

The adult female is about 0.5 mm long. The adult male is about 0.3 mm long. In the summer the adult females are the feeding stages which are yellowish-green, with two pronounced dark spots, one on either side of the body. In males these spots are less conspicuous and the body is smaller and tapered towards the rear. Both sexes possess two reddish eye-spots.

Under artificial long day length or in warm situations such as glass-houses, the mites may continue feeding and reproducing through the winter. Outdoors, when temperatures and day length decrease in autumn, the males die and the adult females stop feeding and change colour to orange-red. They usually migrate to sheltered sites where they congregate to over winter. When the temperatures increase in spring, the females leave their shelters and begin to feed, losing their red colour in the process. They lay up to six eggs a day and usually lay a total of 70 or more. Males develop from unfertilized eggs and females develop from fertilized eggs.

Egg hatch takes from three to 10 days depending on the temperature. The young mites mature in from four to 12 days and may live for about three weeks. In hot weather the complete life cycle may be as short as two weeks and overlapping generations may occur. Damage occurs when nymphs and adults feed by piercing the cells of leaves and sucking the contents, causing collapse and death of the cells. The mite is a profuse web-spinner in its active stages and eggs are laid amongst the webbing. Heavily infested leaves become pallid and may become bronzed and shriveled. Defoliation may occur.

ii) Management: Cutting and burning of severely infested and dried plant parts to be undertaken. Washing of plants with water dislodges the webs. A mixture of dicofol and tetradifon at 300 g/ha or proparigate at 150 g/ha were found effective against adults, eggs and nymphs (Pondolfo, 1984). Spraying of pongamia oil in combination with annona oil at 2% effectively checks mite population (Jhansi Rani and Jagan Mohan, 1997). Spraying of triazophos at 0.06% or ethion at 0.05% or dicofol at 0.05% at 15 days interval is also effective.

Carnation cvs. Isaq, Alcardi and Sunrise are reported resistant to mite attack.

4.3. GERBERA

Gerbera (*Gerbera jamesonii*), commonly known as Transvaal daisy, Barberton daisy or African daisy, is an important flower grown throughout the world for beds, borders, pots and rock gardens. The flowers are of various colours and suit very well for floral arrangements. The cut blooms also have long vase life. In India, it is distributed in the temperate Himalayas from Kashmir to Nepal at altitudes of 1300 to 3200 meters above sea level. Gerbera belongs to the family Asteraceae and is native to South African and Asiatic regions. To meet the quality standards of export market, it has to be grown under naturally ventilated low cost polyhouses. In India, gerbera is commercially grown mostly under cover around big cites like Solan, Shimla, Mandi, Kullu, Chandigarh, Ludhiana, Guragaon, Kalimpong, Pune, Bangalore and Delhi.

4.3.1. Insect Pests

4.3.1.1. White Fly, *Bemisia tabaci*

i) Damage: The pest is active during moderately hot and humid conditions (February - July). Small and fragile white adults lay eggs on lower surface of young leaves. Nymphs and adults suck cell sap from lower side of leaves resulting in discolouration, wilting and drying of matured leaves and curling of young leaves. In case of severe infestation, sooty mould develops on leaves on honey dew secreted by nymphs. Affected plants are retarded in growth and produce deformed flowers with short stalks.

ii) Management

- Remove and burn the heavily infested leaves to check pest build up.
- Spray triazophos 40 EC at 1.5ml/l or methomyl 40 SP at 1.5g/l or acephate 75 SP at 1.5 g/l alternating with pongamia oil 1% for management of nymphs.
- Spray 0.05% dichlorovos or deltamethrin against adults.
- Spraying of permethrin, fenvalerate and amitraz each at 0.1% reduced white fly population by more than 90% (Thachuk *et al.*, 1983).
- Install yellow sticky traps for monitoring and trapping adults.
- Excessive dampness and dark conditions should be avoided.
- Clean cultivation and use of insect proof nets are essential.
- Oil sprays are to be given during evening inside polyhouse for better efficacy. Annona and pongamia oil at 2% were found very effective with more than 90% mortality of nymphs (Jhansi Rani and Jagan Mohan, 1997).
- Release of parasitoid, *Encarsia formosa* at 6 adults/m^2 is recommended for control of white flies (Nucifera and Calabretta, 1986).

4.3.1.2. Leaf Miner, *Liriomyza trifolii*

i) Damage: Attack of leaf miner starts when plants are at 3 to 4 leaf stage. The incidence of leaf miner is severe in March-June. Adults lay eggs by puncturing the tender leaves. Maggots mine the leaves by feeding on mesophyll by leaving characteristic serpent shaped mines (Fig. 4.8). Severely damaged leaves show extensive mining, become brittle, turn brown and dry up. The affected plants become stunted in growth and produce weak and small flowers.

ii) Management

- Remove and destroy severely mined leaves.
- Spray triazophos 40 EC at 1.5 ml/l or abamectin 1.9 EC at 0.4 ml/l against larvae.

- Spray with dichlorvos 76 EC at 1ml/l or decamethrin 2.8 EC at 1ml/l for managing adults.
- Installing yellow sticky traps coated with chlorpyriphos 20 EC at 2.5 ml/l gives considerable protection and encourages parasitization by *Diglyphus isaea* (Calabaretta and Nucifera, 1985).
- Spraying of 1% pongamia oil or 4% neem seed kernel extract is very effective.
- Intercropping with field bean (*Vicia faba*) which acts as trap crop is effective.
- Release of eulophid parasitoid, *Diglyphus intermedius* at 1000 at weekly intervals gives considerable reduction in leaf miner population.

Fig. 4.8. Leaf miner infestation on gerbera.

4.3.1.3. Thrips, *Thrips palmi*

i) Damage: Heavy infestation of thrips is observed on gerbera during September-November and March-May. Adults and nymphs colonize along midrib on upper surface of leaves and petals of growing flowers and suck the sap. Thrips cause damage by sucking sap from leaves, buds and flowers. Damaged leaves become brittle and curl up. Affected buds and flowers get discoloured and distorted in shape (Fig. 4.9). Discolouration, development of brown streaks and severe distortion of petals are seen on the damaged flowers.

Fig. 4.9. Thrip infestation on flowers of gerbera.

ii) Management

- Spray acephate 75 SP at 1.5 g/l or dimethoate 30 EC at 2ml/l along with 0.5% pongamia oil 2-3 times at 15 days interval with onset of new flush.
- Apply fipronil 5 SC at 1.5 ml/l in case of severe infestations.
- Drench the soil with chlorpyriphos 20 EC at 5ml/l for killing pupae in the soil.

4.3.1.4. Aphid, *Myzus persicae*

i) Damage: Pest incidence is found during cool and cloudy months (November-January). Aphids mainly attack growing flowers. Both nymphs and adults colonize on lower side of petals and spread to entire flower. They suck the sap at the feeding site resulting in discolouration, distortion and wilting of flowers. Development of sooty mould on honey dew secreted by aphids makes flowers look sickly with black dots.

ii) Management

- Spray dimethoate 30 EC at 2ml/l or oxydemeton methyl 25 EC at 2ml/l at 10-15 days interval.
- Effective control of the aphid can be obtained by spraying dichlorvos or methomyl each at 0.1% or cypermethrin at 0.04% (Tusandi, 1984). New insecticides like 0.07% imidachloprid or 0.05% cartap hydrochloride also give very good control of the pest.

- Spray pongamia or neem oil at 1%.
- If the incidence is severe, spray imidacloprid 200 SL at 0.4 ml/l.

4.3.1.5. Bud Caterpillar, *Spodoptera litura*

i) Damage: Larvae cause severe damage to flower buds before they open as they are protected by bracts on both open and greenhouse grown gerbera plants. Damage results in petal drop, reduces quality of flowers due to excreta and cause severe flower loss.

ii) Management: A mixture of deltamethrin at 15 g and triazaphos at 30 g/10 litre spray applied at 150 g formulation/750 m gave 97% mortality and treatment with *Bacillus thuringiensis* offers effective control (Inserra and calabretta, 1985). Methyl parathion, quinalphos and methomyl each at 0.05% sprays protect plants from damage by *S. litura*. Spreading of poison bait (10 kg rice bran + 1 kg jaggery + 0.5 kg chlorpyriphos) where damage is noticed and installing pheromone traps after first showers is quite helpful in bringing down the population of *S. litura*.

4.3.2. Mite Pests

4.3.2.1. Red Spider Mite, *Tetranychus ludeni*

i) Damage: Severe attack of the mite is observed during February-April on gerbera. Colonies of red coloured mites are seen on both sides of the matured leaves and spread to flowers in case of heavy infestation. Nymphs and adults feed on the leaf sap resulting in bronzing and drying of damaged leaves. Severely damaged plants produce few and very small flowers which dry up consequently.

ii) Management: Thinning of old and heavily infested leaves prevents severe attack. Weekly sprays of 0.15% dichlorvos or 0.1% methomyl are recommended (Kerenyi-Nemestothy and Tusandi, 1985). Recent acaricides like 0.0025% vertimec or 0.025% diafenthiuron or 0.05% amitraz or 0.015% flufenuxuron provide highly effective control of the mite (Jhansi Rani and Jagan Mohan, 1997). Annona, pongamia or jatropa oil all at 1% cause considerable mortality of mites.

4.4. CHRYSANTHEMUM

Garden chrysanthemum (*Dandrathema grandiflora*), also known as 'Queen of the East', is a highly versatile and accommodating ornamental with a wide range of type, size and colour in flowers. In India, large flowered varieties are grown for cut flower, making garland, wreaths and 'veni', religious offerings and for bedding and potting purposes. Its commercial cultivation is being done in Maharashtra, Rajasthan, Madhya Pradesh and Bihar. Small flowered varieties are commercially grown for loose flowers in Tamil Nadu, Karnataka and Maharashtra. In India, chrysanthemum is grown in about 4,000 ha.

4.4.1. Insect Pests

4.4.1.1. Aphid, *Myzus persicae, Macrosiphoniella sanborni*

i) Damage: The most common aphids infesting chrysanthemums are the green peach aphid, *M. persicae*, and the chrysanthemum aphid, *M. sanborni*. The green peach aphid is a green to greenish-yellow in colour and the chrysanthemum aphid is a shiny mahogany in colour. Greenish-black nymphs and chocolate-brown adults suck the cell sap from growing shoots and lower surface of leaves (Fig. 4.10). Damage results in loss of vigour, yellowing and premature leaf fall and stunted growth of plant. On heavily infested leaves, honey dew secreted by aphids favours development of black sooty mould which is unsightly and inhibits photosynthesis by reducing the amount of light reaching the leaves. In severely infested plants, flower buds are reduced in size and remain unopened.

Fig. 4.10. Aphid infestation on stems and flower of chrysanthemum.

Only pregnant females survive after winter. In spring, females reproduce without mating and produce nymphs (immature stages) which form social broods and feed near their mothers. Winged females develop and disperse and aphid populations build up rapidly with spring growth.

ii) Management

- Spray pongamia or neem oil at 20 ml/l or 4% pongamia kernel extract at 10-15 days interval.
- Spray dimethoate 30 EC at 2 ml/l or oxydemeton methyl 25 EC at 2ml/l or imidacloprid 200SL at 0.4 ml/l or acephate 75 SP at 1.5 g/l.
- Application of *Verticillium lecanii* (Microgermin) at 3g/l.
- Release of 1 day old *Chrysoperla carnea* larvae at 50 : 1 ratio (aphid: chrysopid) eliminated aphid population (Scopes, 1969).

4.4.1.2. Thrips, *Microcephalothrips abdominalis; Frankiniella* spp., *Thrips* spp.

Thrips have been found in all chrysanthemum growing areas. This major insect pest is responsible for millions of dollars worth of crop damage world-wide due to its resistance to many insecticides and efficiency at spreading tomato spotted wilt virus. More than seven species of thrips are reported on chrysanthemum of which the above mentioned species are common.

i) Damage: Thrips usually live and feed on flowers or new plant growth, such as buds and young leaves. They have rasping and suckling mouth parts which cannot penetrate deep into plant tissue, so they usually feed on soft-recent growth or flowers by rasping or scraping surface cells and sucking the cell contents. Damage is not always obvious after feeding but becomes more obvious as the affected flowers, leaves or fruit grow and distort. Damaged flowers look discoloured, withered and dried due to scorching (Fig. 4.11).

Fig. 4.11. Thrip damage on flowers of chtysanthemum.

The thrips life cycle is mostly continuous and all stages of thrips can be found throughout the year. Female thrips live for 30 to 45 days and can produce around 150 to 300 eggs.

ii) Management

- Spray dimethoate 30 EC at 2ml/l at 15 days interval.
- Drench the soil with chlorpyriphos 20 EC at 5 ml/l.
- If the incidence is severe, spray ethofenprox 10 EC at 1ml/l or fipronil 5 SC at 1.5 ml/l.

- Covering of young buds with polythene bags with holes prior to blooming protects flowers from attack by thrips (Wang and Lin, 1984).

4.4.1.3. Leaf Folder, *Hedylepta indicata*

i) Damage: This is a serious pest of chrysanthemum under field conditions. Pale white eggs are laid in groups on lower side of leaves which hatch in 10 days. There are 6 larval instars. Green coloured larvae with white stripes and brown head fold the leaves and feed on chlorophyll resulting in skeletonization and drying. Pupae are seen inside the rolled leaves. The larvae also damage flowers leading to heavy loss occasionally.

ii) Management: Cutting and burning of rolled leaves with larvae and pupae reduce the pest population. Spraying of 0.01% ethofenprox or 0.01% fenvalerate followed by methyl parathion or quinalphos or profenofos each at 0.05% gives effective control of leaf folder.

4.4.1.4. Leaf Miner, *Liriomyza trifolii*

i) Damage: The incidence of leaf miner is severe in March-June. Eggs are deposited in leaf tissue and larvae mine leaves by feeding on mesophyll leaving characteristic trails (Fig. 4.12). Severely damaged leaves show extensive mining which turn brown and dry up.

Fig. 4.12. Leaf miner infestation on chrysanthemum.

ii) Management

- Remove and destroy severely mined leaves to reduce further damage.
- Spray triazophos 40 EC at 1.5 ml/l or methomyl 40 SP at 1.5g/l or abamectin 1.9 EC at 0.4 ml/l against larvae.
- Spray dichlorvos 76 EC at 1ml/l or decamethrin 2.8 EC at 1ml/l for managing adults.

- Install yellow sticky traps coated with chlopyriphos 20 EC at 2.5 ml/l to trap adult population.
- Dipping of chrysanthemum cuttings in 3% margosan (neem formulation) for 2 hr and foliar spray of the same botanical at 0.4% reduce the number of pupae and adults (Lindquest *et al.*, 1986).
- Insect growth regulatory compounds like cyromazine, methoprene, fenoxycarb and avermectin reduce leaf miner population (Parella *et al.*, 1983).
- Removal and destruction of weeds which act as alternate hosts is very important to protect the crop from leaf miner damage.
- The parasitoid, *Diglyphus* sp. at 1000/week effectively checks larval stages (Jones and Parella, 1986).

4.4.1.5. Bud Borer, *Helicoverpa armigera*

i) Damage: Larvae tunnels its way into the buds and feeds on developing flower parts. The pest can be detected by the presence of holes on buds and fecal matter at entrance of such holes (Fig. 4.13).

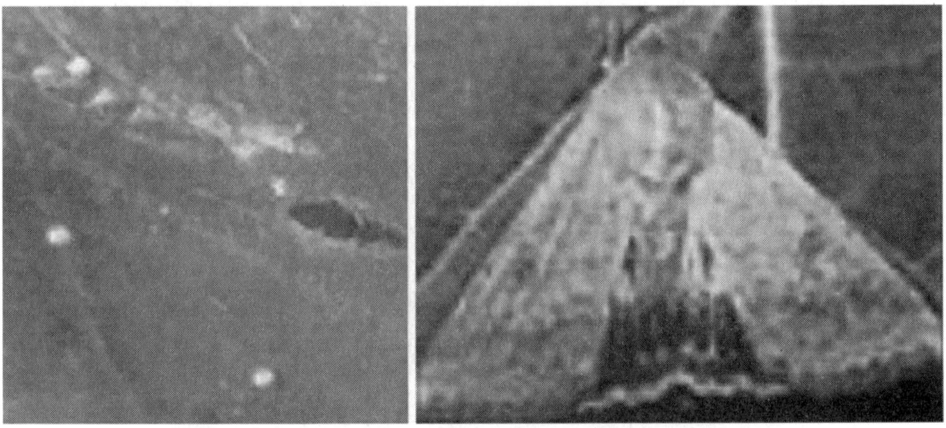

Fig. 4.13. Bud borer infestation on chrysanthemum.

Fully-grown caterpillars are about 40 mm long and may be various shades of yellow, green, pink or brown, with dark flecking and dark markings or stripes along their bodies. Caterpillars feed for two to three weeks in warm weather. When they have eaten enough, they leave the plant and burrow about 100 mm into the soil where they pupate. Adult moths have a wingspan of about 40 mm and have grey-brown to reddish-brown fore wings with darker markings. The hind wings are pale at the base and dark at the tips. Moths are inactive during the day and fly at night to lay eggs and feed on nectar. Each female can lay up to 1000 eggs which are deposited singly on new leaves.

ii) Management: NSKE 4% sprays offers protection to the crop from the borer damage. Other control measures are same as in rose bud borer.

4.4.2. Mite Pests

4.4.2.1. Red Spider Mite, *Tetranychus urticae*

i) Damage: Mite colonies are usually seen on the upper surface of leaves and population spreads to entire plant in severe cases of attack. Damage starts in March and reaches a peak during April-July. Chrysanthemums are damaged when nymphs and adults pierce the leaves and suck the contents of cells causing them to collapse and die. Heavily-infested leaves become lifeless and may become bronzed and shriveled. Plants may lose their leaves and die.

Females lay up to six eggs a day, totaling 70 or more. Eggs are minute, globular and almost transparent pale-yellow-white. Males develop from unfertilized eggs and females develop from fertilized eggs. Egg hatch takes from three to 10 days depending on the surrounding temperature. Newly-hatched larvae have six legs, are pale yellow-white, minute and oval. When they molt, larvae become eight-legged nymphs which mature after another two molts. Young mites take four to 12 days to mature and live for about three weeks. In hot weather, the complete life cycle may be as short as two weeks and overlapping generations may occur. Adult females are about 0.5 mm long and adult males are about 0.3 mm long. When feeding in summer, adult females are a yellowish-green with two pronounced dark spots, one on each side of the body. In males, these spots are less obvious and the body is smaller and tapered towards the rear. Both sexes also have two reddish eye-spots on the body.

Red spider mite is a profuse web-spinner in its active stages. Under artificial long-day length or in warm situations such as glasshouses, spider mite may continue to feed and reproduce through winter. Outdoors, when temperatures and day length decrease in autumn, males die and adult females stop feeding and change colour to orange-red. They usually migrate to sheltered sites where they assemble for over wintering. When temperatures increase in spring, females leave their shelters, begin to feed, and lose their red colour.

ii) Management

- Spray of dicofol 18.5 EC at 2.5 ml/l or dimethoate 30 EC at 2ml/l or triazophos 40 EC at 1.5 ml/l or profenphos 50 EC at 1ml/l.
- After 15 days repeat the spray with pongamia or neem oil at 1%.
- Removing and burning of heavily infested shoots checks further build up of mites.

4.5. GLADIOLUS

Gladiolus (*Gladiolus* spp.) is very much liked for its majestic spikes containing attractive, elegant and delicate florets. These florets open in sequence over a longer duration and hence have a good keeping quality of cut spikes. It is cultivated for cut flowers, garden and interior decoration, making bouquets, bedding, rockeries and pots. West Bengal, Maharashtra, Uttar

Pradesh, Punjab, Haryana and Andhra Pradesh are the major gladiolus-growing states.

4.5.1. Insect Pests

4.5.1.1. Thrips, *Taeniothrips simplex*

i) Damage: Thrips are minute insects that feed by rasping plant tissues and sucking the sap. Gladiolus thrips cause serious damage to late flowering plants and to stored gladiolus corms. The leaves of infested plants develop a bleached and silvered appearance (Fig. 4.14), the flower spikes are stunted, flowers may not open, or are distorted and the petals are marked with small pale flecks. Damage to dark-coloured varieties is most noticeable and even a few flecks are enough to considerably reduce the value of the flower (Fig. 4.14).

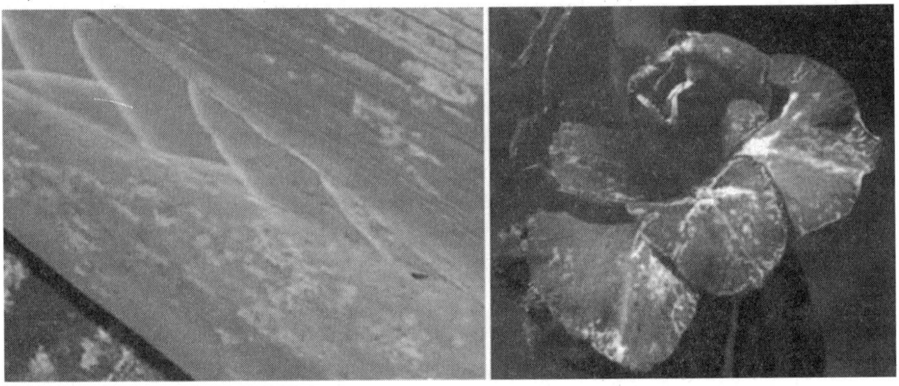

Fig. 4.14. Thrip damage on leaf and flower of gladiolus.

Young nymphs feed inside the leaf tubes and buds. If the plants are left in the field for a long time after flowering, the thrips enter the corms and continue feeding. Stored corms are attacked by thrips, which cause a sticky exudate. The infested corms later become hard and scabby. When the damaged corms are planted out, the thrips begin feeding on the young roots, and growth from the corm may be seriously affected.

Thrips can over winter at any stage of their life cycle on stored corms or on plants growing in greenhouses. Although the gladiolus thrips can maintain a population outside during the summer, they cannot over winter outside in areas where the temperature consistently falls below 10°C as a result the monitoring of corms in storage is extremely important.

The eggs are opaque, white, kidney-shaped, and about 0.34 mm long and 0.2 mm in diameter. Gladiolus thrips emerge as creamy yellow larvae, but soon turn dark brown, except for the tips of the legs which are lighter. The wings are brown above and lighter below and appear to be darker with a grey band when folded over the back. Gladiolus thrips are about 1.5 mm long and extremely slim. Males are smaller and lighter in colour than the females. The life cycle of the thrip is very much determined by the weather and in warmer conditions several generations can be produced in a short time hence the apparent sudden epidemic.

ii) Management

- Spraying 0.1% acephate or 0.5% dimethoate or oxydemeton methyl when initial symptoms are noticed and at 2 weeks interval checks damage.
- Storing of infested corms at 2°C for 6 weeks and later treating them in hot water at 46°C, completely kills thrips on corms (Conijin and Goren, 1990).
- Acephate at 0.15%, primophos methyl at 0.1% and cypermethrin at 0.05% were found effective as dipping agents for the control of thrips on corms (Labnovisky and Mynett, 1990).

4.5.1.2. Cut Worm, *Agrotis segetum*

i) Damage: The pest attack is severe in hilly regions. Female moth lays eggs near ground level on plant parts. Larvae feed on emerging shoots at night. Grown up clay coloured larvae cut the plants at ground level. Plants are vulnerable to attack up to 3-leaf stage. Cut worms also damage underground corms and developing spikes in severe cases.

ii) Management

- Deep ploughing during summer exposes pupae to predators.
- Spread the poison bait consisting of rice bran, jaggery and chlorpyriphos (10: 1: 0.5).
- Spray methyl parathion 50 EC at 1ml/l or quinalphos 25 EC at 2ml/l or acephate 75 SP at 1 g/l.

4.5.1.3. Leaf Caterpillar, *Spodoptera litura*

i) Damage: Caterpillars (Fig. 4.15) can also be a pest and when present will often attack flower buds chewing a hole through the entire bud and thus the petals are held inside. In addition they may also emerge from the tips of a spike in tight buds generating further holes which will detract from the beauty of the plant.

Ash coloured eggs are laid in groups of 200-300 on lower side of leaves. Early instar larvae feed on lower surface of leaves by scraping and skeletonizes them. Greenish brown mature larvae feed voraciously during nights on leaves, flowers and spikes. Pupation takes place in the soil.

ii) Management

- Collect and destroy egg masses and leaves with early larval instars.
- Deep ploughing in summer for exposing pupae to predators.
- Spray quinalphos 25 EC at 2ml/l or chlorpyriphos 20 EC at 2.5 ml/l in combination with diflubenzuron 25 WP at 2g/l.
- Spraying 4% NSKE protects plants from damage.

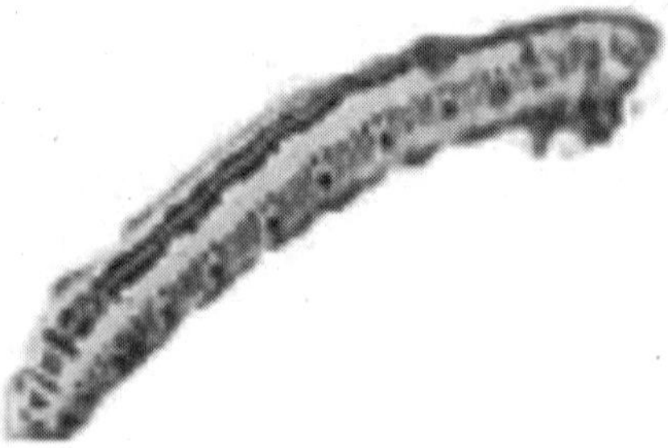

Fig. 4.15. Caterpillar pest on gladiolus.

4.5.1.4. Mealy Bug, *Ferrisia virgata*

i) Damage: The infestation begins in the field on underground corms during dry conditions and is carried on to the storage. Nymphs and adults damage corms by sucking the sap resulting in shriveling and drying of affected corms.

ii) Management

- Spray methyl parathion 50 EC at 1ml/l (Krishnaiah *et al.*, 1975) or dimethoate 30 EC at 2ml/l or acephate 75 SP at 1.5 ml/l at fortnightly intervals.
- Spray 2% pongamia or neem oil.
- Spray *Verticillium lecanii* at 3 g /l .

4.5.1.5. Slugs and Snails

i) Damage: Both slugs and snails can cause damage but in most cases it is purely cosmetic. However, a severe attack on the leaves (Fig. 4.16) can affect the plants ability to produce a new corm for the next season. Slugs have the potential to destroy the growing shoot of a newly planted corm but once established the plant should be safe from their clutches. Snails can affect the florets and make them unsuitable for exhibition.

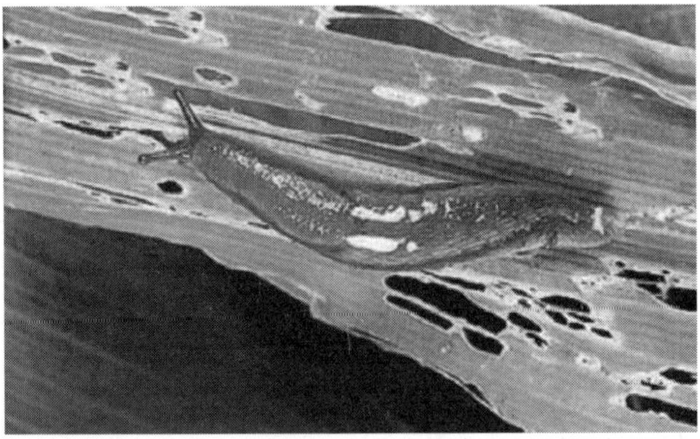

Fig. 4.16. Snail damage on gladiolus leaf.

4.5.2. Mite Pests

4.5.2.1. Mite, *Tetranychus equatorius*

i) Damage: Colonies of mite are seen on undersurface of leaves. Eggs are spherical and translucent in colour. Protonymph and deutonymph are green with dark specks on the dorsum while adults are carmine coloured. Nymphs and adults damage tissue and suck the oozing sap from leaves resulting in white spots. Damaged leaves look discoloured, gradually wilt and finally drop. Mites also infest bulbs during storage.

ii) Management

- Spraying of dicofol or frofenphos each at 0.05% or 0.06% triazophos causes significant mortality of the mites.
- Spraying of 1% neem or pongamia oil also gave considerable control of mites.

4.5.2.2. Bulb Mite, *Rhizoglyphus echinopus*

i) Damage: This slow moving mite is about 0.5 mm long, globular, and yellowish-white with brownish legs. Infested corms produce stunted plants with yellow and distorted leaves. Early infestations are found around the basal plate of the old corm. Roots are destroyed first and stems are attacked later. Corms can be completely destroyed by the combined action of the mites and micro-organisms that invade the damaged tissue.

ii) Management: Hot water treatment will kill the mites, but good sanitation is very important when digging up, storing, or planting out corms.

4.6. TUBEROSE

Tuberose (*Polianthes tuberosa*) is one of the most important bulbous ornamentals grown both for cut and loose flowers, and also for the extraction of its highly valued natural flower oil. It is also grown for garden decoration in pots, beds and borders. It is commercially grown in Karnataka, West Bengal, Maharashtra, Punjab, Andhra Pradesh and Tamil Nadu. At present, the total area under tuberose cultivation in the country is estimated at about 3,000 ha.

4.6.1. Insect Pests

4.6.1.1. Aphids, *Aphis* spp.

i) Damage: Black coloured aphids colonize on growing spikes at base of the buds and suck the sap. Damaged buds and flowers fall off while affected spikes look withered and droop. Black sooty mold develops on the buds in case of heavy infestations.

ii) Management

- Spray dimethoate 30 EC at 2ml/l or oxydemeton methyl 25 EC at 2ml/l at 10-15 days interval.

- Spray imidacloprid 200 SL at 0.4 ml/l or cartap hydrochloride 50 SP 1g/l, if the incidence is serious.
- Pongamia or neem oil sprays at 1% at 10-15 days interval are effective against the pest.
- Use of commercial formulations of *Verticillium lecanii* at 15 days interval reduces aphid population.

4.6.1.2. Thrips, *Thrips* sp.

i) Damage: Thrips cause damage by sucking sap from leaves and growing buds resulting in browning and distortion of damaged leaves. Damaged leaves become brittle and develop brown streaks while affected buds get distorted in shape and dry up. Severely damaged plants get retarded and do not produce flower spikes. Sometimes the pest is associated with transmission of bunchy top disease in which inflorescence gets malformed.

ii) Management

- Spray oxydemeton methyl 25 EC at 2ml/l or dimethoate 30 EC at 2ml/l or acephate 75 SP at 1g/l along with 0.5% pongamia oil at 15 days interval at appearance of early symptoms

4.6.1.3. Weevil, *Myllocerus* spp.

i) Damage: Ash coloured adult weevils feed on leaves from edges and produce characteristic holes and also damage tender shoots. Grubs feed on roots and tunnel into bulbs causing wilting of attacked plants. Full grown grubs pupate in earthen cells below the ground level.

ii) Management

- Treatment of soil with 0.1% carbaryl or malathion dust before planting control grubs.
- Sprays of fenthion or carbaryl at 0.1% protect the foliage from adults damage.
- Neem oil at 1% effectively checks feeding by the weevils.

4.6.1.4. Bud Caterpillar, *Helicoverpa armigera*

i) Damage: Cream coloured eggs are deposited on leaves, buds and flowers. Larvae bore into buds and feed on flowers resulting in considerable flower loss.

ii) Management

Effective control of the caterpillar can be obtained by a spray of 0.01% fenvalerate or 0.05% ethofenprox followed by 0.07% endosulfan or 0.05% methyl parathion at 10 days interval (Jhansi Rani and Jagan Mohan, 1997).

4.6.2. Mite Pests

4.6.2.1. Red Spider Mite, *Tetranychus sp.*

i) Damage: Mite colonies are found on lower surface of leaves which are covered by dirty webs. Affected leaves get discoloured and drop. Attacked plants lose vigour, get stunted in growth, wilt and dry if damage is severe. Damage will be severe if the plants are affected in the early stage.

ii) Management

- Cutting and burning of severely infested plant parts.
- Thorough spraying of water on the plant dislodges the webs.
- Spray dicofol 18.5 EC at 2.5 ml/l or ethion 50 EC at 1ml/l or profenphos 50 EC at 1ml/l followed by a spray of pongamia oil 1% is also effective.

4.7. CROSSANDRA

Crossandra (*Crossandra undulaefolia*) is mainly cultivated commercially for its loose flowers in Tamil Nadu, Karnataka and Andhra Pradesh. It is also grown in Sri Lanka, Tropical Africa and Madagascar. It is a perennial herb and belongs to the family Acanthaceae.

4.7.1. Insect Pests

4.7.1.1. Brown Scale, *Saissetia nigra*; White Scale, *Orthezia insignis*

i) Damage: Yellowish nymphs and dark brown adult scales are seen in large numbers under leaves, on shoots, petioles and stem (Fig. 4.17). Severe infestation results in stunted growth and leaves turn yellow and drop. Honey dew secreted by *Orthezia* sp. favours development of sooty mould which harms the plant growth.

Fig. 4.17. Scale infestation on leaves of crossandra.

ii) Management

- Removal and burning of infested portion of plants prevents further spread.
- Application of carbofuran at 1.0 kg a.i./ha and spraying of 1% pongamia oil in combination with 0.05% methomyl or 0.01% acephate or fenthion provide effective control of scales on crossandra.
- Spray ethion 50 EC at 1ml/l or chlorpyriphos 20 EC at 2.5 ml/l or dimethoate 30 EC at 2ml/l in combination with malathion 50 EC at 1ml/l followed by 1% pongamia oil.

4.7.1.2. Spike Borer, *Helicoverpa armigera, Archips epicyrta*

i) Damage: Spike borer attack is seen from October to February. Eggs are laid singly on tender spikes. Hatched larvae feed on spikelets, flowers and developing seeds. Early instar larvae bite holes into spikelets. Matured larvae feed on spikelets, developing buds and tender seed. *Archips epicyrta* often make silken webs on spikelets. Pupae are formed outside the spikes and brown adult emerges in 10 days (Vadivelu *et al.*, 1976).

ii) Management

- Collection and destruction of grown up larvae and webbed spikes reduces borer population.
- Spray endosulfan 35 EC at 2ml/l or methyl parathion 50 EC at 1ml/l in combination with diflubenzuron 25 WP at 2g/l two to three times at 15 days interval for protecting spikes from borer.

4.7.1.3. Crossandra Bug, *Gynencia affinis*

i) Damage: Crossandra bug is very specific on this crop. Bugs are small, greenish with brown markings. The entire development from egg to adult takes 23-28 days. Both nymphs and adults suck sap resulting in brown lesions. Severely infested plants fail to produce flowers. Population is high during winter although the pest occurs through-out the year.

ii) Management

- Spray quinalphos 25 EC at 2ml/l or acephate 75 SP at 1.5 g/l at or carbaryl 50 WP at 2g/l
- Give subsequent spray with 1% neem oil to protect foliage from bug damage.

4.7.1.4. Midge, *Contarinia sp.*

i) Damage: Attack of the midge is more during July-November. Black coloured adult fly lays eggs singly within buds of the spikes. Orange maggots feed on ovaries of developing buds and pupate inside. Due to damage, the buds fail to develop and rot in severe cases.

ii) Management

- Soil application of carbofuran at 1 kg a.i./ha.
- Spraying of triazophos 40 EC at 1.5ml/l or ethion 50 EC at 1ml/l or acephate 75 SP at 1.5 ml/l.

4.8. CHINA ASTER

China aster (*Callistephus chinensis*) is one of the important annual flower crops. It ranks next to chrysanthemum and marigold among traditional flowers. In recent years, it has gained more popularity due to its multifarious uses. The flowers fetch very good price and returns to the farmer when their production coincides with their demand during particular season/ festival. It is more popular among small and marginal farmers due to its easy cultivation. It is grown for cut flowers, herbaceous borders, bedding, bouquet and pots and cultivated in most parts of the country. It is commercially grown in Karnataka, Tamil Nadu, Andhra Pradesh, Maharashtra (Pune) and West Bengal. In Bangalore and Pune alone it is being grown in an area of 500 and 400 ha, respectively.

4.8.1. Insect Pests

4.8.1.1. Flower Caterpillars, *Helicoverpa armigera, Phycita* sp.

i) Damage: Caterpillars feed on the floral parts damaging the flowers by making groves in the receptacles. They feed on ovaries and stamens of flower buds.

ii) Management

- Spray endosulfan 35 EC at 2ml/l or methyl parathion 50 EC at 1ml/l or phenthoate 50 EC at 1ml/l or fenvalerate 20 EC at 0.5ml/l or 4% neem seed kernel extract.

4.8.1.2. Stem Borer, *Platyptilia molopias*

i) Damage: The adult female plume moths lay white elongate eggs on tender leaves and tips of shoots. Hatched larvae bore into stem and side branches and their feeding results in hollowness and shoot wilt. Caterpillars are white with black head, which feed inside shoots and make them hollow. Damaged shoots bear black excreta on the top. Affected shoots wilt and dry. Larvae pupate inside the stem. Pupae are green in colour (Udayagiri, 1985).

ii) Management

- Collection and destruction of infested shoots.
- Spray profenphos 50 EC at 1ml/l or endosulfan 35 EC at 2ml/l or ethofenprox 10 EC at 1ml/l and application of carbofuran granules at 1 kg.a.i./ha.

4.8.1.3. Aphids, *Aphis gossypii*

i) Damage: Aphids mainly infest lower side of flowers and base of petals (Fig. 4.18). Nymphs and adults suck the sap from the flowers causing discolouration, deformation and withering.

Fig. 4.18. Aphid infestation on flower of China aster.

ii) Management

- Spray oxydemeton methyl 25 EC at 2ml/l or dimethoate 30 EC at 2ml/l.
- Spray 1% neem or pongamia oil if necessary.

4.8.1.4. Red Pumpkin Beetle, *Aulacophora foveicallis*

i) Damage: Beetles attack China aster plants soon after transplantation. Adults feed extensively on leaves by making holes and also cut tender shoots thereby affecting the growth of the plants. Grubs feed on underground roots and stem leading to wilting and drying of attacked plants.

ii) Management

- Neem oil 1% or NSKE 4% reduces damage to foliage by beetles.
- Two or three sprays of quinalphos or methyl parathion each at 0.05% or 0.1% carbaryl give good control of the beetle.
- Soil application of phorate granules at 0.5 kg a.i./ha control grubs.

4.8.1.5. Semilooper, *Ctenoplusia albistriata*

i) Damage: The attack of semilooper is observed during June-July on young plants. Eggs are deposited on leaves. Green coloured semilooper with lines on dorsal and lateral sides feed voraciously on leaves and pupate on the plant causing severe defoliation.

ii) Management: Collecting and killing of grown up larvae reduces further damage. Spraying of quinalphos or methyl parathion each at 0.05% or endosulfan at 0.07% controls semilooper effectively.

4.9. MARIGOLD

4.9.1. Insect Pests

4.9.1.1. Aphid, *Aphis gossypii*

i) Damage: Aphids mainly infest growing tips and floral parts. Nymphs and adults suck the sap from growing parts and the plants give a sickly appearance.

ii) Management

- Spray oxydemeton methyl 25 EC at 2ml/l or dimethoate 30 EC at 2ml/l.
- Spraying of 1% neem or pongamia oil cause considerable reduction in aphid population.

4.9.1.2. Bud Caterpillars, *Helicoverpa armigera, Phycita* sp.

i) Damage: Marigold is one of the preferred hosts of *H. armigera* and an average of 4-5 larvae are seen per plant (Joginder Singh and Sohi, 1990). Moth lays eggs singly on young buds. Hatched larvae bore into buds and growing flowers and feed on floral parts damaging florets and petals (Fig. 4.19).

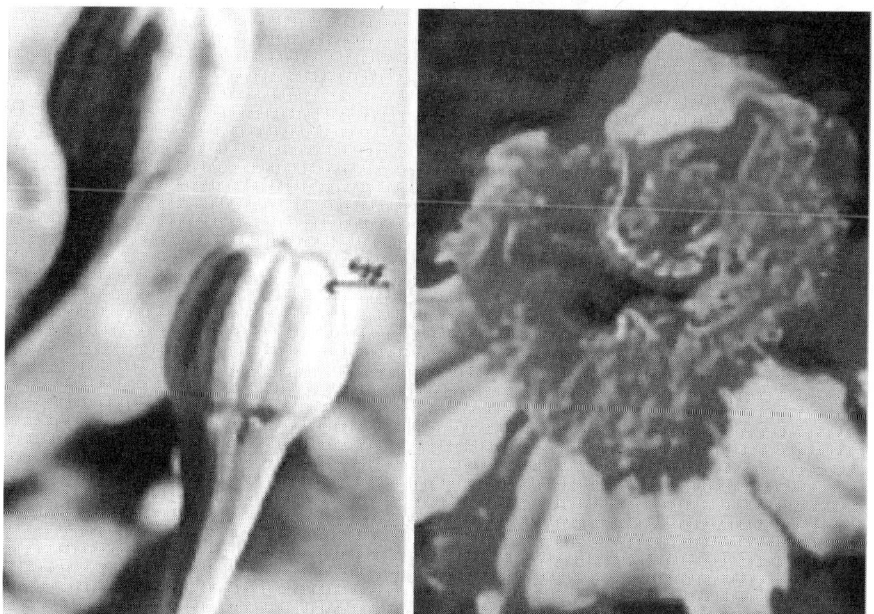

Fig. 4.19. Bud caterpillar on marigold bud and flower

Larvae of *Phycita* sp. damage flowers by making groove on the central receptacle and feed on ovaries and stamen filaments. Excreta of the caterpillars are seen on damaged flowers.

ii) Management

- Collection and destruction of infested buds, flowers and mature larvae to reduce further damage.
- Spraying of endosulfan 30 EC at 2ml/l or methyl parathion 50 EC at 1ml/l or fenvalerate 20EC at 0.5 ml/l in combination with difluben-zuron 25 WP at 2g/l at appearance of eggs on tender foliage checks borer damage.
- Spray 4% neem seed kernel extract.
- Spray *Ha* NPV at 250 LE/ha.

4.9.2. Mite Pests

4.9.2.1. Red Spider Mite, *Tetranychus urticae*

i) Damage: Mite attack on marigold is observed during August-September. Colonies of mites are found on lower surface of leaves which are covered by dirty webs (Fig. 4.20). Both nymphal and adult stages feed on the sap. Affected leaves get discoloured and drop. Attacked plants get stunted in growth, wilt and dry if damage is severe.

Fig. 4.20. Red spider mite damage on marigold.

ii) Management

- Cut and burn the severely infested plant parts.
- Thoroughly spray water on the plant for dislodging the webs.
- Spray dicofol 18.5 EC at 2.5 ml/l or wettable sulphur 80 WP at 3g/l or ethion 50 EC at 1ml/l or profenophos 50 EC at 1ml/ followed by spray of pongamia oil 1%.

4.10. NARCISSUS (DAFFODIL)

4.10.1. Insect Pests

4.10.1.1. Bulb Flies, *Merodon equestris, Eumerus tuberculatus*

i) Damage: The adult narcissus bulb fly is about 0.5 inch long and closely resembles a small bumblebee. The larva is yellowish-white and 0.75 inch long. Larvae feed in bulbs, causing yellow foliage and stunted plants. If infested, large bulbs produce fewer leaves and distorted growth. Instead of a large shoot, small leaves may emerge in a ring around the central growing point that has been killed by a larva. Blooms and small bulbs often die. *Merodon equestris* attacks healthy plants; *Eumerus* spp. prefer bulbs and roots that are already diseased or decayed. The narcissus bulb fly attacks amaryllis, daffodil, *Galtonia,* hyacinth, iris, lily, *Leucojum, Narcissus, Scilla* , tulip, and *Vallota.*

ii) Biology: The female lays eggs near the bulbs during May or early June. The eggs hatch into grubs that move down into the soil and burrow into the base of the bulb to feed there, reducing center of the bulb to a rotten mass. In the process, the embryonic flowers that would have unfolded the following spring are destroyed. After feeding, larvae leave the bulb to pupate in the soil. In the spring, the grubs emerge as adult flies to mate and lay eggs.

iii) Management : Prevention is the most effective bulb fly control strategy. Plant only pest-free bulbs. Handle bulbs carefully to avoid injury. Remove and destroy infested bulbs and plants as soon as you find them.

Narcissus fly can be evaded by deep planting (10 inches). Plant in open areas exposed to wind. Bulb flies are usually less of a problem in such sites. Covering foliage of plants with a row cover after bloom prevents egg-laying. Plant only firm, healthy bulbs. Soft bulbs may be infested from the previous season. Destroy any soft bulbs to prevent emergence of adult bulb flies.

4.10.2. Mite Pests

4.10.2.1. Bulb Mite, *Rhizoglyphus echinopus*

i) Damage: This slow-moving mite is about 0.5 mm long, globular, and yellow-white with brownish legs. Infested bulbs produce stunted plants with yellow and distorted leaves. Early infestations are found around the basal plate of the bulb. Roots are destroyed first and bulb scales and stems are attacked later. Bulbs can be completely destroyed by the combined action of the mites and micro-organisms that invade the damaged bulbs.

ii) Management: Hot water treatment will kill the mites but good sanitation is very important when digging up or planting out bulbs.

4.11. ORCHID

4.11.1. Insect Pests

4.11.1.1. Armoured Scale, *Diaspis boisduvalli*

i) Damage: This is a serious and persistent pest on orchids. Scales colonize on the lower side of leaves, stems and shoots. They suck the plant juice and create toxic effects on plants resulting in burning and drying of damaged parts.

ii) Management: Removal of heavily infested leaves avoids spread of scales. Spraying of dimethoate or oxydemeton methyl or ethion each at 0.05% followed by spray of 1% pongamia oil controls the scale. Spray of fish oil rosin soap 0.05 kg in 30 litres of water is recommended for controlling scale population (Abraham and Vatsala, 1981).

4.11.1.2. Mealy Bugs, *Pseudococcus maritimus, Ferrisia virgata*

i) Damage: These two species of mealy bugs are reported to occur commonly on orchids in India. They have soft and filamentous yellow or pink coloured body covered with a white powdery wax. They suck plant sap and cause deformation of leaves and retardation in growth of the plant. They excrete honey dew which attracts ants and sooty mold develops in case of severe infestation.

ii) Management : Prompt collection and destruction of heavily infested plant parts reduce further spread of the pest. Spraying of methyl parathion or methomyl each at 0.05% or 0.1% acephate, when crawling of ants on plant is noticed, provide effective control of mealy bug. Pongamia oil 2% sprays check mealy bug infestation. Release of coccinellid predator, *Cryptolaemus montrouzieri* and a parasitoid, *Leptomastrida abnormis* reduces the mealy bug population.

4.11.1.3. Thrips, *Anaphorathrips orchidii*

i) Damage: Both nymphs and adult stages suck the sap from tender leaves, growing buds and flowers. Affected leaves develop brown streaks while buds and flowers are distorted and discoloured causing flower loss.

ii) Management: Spraying of oxydemeton methyl or dimethoate or cartap hydrochloride or profenophos each at 0.05% or imidachloprid at 0.007% at 10 days interval, when early symptoms are noticed, effectively checks thrips damage. Misting using deltamethrin at 0.0015% or acephate at 0.05% controls thrips in glasshouse (Timmerman *et al.*, 1982).

4.11.1.4. Snail, *Achatina fullica*

i) Damage: Attack of the snails is observed during monsoon season on orchids. Adult and young stages devour the plants by feeding on leaves at night. They scrape leaves leaving only veins and cut tender shoots (Fig. 4.21).

Fig. 4.21. Snail damage on leaves of orchid.

ii) Management: Hand picking of grown up stages and killing them by dropping in 5% salt solution reduce the snail population. Spreading of 3% metaldehyde pellets (Snail Kill) in the field or application of 5% metaldehyde dust on plants gives effective control of the pest. Neem oil 1% or soapnut extract 6% sprays protect foliage from damage.

4.12. ANTHURIUM

4.12.1. Insect Pests

4.12.1.1. White Fly, *Aleurotulus anthuricola*

i) Damage: Nymphs and adults colonize under flowers, petiole sheath and leaves; and secrete white, powdery, waxy material. As a result of sucking, the infested parts turn yellow, deform and dry in severe cases. The life cycle of white fly is completed in 35 days on anthurium sheaths.

ii) Management: Removal and burning of old and heavily infested leaves check pest build up. Spraying of methomyl or acephate each at 0.05% significantly reduced the number of nymphs and pupae (Hat and Hara, 1992). A holistic spray schedule of methomyl or acephate at 0.05% for adults, 0.06%

triazophos or methomyl or profenophos each at 0.05% followed by a spray of 1% pongamia oil for nymphs and pupae effectively controls white fly menace on anthurium.

4.12.1.2. Thrips, *Chactanaphothrips orchidii*

i) Damage: This is a serious pest of anthurium which causes white streaks and distortion of spathes sucking the sap. Severe injury to flowers is caused during September. Life cycle of thrips is completed in about 27 days. Matured nymphs migrate and pupate in the growing medium.

ii) Management: Hara *et al.* (1987) recommended 0.01% oxydemeton methyl, dimethoate and chlorpyriphos each at 0.06% for the control of thrips. Spraying of new chemicals like profenophos and cartap hydrochloride each at 0.05% or acephate at 0.01% also provides effective control.

4.12.2. Mite Pests

4.12.2.1. Mite, *Brevipalpus* sp.

i) Damage: Colonies of the mite are seen on the lower side of leaves and inside spathes. Entire plant is covered with mites in case of severe attacks. Feeding by mites causes mottling and wilting of affected parts.

ii) Management: Successful control of mites can be obtained when 6% flower injury (ETL) is noticed. Foliar application of 0.0025% vertimec or 0.025% diafenthuron followed by profenophos or amitraz each at 0.05% reduces mite population effectively.

Chapter 5

MEDICINAL PLANTS

In medicinal plant sector, the WHO has estimated that about 80% of the population in developing countries rely on traditional medicines mostly plant drugs for primary health-care needs. Even modern medicines contain about 25% drugs derived from plants. A large number of people are earning their livelihood from different activities of medicinal plants such as collection, cultivation, marketing, processing, etc. The area under cultivation of the above crops is low, but the cash returns are high because of intensive cropping.

In recent times, due to increasing realization of health hazards and toxicity caused by synthetic drugs (side effects), there has been a renewal of interest in the use of plant based drugs throughout the world. A campaign is going on worldwide to utilize more and more plant derived chemicals in the human health care/personal care system. As a result, the demand for plant-based raw materials has increased enormously in both national and international markets. Herbal plants are finding diverse uses as raw materials, not only for medicines but also as biopesticides.

Medicinal crops form one of the important groups due to their demand in various pharmaceutical industries and also to earn foreign exchange by way of export. The potential for earning foreign exchange by India from the exports of medicinal and aromatic plants is estimated over US$ 3,000 million per annum. The export of these plants and their products has a tremendous potential in advanced countries like Europe, USA and Japan. The international market for medicinal plants related trade is estimated at US$ 60 billion/year having a growth rate of 7% per annum. Destination of foreign tourists is now aimed at India for health treatment. Kerala and Gujarat have become their favourite destinations since Ayurvedic treatment is very popular in these two states.

Unless we start cultivating the important medicinal plants, it will not be possible to meet the increased demands. However, cultivation of medicinal plants has to be taken up in polyculture models rather than in monoculture models. There is a huge task to be carried out at every level to make the Indian system of medicine a viable, sustainable and modern. For this task, urgency is felt which can only be established through regular and systematic scientific research on collection, conservation, evaluation and development of good agricultural practices to compete in the world market.

5.1. SOLANUM

Solanum (*Solanum viarum*), also called as Khasi Kateri, is a steroid-bearing perennial, tall bush distributed all over Assam, Sikkim and Manipur between 1,600-2,000m elevations. Its berry pulp is rich in Solasodine alkaloid, which is a starting chemical for production of steroids. About 22 species including *Solanum viarum* contain an alkaloid Solasodine in their berries. Solasodine serves as an excellent raw material for the manufacture of Cortisone and Cortisonal hormones. It is used in the production of contraceptive pills, corticosteroids and sex hormones. It is cultivated in 3,000-5,000 ha, mainly in Maharashtra and Andhra Pradesh. It is also cultivated in West Bengal, Assam, Madhya Pradesh and Gujarat on a small scale.

5.1.1. Insect Pests

5.1.1.1. Fruit Borers, *Leucinodes orbonalis, Helicoverpa armigera*

i) Damage: Female moth lay eggs on tender leaves, buds and fruits. *Helicoverpa armigera* larvae feed on foliage initially and later bore into developing fruits by making holes and feed on internal contents. Early instars of *Leucinodes orbonalis* initially bore into tender shoots resulting in wilting and drying of damaged shoots. Grown up stages bore into fruits and feed inside.

ii) Management: Collection and destruction of bored fruits reduce larval population. Spraying of 0.05% cypermethrin or 0.0015% deltamethrin followed by a spray of 0.07% endosulfan or 0.05% quinalphos or monocrotophos at 0.5 ml/l beginning with flowering stage provides protection to fruits from borer damage.

5.1.1.2.Fruit Fly, *Dacus latiformis*

i) Damage: Adult flies lay eggs inside young fruits by puncturing. Hatched grubs enter into fruit and feed on the pulp. Damaged fruits rot, turn black and drop causing severe fruit loss. Loss is high in rainy season.

ii) Management: Collection and destruction of fallen fruits reduce fruifly population. Spraying of 0.005% cypermethrin or 0.0015% deltamethrin or 1% neem oil kills the eggs and effectively controls fruit fly damage.

5.1.1.3. Foliage Caterpillars, *Spodoptera litura, Spilosoma obliqua*

i) Damage: Eggs are laid in clusters on lower surface of leaves. Hatched larvae are gregarious in nature and they damage leaves by scraping resulting in skeletonization and drying. Later instars feed voraciously on leaves causing severe defoliation.

ii) Management: Collection of egg masses and early instar larvae and killing them effectively reduce the pest population. Use of poison baits consisting of rice bran, jaggery and chlorpyriphos (in 10: 1: 0.5 ratio) kills grown up stages. Spraying of methyl parathion or phenthoate each at 0.05% or acephate at 0.01% or quinalphos at 2.0 ml/l gives good control of the caterpillars.

5.2. PERIWINKLE

Periwinkle (*Catharanthus roseus*), commonly known as vinca, is a common bedding plant. It is popular in container gardens and new flower colours produced on compact cultivars, have increased its use in many public and home gardens. It contains several alkaloids found to be effective in the treatment of leukemia and other diseases such as diabetes, hypertension, etc. The more important alkaloids obtained from Vinca are Leurosine, Vinblastin, Vincristin, etc.

5.2.1. Insect Pests

5.2.1.1. Sphingid Larvae/Oleander Hawk moth, *Dilephila nerii*

i) Damage: Smooth, green, stout larvae with horn on dorsal side cause damage to plants by feeding extensively on leaves, buds and flowers resulting in severe defoliation and yield loss. Grown up larvae pupate in soil in earthen cocoons.

ii) Management: Deep ploughing before planting expose pupae to predators. Collection and destruction of grown up larvae check further attack. Spraying of carbaryl at 0.2% or methyl parathion or chlorpyriphos each at 0.05% in combination with 0.5% neem oil controls damage by larvae.

5.2.1.2. Leaf Eating Caterpillar, *Anomis flava*

i) Damage: Eggs are deposited on tender foliage. Larvae feed voraciously on leaves and cause severe damage to foliage.

ii) Management: Spraying of phenthoate or quinalphos each at 0.05% or acephate at 0.01% protects foliage from damage by caterpillars. Reddy *et al.* (1981) found that sprays of 0.07% endosulfan and 0.2% carbaryl were effective against the caterpillars.

5.2.1.3. Blister Beetle, *Mylabris pustulata*

i) Damage: Adult and young stages of blister beetle attack plants in large numbers during the day and feed on the flowers and buds causing severe flower loss.

ii) Management: Collection and destruction of beetles by tapping them into insecticidal solution reduce the pest population. Spraying of 0.2% carbaryl or 0.1% malathion provides protection to plants from beetle attack.

5.3. ISABGOL

Isabgol (*Plantago ovata*) is an important cash crop of Gujarat (Ahmedabad, Banaskantha, Bhavnagar, Jamnagar, Junagadh, Khaira, Kutch, Mehsana, Rajkot) where it is cultivated in 33,600 ha area producing 30.742 tonnes of seed. Isabgol is also grown in Bihar (Sasaram), Haryana (Rewari), Madhya Pradesh, Rajasthan (Sirohi) and Uttar Pradesh.

The husk contains xylose, arabinose, galacturonic acid, rhamnose and galactose. Husk is a safe laxative, beneficial in habitual constipation, chronic dysentery and diarrhea. The seed and husk are used to cure inflammation of the nervous membrane of gastrointestinal and genito-urinary tracts, duodenal ulcer, gonorrhea and piles. It can also be used as cervical dilator for termination of pregnancy. Mucilaginous husk absorbs toxins from the gut and helps in excreting them from the body. It acts as a demulcent and anti-irritant in soothing the intestinal ulcers and reduces the serum cholesterol. Seed contains protein (17-19%) and yellow semi-drying oil (5%) containing linoleinic acid (0.2%), linoleic acid (47.9%), oleic acid (36.7%), palmitic acid (3-7%), stearic acid (6.9%) and liguoceric acid (0.8%).

5.3.1. Insect Pests

5.3.1.1. Aphid, *Aphis gossypii*

i) Damage: Yellowish green and brown coloured aphids colonize the plants on tender shoots and lower surface of leaves. Both adult and young stages suck the sap causing curling, withering and retardation of growth of the infested plant parts (Fig. 5.1).

Fig. 5.1. Aphid infestation on isabgol.

ii) Management: Spraying 1% neem or pongamia oil effectively reduces aphid population. Dimethoate or oxydemeton methyl or fenthion each at 0.05% were found effective against the aphids (Prem Sagar, 1992).

Oxydemeton-methyl at 0.025% was most effective with a higher benefit cost ratio of (16.9: 1). Among neem products 2% crude extract treatment resulted in the next highest benefit cost ratio (13.4: 1).

5.3.1.2. Seed Beetle, *Lasioderma serricorne*

i) Damage: This is a serious pest of seeds in storage. Both grubs and adults cause damage by feeding on the internal contents of the seed.

ii) Management: Seed treatment with 2% neem/pongamia oil protects them from beetle damage. Seed treatment with 0.05% chlorpyriphos or 0.07% endosulfan also protects seed from beetle damage.

5.3.1.3. White Grubs and Termites

i) Damage: Sometimes, white grubs and termites damage the isabgol crop by cutting the roots.

ii) Management: These pests can be controlled by broadcasting 65% lindane at 1.25 kg/ha before the last ploughing.

5.4. ASWAGANDHA

Ashwagandha (*Withania somnifera*) is a plant of immense medicinal importance growing all over North-Western and Central India. Roots of this plant are the major source of alkaloids and steroidal lactones (withanoids), which are intensively used in various pharmaceutical industries. It has adaptogenic, immuno-modulator, aphrodisiac, anti-stress and mildly sedative properties. Its roots yield valuable drugs which are employed in rheumatic pain, inflammation of joints, nervous disorders, cough, cold, epilepsy, and cardio and nerve tonic. At present, 4,000-5,000 ha area is under its cultivation mainly in Madhya Pradesh (Manasa Neemuch and Jawad tehsil of Mandsaur district) and neighbouring districts of Rajasthan (Kota and Charu).

5.4.1. Insect Pests

5.4.1.1. Epilachna Beetle, *Henosepilachna vigntioctopunctata*

i) Damage: Clusters of eggs are laid on lower surface of leaves. Grubs and adults fed on leaves by scraping resulting in skeletonization, drying of leaves and defoliation of attacked plants.

ii) Management: Collection and destruction of eggs, grubs and adults bring down the population. Spraying of 0.2% carbaryl or 0.1% malathion or 0.01% fenthion protects foliage from the beetle damage. Spraying 2% neem oil effectively prevents feeding of the beetle.

5.4.1.2. Mealy Bug, *Coccidohysterix insolitus*

i) Damage: White and wax coated nymphs and adults infest lower side of leaves in colonies and suck the sap. Affected leaves get deformed; turn yellow and drop-off in severe cases. Honey dew excreted by bugs favours development of sooty mold which affects the plant growth.

ii) Management: Removal and destruction of severely infested plant parts avoid further spread. Spraying of 0.1% acephate or methomyl or fenthion each at 0.05% followed by a spray of 2% pongamia oil checks mealy bug infestation.

5.4.1.3. Ash Weevil, *Myllocerus viridanus*

i) Damage: Adult weevils cause damage to leaves by eating from edges in a characteristic manner resulting in severe foliage loss if populations are very high. Grubs feed on roots leading to wilting and drying of the plants.

ii) Management: Tapping of adults into insecticide solution reduces their population. Spraying of 0.2% carbaryl 0.05% fenthion protects foliage from weevil attack. Soil application of malathion dust or drenching of soil with 0.1% chlorpyriphos kills the grubs.

5.5. COLEUS

Coleus (*Coleus forskohlii*) belongs to the family Lamiaceae and is indigenous to India. It is used in Ayurvedic medicines and as a condiment. It shot into prominence in modern medicine with the isolation of the diterpinoid, forskolin, from its tubers. The therapeutical properties of forskolin in the treatment of glaucoma, congestive cardiomyopathy, asthma and certain cancers combined with its use in cosmetics has enhanced importance of coleus in modern medicine. Coleus proved to be the exclusive source of forskolin and therefore, the sudden spurt in the demand for its tubers led to exploitative collection from natural stand. The drug is claimed to improve appetite, facilitate digestion, increase vitality and is useful in anemia and inflammation. Recently it is also being used in the treatment of alcoholic addiction. Forskolin has a multiple biological activities like positive iontropic, antihypertensive, branchospasomolytic, antithrombotic, platelet aggregation inhibiting and adenylate cyclase stimulation. The species is now considered endangered and a prime candidate for cultivation, more especially captive cultivation.

5.5.1. Insect Pests

5.5.1.1. Tinged Bug, *Monanthia globulifera*

i) Damage: Female bug inserts eggs on leaves and tender foliage. Black, spiny nymphs and lace winged adults cause damage by sucking the sap from the leaves. Attacked leaves develop yellow patches, become brown, shrivel and dry up from margins in case of heavy infestations.

ii) Management: Removal and destruction of heavily infested leaves reduces the bug population. Acephate or carbaryl or methyl parathion or phenthoate each at 0.05% sprays offer effective control of the bug.

5.5.1.2. Spike Borer, *Helicoverpa armigera*

i) Damage: Cream coloured, round eggs are seen on tender foliage. Larvae initially feed on tender leaves and later bore into developing spikes by making holes and feeding on buds and flowers.

ii) Management: Two or three sprays of 0.07% endosulfan or methyl parathion or ethofenoprox each at 0.05% at 15 days interval controls the pest.

5.5.1.3. Grass Hopper, *Chrotogonus* sp.

i) Damage: Nymphs and adult grass hoppers attack the crop during the day. They cause damage by making irregular holes on leaves and cutting tender shoots.

ii) Management: Malathion at 0.1% or carbaryl at 0.2% sprays offer protection to foliage from grass hopper damage. Spraying 2% neem oil checks attack of the pest.

5.6. OPIUM POPPY

Opium poppy (*Papaver somniferum*) is grown in Banswara, Bara, Bhilwara, Bundi, Chittorgarh, Jhalawar, Kota and Udaipur districts of Rajasthan; Faizabad, Barabanki, Bareilly and Shahjehanpur in Uttar Pradesh and Neemuch, Mandsaur and Ratlam in Madhya Pradesh. It is a source of raw opium which is used in narcotic, veterinary and pharmaceutical industries. It contains alkaloids like Morphine, Codeine, Narcotine, Papaverine and Thebain. Poppy seed is used for culinary purpose - breads, sweets and confectioneries. Capsule husk is brewed with tea (Bonda chai).

5.6.1. Insect Pests

5.6.1.1. Cut Worm, *Agrotis ipsilon*

i) Damage: Female moths lay eggs on lower surface of leaves or on the moist soil. Dark brown larvae cut young plants at ground level and damage shoots of grown up plants after dusk and hide during day time. Attack by the pest leads to retardation in plant growth and yield loss.

ii) Management: Deep ploughing during summer exposes pupae. Flooding of infested field with water mixed with 0.1% chlorpyriphos and application of 0.1% malathion or 0.2% carbaryl protect newly planted crop from cut worm. Spreading of poison bait made using rice bran, jaggery and chlorpyriphos (in 10: 1: 0.5 ratio) in the field kills larvae. Spraying of methyl parathion or phenthoate each at 0.05% also controls the pest effectively.

5.6.1.2. Capsule Borer, *Helicoverpa armigera*

i) Damage: Spherical, cream coloured eggs are laid on tender leaves and later bore into developing spikes and capsules by making holes.

ii) Management: Spraying 0.01% fenvalerate or 0.05% methyl parathion or quinalphos or 0.07% endosulfan, when larvae are noticed, controls the borer damage. Spraying of 5% NSKE or 1% neem oil checks feeding by the larvae.

5.6.1.3. Aphid, *Myzus persicae*

i) Damage: Nymphs and adults colonize on growing shoots and cause damage by sucking the sap. Affected shoots get deformed, wither and retarded in growth.

ii) Management: Spraying 2% neem/pongamia oil at 10 days interval controls aphids effectively. Dimethoate or phenthoate or profenphos each at 0.05% sprays also reduces aphid population considerably.

5.7. SARPAGANDHA

Sarpagandha (*Rauvolfia serpentina*) is a perennial native Indian herb. Its roots are used for controlling high blood pressure and certain forms of insanity in Ayurvedic system of medicine since ancient times. It has received worldwide recognition after isolation of its bioactive reserpine alkaloid in allopathic medicines. In addition, its sedative property is utilized by Ayurvedic physicians in treatment of insomania, epilepsy and asthma. It is commercially cultivated in Madhya Pradesh, Uttar Pradesh, West Bengal, Assam and Orissa. The root is its economic part, containing 55 alkaloids of which reserpine, deserpidine, ajmalcine, serpentine and yohimbine being pharmacologically active and important. Of these, ajmalicine, reserpine and serpentine are highly potent in medicine. The commercial root crop contains 1.4-3.0% of total alkaloids.

5.7.1. Insect Pests

5.7.1.1. Scale, *Saissetia coffeae*

i) Damage: Brownish black scales infests lower surface of leaves along midrib and veins. They also infest petioles and tender shoots and deplete the sap from plant. Attacked leaves turn yellow and drop off while plants lose vigour and become stunted in growth.

ii) Management: Removal and burning of heavily infested plant parts bring down scale population. Spraying of ethion or chlorpyriphos each at 0.05% or acephate at 0.1% effectively controls the scale attack.

5.7.1.2. Grass Hopper, *Orthacric simulans*

i) Damage: Both young and adult stages damage the crop by making irregular holes on leaves and cutting tender shoots. Attack on young plant result in plant loss.

ii) Management: Spraying of 4% NSKE or 1% neem oil prevents feeding by the grass hoppers. Foliar sprays of carbaryl or malathion each at 0.1% effectively checks damage by the pest.

5.7.1.3. Leaf Folder, *Glyphodes vertumalis*

i) Damage: Eggs are deposited on tender foliage. Larvae fold leaves together and feed on green tissue causing drying of leaves and stunted growth of the plant in case of severe attack.

ii) Management: Collection and destruction of dried and infested plant parts reduce the pest population. Spraying of profenophos or phenthoate or ethofenoprox each at 0.05% effectively controls the larval damage.

5.8. SENNA

Senna (*Cassia angustifolia, C. acutifolia*) is a well known drug made from leaves and pods. It is grown principally under dry land conditions and usually as a semi-arid irrigated crop. Leaves and pods of senna contain two glycosides viz., Sennoside A and Sennoside B which are used as laxatives. It is also used in medicine as a cathartic. It is used as an ingredient of herbal tea in Europe. The major senna growing areas include Madurai, Tirunelveli and Ramanathapuram in Tamil Nadu; Pune in Maharashtra; Anand and Mehsana in Gujarat and Jodhpur in Rajasthan.

5.8.1. Insect Pests

5.8.1.1. Green Leaf Eating Caterpillar

i) Damage: The pest feeds on young leaves.

ii) Management : Spraying endosulfan at 2 ml/litre of water is effective.

5.8.1.2. Pod Borer

i) Damage: The pod borer bores into pods and feeds on the seed.

ii) Management : Spraying endosulfan at 2 ml/litre of water is found effective.

5.8.1.3. Aphids

i) Damage: Aphids suck the sap and devitalize the plant.

ii) Management : Spraying endosulfan at 2 ml/litre of water gave effective control of aphids.

5.9. LIQUORICE

Liquorice (*Glycerrhiza glabra*) is grown in Gujarat, Haryana and Rajasthan. Dried peeled or unpeeled underground stem and roots contain glycyrrhizin, which is 5 times sweeter than sucrose. It is used as flavouring agent in pharmaceutical industry for manufacturing cough syrups.

5.9.1. Insect Pests

5.9.1.1. Termites

i) Damage: Termites live in above-ground and under-ground mud galleries and colonize the dry parts of plants including roots. Plants are easily uprooted.

ii) Management: Destroy termite mound near the vicinity. Remove above-ground mud galleries. Drench with chlorpyriphos at 7 ml/l of water around soil near root zone.

5.9.1.2. Cotton Ash Weevil, *Myllocerus undecimpunctatus*

i) Damage: The ash weevil causes injury to leaves.

ii) Management: Foliar spraying of quinalphos at 2 ml/l of water is found effective. Application of neem cake at 0.5 t/ha at planting also gave effective control of the pest.

5.10. BELLADONNA

Indian belladonna (*Atropa accuminata*) and European belladonna (*Atropa belladonna*) are the source for tropane alkaloids which include hyoscine, hyoscyamine and atropine (0.13-0.7%) which are used in the manufacture of tinctures, treatment of whooping cough, epilepsy, Parkinson's disease, renal and bilary colic. Indian belladonna is grown in the farms of Kashmir State Forest Department which is a remunerative crop for North-Western hilly region of India.

5.10.1. Insect Pests

5.10.1.1. Cut Worm, *Agrotis flammetra*

i) Damage: The cut worm attacks tender seedlings during early summer.

ii) Management: Spraying endosulfan at 2 ml/litre of water is effective.

5.10.1.2. Leaf Defoliator, *Archips imcaceana*

i) Damage: The pest defoliates the leaves.

ii) Management: Spraying endosulfan at 2 ml/litre of water is effective.

5.11. GLORIOSA

Gloriosa (*Gloriosa superba*) tubers and seeds contain colchicine at 0.15-0.3% and 0.7-0.9%, respectively. Colchicine is used in treatment of gout.

5.11.1. Insect Pests

5.11.1.1. Lily Caterpillar, *Polytela gloriosae*

i) Damage: The lily caterpillar is a voracious feeder of leaves and growing point.

ii) Management : Spraying carbaryl at 3 g/litre of water is effective for the management of lily caterpillar.

5.11.1.2. Green Caterpillar, *Helicoverpa armigera*

i) Damage: The green caterpillar completely eat the flowers.

ii) Management: Spraying carbaryl at 3 g/litre of water gives effective control of green caterpillar.

5.12. GUGGAL

Guggal (*Commiphora mukul*) is grown in Assam, Gujarat, Karnataka, Maharashtra, Rajasthan and Tamil Nadu. It is a source of Indian Bedellium, an oleo-gum resin obtained by incision of the bark. The oleo-gum resins are mixtures of resins (61%), gum (29.3%), volatile oil (1.45%), occasionally combined with other substances, mainly myrecene, dimyrecene and polymyrecene. Resin is used in incense, perfumery and in indigenous medicine as antiseptic, astringent, carminative and digestant. It is used to treat arthritis, pyorrhoea, tonsillitis, obesity etc.

5.12.1. Insect Pests

5.12.1.1. Leaf Eating Caterpillar, *Euproctis lanata*

i) Damage: The pest is voracious leaf feeder which causes extensive defoliation.

ii) Management: Spraying monocrotophos at 2 ml/litre of water is effective in managing the pest.

5.12.1.2. White Fly, *Bemisia tabaci*

i) Damage: White flies sucks sap from leaves which turn yellow and fall.

ii) Management: Spraying monocrotophos at 2 ml/litre of water gives effective control of whiteflies.

5.13. LONG PEPPER

Long pepper (*Piper longum*) is grown in Karnataka, Kerala, Madhya Pradesh, Maharashtra, Tamil Nadu, Eastern Uttar Pradesh and lower hills of West Bengal. Its spikes contain the alkaloid piperine (4-5%) and piplatin, piperolactum A, piperolactum B and piporadione. Roots contain alkaloids piper longuminine (0.2-0.25%) and piper longumine (0.02%). Further purification of piper longumine yields alkaloids like cepharadione A, cepharadione B, cepharanone B and aristolactum A11. Dried immature roots and fruits find extensive use in Ayurveda and Unani medicine in treatment of chronic bronchitis, cough and cold.

5.13.1. Insect Pests

5.13.1.1. Mealy Bugs

i) Damage: Mealy bugs infest roots of plant and suck the sap.

ii) Management: Soil drenching with chlorpyriphos at 3 ml/litre of water around the base of plants is effective in managing the disease.

5.13.1.2. Tea Mosquito Bug, *Helopeltis theivora*

i) Damage: Adults and nymphs feed on tender foliage causing necrotic lesions around feeding punctures and leave shot holes on leaf lamina.

ii) Management : Spray any commonly available systemic insecticide to control the pest.

5.14. AMBRETTE

Ambrette (*Abelmoschus moschatus*), also called as Muskdana and Kasturi Bhendi, grows wild all over the Deccan plateau and Karnataka, in the hilly region and at the foot hills of the Himalayas. It is raised as a *kharif* crop in Punjab, Nagaland, Arunachal Pradesh, Bihar, Terai area of Kumaon, adjoining areas of Lucknow, Mahmudabad and Sitapur (Uttar Pradesh). Major constituent of oil is Ambrettolic acid and its lactone - 'ambrettolid'. The volatile oil finds use in perfumes, cosmetics, talcum powder, toiletries and confectionaries. Various parts of the plant find extensive use in Ayurvedic medicine. It is also used in manufacture of 'zarda', a flavoured chewing tobacco.

5.14.1. Insect Pests

5.14.1.1. Aphids and Thrips

i) Damage: The early season sucking pests i.e. aphids and thrips cause considerable damage to the crop in early stage of its growth, particularly in the early sown summer crop.

ii) Management: The foliar spray of dimethoate, endosulfan, malathion, phosphamidon and vamidothion reduces the risk of aphids and thrips.

5.14.1.2. Spotted Boll Worm

i) Damage: The spotted boll worm attacks succulent shoot, vegetative buds, flowers and pods during vegetative growth and fruiting stage. The infested shoots above the point of infestation become brown, bend down and die. The fruits are also infested which causes damage to seeds in pods.

ii) Management: Endosulfan at 0.07% can be sprayed at 10-15 days interval right from the stage of one month old crop to the last picking to control the pest. To eradicate spotted boll worm, methyl parathion, quinalphos, toxaphene, dimethoate and phosphamidon can be applied.

5.14.1.3. Pink Boll Worm

i) Damage: The deep pink coloured caterpillar of pink boll worm also infests the crop. The caterpillar bore into the pods and seeds.

ii) Management: Healthy seeds free from pink bollworm infestation should be sown as precautionary measure. Chlorpyriphos + cypermethrin each at 0.05% can be sprayed for the control of boll worms.

5.14.1.4. Red Cotton Bug

i) Damage: The red cotton bug feeds on the surface of hairy leaves.

ii) Management: Spray cypermethrin or fenvalerate each at 0.01% to control the red cotton bug.

5.14.2. Mite Pests

5.14.2.1. Red Spider Mite

i) Damage: The web forming red spider mite feeds on the under surface of leaves and produces white specks. The infested plants become yellow, weak and shed leaves.

ii) Management : Dimethoate, phosphamidon, folidol each at 0.05% can be sprayed once on the infested plants to ward off red spider mites.

Chapter 6

AROMATIC CROPS

In recent years, there has been an increased interest in the cultivation of aromatic plants to meet the requirements of cosmetic, flavouring and perfumery industries. The aromatic plants provide raw material for the production of flavours, herbal cosmetics, perfumery etc. Aromatic plants and their products, particularly the essential oils, are now becoming one of the more important export items from many developing countries of Asia. Chemically essential oils are terpenes, which act as carriers of the aromatic substances. Most essential oils also contain camphors, and the more odiferous compounds present in them consist of oxygen derivatives of terpenes, alcohols, esters, aldehydes and ketones. India has enjoyed a prominent position in the manufacture of superior perfumes and aromatics by using essential oils. The important aromatic plants are lemon grass, vetiver, patchouli, palmarosa, citronella, mints, geranium, lavender, basil, jasmine etc.

In the world market, India stands in second position. Most of the essential oils produced though marketed within the country, but a sizeable amount has also been exported. Among the essential oils exported from India are Japanese mint oil, ambrette seed essence, sandal wood oil, citronella oil, lemon grass oil, palmarosa oil, ginger oil, clove oil, eucalyptus oil, vetiver oil, pepper oil, etc. The annual exports of the derivatives of aromatic plants are to the tune of Rs. 600-700 million.

6.1. JASMINE

Jasmine (*Jasminum* spp.) is one of the leading traditional flowers of India. Its flowers are used for making garlands, adorning hairs of women, in religious and ceremonial functions, and for producing perfumery oil. It is commercially cultivated in Karnataka, Tamil Nadu, Andhra Pradesh, Maharashtra, Gujarat, Haryana, Punjab, West Bengal, Rajasthan, Assam and Uttar Pradesh. Apart from internal trade, fresh flowers of jasmine are exported to Malaysia, Singapore and Sri Lanka.

6.1.1. Insect Pests

6.1.1.1. Bud-Worms, *Hendecasis duplifascialis*

i) Damage: This is a serious pest of *Jasminum sambac* and *J. pubescens*. The pest appears in severe form during post-monsoon months. The eggs are laid on flower clusters. The caterpillars are green with a black or brown head and feed on fresh leaves and floral parts and leave behind fecal pellets entangled in the silk strands (Fig. 6.1). Webbing is seen in severe infestation and pupation takes place in soil or in the webs. Severely affected bushes show burnt appearance. Pupation takes place in the soil or inside galleries. The moth is small and white in colour (Prasad *et al.*, 1983).

Fig. 6.1. Bud worm on jasmine.

ii) Management

- Collection and destruction of infested plant parts.
- Hand picking of larvae and pupae to reduce the insect population.
- Spray acephate 75 SP at 1.5 g/l or cypermethrin 25 EC at 1 ml/l or quinalphos 25 EC at 2ml/l or profenphos 50 EC at 1ml/l.
- Botanicals like 5% neem seed kernel extract or 1% neem oil are also effective against the pest (Nelson *et al.*, 1993).
- This pest is highly susceptible to *Bacillus thuringiensis* (Sundarababu, 1969).

6.1.1.2. Web Worm, *Nausinoe geometralis*

i) Damage: This is a serious pest of *J. sambac*. Eggs are seen on matured leaves. Hatched larvae scrape leaf tissue initially and web leaves together with silken threads. Green coloured caterpillar with dark warts and hairs on the body feed gregariously on leaves resulting in skeletonization and drying.

ii) Management: Collection and destruction of webs having larvae and pupae reduce the damage. Spraying of 0.1% carbaryl, 0.05% quinalphos or 0.01% fevalerate was observed to be effective against the web worm (Sandhu and Shukla, 1984). Methyl parathion or phenthoate each at 0.05% also provide good control of the pest.

6.1.1.3. Blossom Midge, *Cantarinia maculipennis*

i) Damage: The maggot attacks *J. sambac* and *J. auriculatum* petals and the flower shows the signs of purple discolouration 4-5 days after oviposition. These maggots are dull white and feed on inner petals, anthers and stigma. Mature yellowish orange maggots fall down to the soil and pupate. Pupation occurs in a thin papery white case in the top layer of the soil. The life cycle is completed in 13-18 days.

ii) Management

- Soil application of carbofuran 3G at 40 g per bush or foliar application of cypermethrin 25 EC at 1 ml/l (David and Natarajan, 1988) or triazophos 40 EC at 1.5 ml/l or profenophos 50 EC at 1ml/l at fortnightly interval.
- Apply neem cake at 250 kg/ha as soil treatment.
- Spraying 2% neem oil mixed with teepol at 0.05% effectively reduces the damage (David *et al.*, 1992).

6.1.1.4. Gallery Worm, *Elasmopalus jasminophagus*

i) Damage: The pest attacks all *Jasminum* species. The larvae are green in colour with red head and brown streaks on the body. It builds galleries of fecal matter and silk on the terminal leaves, shoots and flower heads and feeds inside. Pupation takes place inside the galleries and the adult is dark grey in colour.

ii) Management : Collection and destruction of infested plant parts reduce the pest population. Cypermethrin at 0.05% or acephate at 0.01% or methyl parathion at 0.05% sprays offer very effective control of gallery worm.

6.1.1.5. Shoot Borer, *Sycophila* sp.

i) Damage: The pest is observed on *J. grandiflorum*. Larvae tunnel shoots, feed inside by making grooves and pupate inside the shoots. Series of holes are seen on damaged shoots.

ii) Management: Application of carbofuran at 1.0 kg a.i./ha or spraying of 0.06% triazophos or 0.05% chlorpyriphos at fortnightly intervals check shoot borer attack.

6.1.2. Mite Pests

6.1.2.1. Gall Mite, *Aceria jasmini;* Carmine Mite, *Tetranychus cinnabarinus*

i) Damage: Gall mite is a major pest of *J. auriculatum* and *J. pubescens.* Colonization and feeding by the pest results in the formation of felty outgrowths resulting in malformation of vegetative and floral parts (Fig. 6.2) and consequently reduces crop yield.

Carmine mite attacks jasmine leaves during summer resulting in blotching, browning and falling of leaves.

Fig. 6.2. Gall mite infestation on jasmine.

ii) Management

- Spraying of 0.3% wettable sulphur 80 WP at 3 g/l or 0.06% triazophos 40 EC at 1.5ml/l with the onset of monsoon at 14 days interval, repeated 2-3 times, provides effective control of gall mite (Shivaramu and Jhansi Rani, 1996).

- Sprays of 0.05% ethion or 0.01% fenproparthin and 0.02% dicofol are effective in reducing galls by the mite (Mallikarjunappa *et al.*, 1991).
- Karuppuchamy *et al.* (1986) reported successful control of adults and egg hatching of spider mite by spraying 0.05% dicofol or 0.3% wettable sulphur.
- Spray 1% neem or pongamia oil, if necessary.

6.2. MINTS

Mints (*Mentha* spp.) are a group of aromatic herbs belonging to the family Lamiaceae, which are considered to be the most important cash crops in Indo-Gangetic plains. Cultivation of mints has been done in a large scale in many tropical and sub-tropical countries of the world including India, China, Brazil, Japan and USA. Indian farmer grow it as a bonus crop on an area of more than 1,50,000 ha in Central and Northern parts of Indo-Gangetic plains. The crop fit well in the traditional food based cropping system with other crops like paddy, wheat, potato, mustard, maize, okra, carrot, onion, spinach, pigeon pea, cowpea, etc. Similarly it can be grown as an intercrop with sugarcane and some legumes. In India, the highest area under commercial cultivation of mint has been in Uttar Pradesh (Terai region), but during last few years it has spread through out the country, especially in North and North-Western states including Madhya Pradesh.

Among different oil yielding species of mint, Japanese mint (*M. arvensis*) occupies the highest area under mint cultivation followed by Peppermint (*Mentha spicata*), Spear mint (*M. piperata*), Bergamot mint (*M. citrata*) and Scotch spearmint (*M. cardiaca*). Japanese mint essential oil is used as a source of menthol, menthyl acetate, menthone and terpenes which finds wide use in medicine, perfumes, food and cosmetics. The oil of peppermint, which contains less menthol but has a sweeter aroma and taste, is used in tooth paste, chewing gums, candies, high grade liquors, medicines and other pharmaceutical preparations. Spearmint oil having carvone and limonene, is mainly used for flavouring; whereas, Bergamot mint oil is mainly used in perfumery and flavouring due to higher content of linayl acetate and linalool.

6.2.1. Insect Pests

6.2.1.1. Semilooper, *Thysanoplusia orichalcea*

i) Damage: The pest infests the crop round the year. It has overlapping generations and the life cycle is completed in 27 to 38 days (Prem Sagar and Ramji, 1991). Eggs are found on lower surface of leaves. Larvae devour the leaves and cause severe defoliation.

ii) Management: Spraying of methyl parathion or quinalphos each at 0.05% or fenvalerate at 0.01% provides god control. NSKE 4% or 1% neem oil reduces damage by semilooper to foliage.

6.2.1.2. Leaf Webber, *Syngamia abruptalis*

i) Damage: Infestation of leaf webber is severe during May-June and August-October. Larvae web leaves together and feed on them resulting in skeletonization and drying of damaged leaves. Pupation takes place inside the web. The life cycle is completed in 31 to 43 days in different months.

ii) Management: Clipping and destruction of webs reduce webber population. Spraying of 0.005% cypermethrin followed by 0.05% quinalphos or 0.1% dichlorvos or 0.07% endosulfan gives successful control of leaf webber.

6.2.1.3. Aphid, *Aphis affinis*

i) Damage: This pest is active during March-April. About 15 generations are reported on this crop (Prem Sagar and Singh, 1981). Colonies of the aphid are found on tender shoots and lower side of leaves. They suck the sap resulting in discolouration, deformation and falling of leaves. Affected shoots become stunted in growth and wither.

ii) Management: Spraying of 4% PSKE or 1% neem or pongamia oil reduces aphid infestation. Dimethoate or oxydemeton methyl each at 0.05% also provides good control of the aphids.

6.2.1.4. White Fly

i) Damage: The initial adult population of white fly was found building up from March and was maximum in the month of May.

ii) Management : Among genotypes evaluated for their resistance potential to whitefly population, significantly low number of flies was observed in cv. 'Kushal'.

6.2.1.5. Leaf Roller, *Syngamia abruptalis*

i) Damage: The caterpillar folds the leaf in the form of a roll and feed inside the leaf tissue. The edges of the leaves are held together with silk-like filaments. Leaves get rolled in August-September.

ii) Management : Spraying of 0.1% phosphamidon or 0.07% endosulfan or thiodan at 1.5ml/litre of water gives effective control of leaf roller. Two to three sprayings are sufficient to control the pest.

6.2.1.6. Red Pumpkin Beetle, *Aulacophora foveicallis*

i) Damage: Leaves and buds are eaten away.

ii) Management: Spraying 0.05% malathion twice at 20 days interval is effective for the management of beetles.

6.2.1.7. Hairy Caterpillar, *Diacrisia obliqua*

i) Damage: The caterpillar starts eating the under-surface of the leaves resulting in leaf damage.

ii) Management : Spraying of folidol or malathion each at 0.05% twice at 20 days interval is effective for the control of the pest.

6.2.1.8. Cut Worms, *Agrotis flammatra*

i) Damage: Pest damages the young plants at the collar region during spring season.

ii) Management : Soil treatment with phorate before planting the crop is recommended.

6.3. BASIL

Among several species of basils (*Ocimum* spp.), sweet basil (*O. basilicum*) is considered to be the most important crop for its high quality of essential oil. Major components of its oil are linalool (43-50%), menthyl chavicol (18-33%), eugenol and isoeugenol (5-6%) and the minor constituents are alpha and beta pinene, camphor, geraniol, etc. The oil of sweet basil owes its importance to its extensive use in condimentary products, cosmetics, toiletry, perfumery and confectionary industries, particularly in European countries.

6.3.1. Insect Pests

6.3.1.1. Aphid, *Aphis gossypii*

i) Damage: Dark brown adults and green coloured nymphs colonize tender shoots, leaves and inflorescence. They damage plants by sucking sap resulting in yellowing and deformation of leaves. Attacked buds and flowers drop-off.

ii) Management: Spraying 2% neem/pongamia oil causes considerable mortality of aphids. Dimethoate or oxydemeton methyl each at 0.05% also provides good control.

6.3.1.2. White Fly, *Dialeurodes* sp.

i) Damage: White, elongated eggs are laid singly on lower side of leaves. Nymphs and adults suck sap from leaves causing curling, withering and drying of leaves. Sooty mold develops on heavily infested leaves resulting in leaf drop.

ii) Management: Spraying 0.05% dichlorvos or 0.0015% deltamethrin controls adults. Triazophos or methomyl each at 0.05% sprays cause mortality of nymphs and eggs. Butani (1982) recommended spray of endosulfan or malathion each at 0.05% and application of carbofuran granules at 2-3 g/plant in case of severe infestation.

6.3.1.3. Tinged Bug, *Monanthia globulifera*

i) Damage: Spiny, black coloured nymphs and lace winged adults suck sap from ventral side of leaves resulting in yellowing, curling and drying of damaged parts.

ii) Management: Spraying of methyl parathion or fenthion or phenthoate each at 0.05% provides effective control. Sprays of 1% neem/ pongamia oil cause considerable mortality of bugs.

6.4. LEMON GRASS

Lemon grass (*Cymbopogon flexuosus*), West Indian lemon grass (*C. citratus*) and Jammu lemon grass (*C. pendulus*) are grown in Assam, Kerala, Maharashtra, Tamil Nadu and parts of Uttar Pradesh. The chief constituent of palmarosa oil is Citral. Citral is the starting material for preparation of α-Ionone (used in flavours, cosmetics and perfumes) and β-Ionone (used in the manufacture of synthetic vitamin A).

6.4.1. Insect Pests

6.4.1.1. Stem Boring Caterpillar, *Chilotrea* sp.

i) Damage: The most important pest species infecting lemon grass is *Chilotrea* sp. The caterpillar is white in colour with a black head and black spots on the body. It bores into the stem and remains there, feeding on the shoot. It is usually found at the bottom of the stem. As the grass is cut a few centimeters above the ground, the caterpillar, which is lodged lower down, is not affected. The first symptom of the attack is the drying up of the central leaf. Subsequently, the whole shoot dies, resulting in a significant reduction in the yield of the grass.

ii) Management

- The dry stubbles are set on fire during the off-season in summer. The caterpillars lurking inside the stubbles are thus destroyed.
- The affected shoots are pulled out and destroyed.
- When attack is serious, a spray of Folidol E 605 is used.

6.4.1.2. Shoot Borer, *Chilo infuscatellus*

i) Damage: Female moth lay eggs in batches on lower side of leaves. Larvae cut a hole on stem near ground level and feeds inside the shoot. Attacked shoots wilt and dry consequently.

ii) Management: Cutting and burning of dried shoots and earthing up to a height of 25 cm around stem are effective in preventing further attack. Application of carbofuran granules at 1.5 kg a.i./ha and spraying of chlorpyriphos or phenthoate each at 0.05% or ethofenoprox at 0.01% provide good control of the borer.

6.4.1.3. Spittle Bug, *Clovia bipunctata*

i) Damage: This is a commonly found pest of lemon grass. Eggs are deposited on leaves. Nymphs secrete frothy spittle mass and remain inside. They cause damage by sucking sap resulting in yellowing of leaves.

ii) Management: Collection and destruction of leaves with spittle bugs. Spraying fenthion or methyl parathion each at 0.05% or acephate at 0.1% checks the bug infestation. Saradamma and Dale (1983) recommended spray of 0.05% quinalphos for the control of this pest.

6.4.1.4. Aphid, *Macrosiphum muscanthii*

i) Damage: Infestation of aphid is mainly found on tender growing shoots. They suck the sap and cause deformation and stunted growth of the attacked plant.

ii) Management: Two sprays of 1% neem/pongamia oil at 10 days interval cause considerable mortality of aphids. Dimethoate or oxydemeton methyl or profenophos each at 0.05% checks aphid damage.

6.4.1.5. Termites, *Microtermes obesi*

i) Damage: Attack of termites to the crop is severe during summer when dry soil conditions prevail. Termites damage roots and stem portion by building earthen galleries and feeding inside. Infested plants show wilting and drying of leaves followed by death of the plant.

ii) Management: Avoid dry soil condition by irrigating frequently. Scrape earthen galleries on stems and spray 2% neem oil or 0.1% malathion or 0.07% endosulfan. Drench the soil with 0.1% chlorpyriphos after raking up to kill soil inhabiting stages. Killing of the queen by destroying the colony provides permanent control.

6.5. PATCHOULI

Patchouli (*Pogostemon patchouli*) is a highly aromatic bushy under-shrub. It is cultivated for its highly fragrant leaves which contain a very sweet smelling oil of lasting sticky odour. The crop is grown in small pockets in Karnataka, Kerala and Tamil Nadu. It has a very characteristic aroma and blends well with other essential oils. The oil is used in very low concentration (2 ppm) in scenting soaps, cosmetics, after-shave lotions, detergents and many fancy products. The oil is also used to flavour foods, beverages, candy and baked products. In combination with sandal wood oil, it is used in blending tobacco and making incense sticks.

6.5.1. Insect Pests

6.5.1.1. Leaf Caterpillars, *Pronomis profusalis, Psara stultalis*

i) Damage: *P. profusalis* larvae feed extensively on tender leaves leaving only veins and cause foliage loss. Caterpillars of *P. stultalis* characteristically roll the leaves and feed inside causing browning and drying of attacked leaves. Damage by these caterpillars affects growth of the plant.

ii) Management: Clipping and burning of webs and rolled leaves bring down the pest population. Two to three sprays of 0.01% fenvalerate followed by quinalphos or methyl parathion each at 0.05% are recommended for the control of these pests (Reghupathy *et al.*, 1979).

6.5.1.2. Pentatomid Bug, *Nezara viridula*

i) Damage: Green coloured nymphs and adults suck the sap from tender leaves and shoots. Deformed leaves and stunted growth of the plant are the symptoms of attack.

ii) Management: Collection and destruction of egg masses prevent damage. Spraying of carbaryl or malathion each at 0.1% or fenthion at 0.05% protects foliage from the bug damage.

6.6. DAVANA

Davana (*Artemisia pallens*) is the most important aromatic herb for its high value essential oil used in the perfumery and cosmetic industries. Its foliage and floral tops produce a viscous essential oil, emitting a delicate, persistently fruity fragrance which is used in floral decorations, bouquets and cosmetics. It is also used in flavouring of cakes, pastries, beverages and tobacco products. India is one of the major exporters of davana oil to the rest of the world (mostly to USA). It is grown in 2,000 ha mainly in Karnataka and to a lesser extent in Tamil Nadu and Maharashtra.

6.6.1. Insect Pests

6.6.1.1. Bug, *Nysius ericae*

i) Damage: Small bugs are found mainly at flowering stage of the crop. They suck the sap from leaves resulting in yellowing and drying of leaves.

ii) Management : Collection and destruction of leaves infested with bugs reduces the pest incidence. Spraying malathion or dichlorvos each at 0.1% or quinalphos at 0.05% at the commencement of flowering is effective in managing the pest.

6.6.1.2. Semilooper, *Thysanoplusia orichalcea*

i) Damage: Larvae feed voraciously on leaves and cause defoliation of the plants. Eggs are laid on lower surface of leaves. This pest infests the crop throughout the year. It has four overlapping generations and the life cycle is completed in 27 to 38 days.

ii) Management : Collection and destruction of larvae reduces the damage. Spraying of 0.01% fevalerate or 0.1% malathion or methyl parathion or chlorpyriphos each at 0.05% check the damage by semilooper.

6.7. SCENTED GERANIUM

Scented geranium (*Pelargonium graveolens*) is an aromatic perennial herb. It is grown in cooler, sub-tropical climate of Mysore and Bangalore

(Karnataka), and extended to Hyderabad (Andhra Pradesh). Its oil has refreshingly delicate long-lasting rosy odour with fruity under-note, containing 66-78% primary alcohols (rhodinol). It is used in manufacturing of perfumes, creams, talcum powder and body lotion. It is stable in alkaline medium and is therefore used for scenting soaps.

6.7.1. Insect Pests

6.7.1.1. White Fly

i) Damage: They are commonly found on geraniums. The adults look like tiny moths, white in colour about 1-2 mm long and can be found on the underside of leaves. If not acted upon, small scales or eggs which will be of next generation are also observed. These tiny insects excrete honey dew which they obtain from the plant sap. A black sooty mould starts to grow on these deposits. The adults have a waxy coating which repels water making it hard to use any insecticides.

ii) Management : One way to remove this waxy covering is to spray with a weak solution of a detergent (a few drops in 1 pint of water). This will remove the coating and weakens the insect and then kills it.

The parasitic wasp, *Encaria* sp. which lays its eggs within the eggs of the white fly can be used as a biological control agent.

6.7.1.2. Aphids

i) Damage: Aphid is a small winged insect like the whitefly but is easier to eradicate. The indication of crimped leaves will be the first sign and on the underside of the leaves a small colony of aphids may be observed. These tiny, soft-bodied, pear-shaped insects suck plant juices and can spread viral diseases.

ii) Management : They can be controlled by knocking them off the plants with a strong stream of water, application of insecticidal soap (pay special attention to treat the undersides of leaves) or malathion, spraying with horticultural oil, and releasing beneficial insects, such as parasitic wasps, lady bugs, green lacewings and aphid midges.

6.7.1.3. Caterpillars

i) Damage: Several species of caterpillars chew geranium leaves. Caterpillars in summer can be a nuisance. They are small, green in colour and about 6-8 mm long. The first signs are holes in the leaves where they have been feeding.

ii) Management : Caterpillars can be hand picked and destroyed or they can be controlled with an application of Bt (*Bacillus thuringeinsis*) which kills caterpillars, but does not harm beneficial insects. Cultivating the soil in winter will kill over wintering pupae. Attracting or releasing beneficial insects such as lady bugs and parasitic wasps will also help to control caterpillars.

6.7.1.4. Thrips

i) Damage: Thrips can be a problem in late summer. Small grayish marks on the top side of the leaves is an indication of this pest. They are grey-green in colour, 3 mm long and jump when disturbed.

ii) Management: Remove by hand or spray insecticide.

6.7.1.5. Vine Weevil

i) Damage: The main damage is done by the larva which lives in the soil and chews on the root system of the plant. First indication will be a sudden wilting of the plant. By uprooting the plant, white fat maggots with a brown head and curled within the compost can be seen.

ii) Management : Treat with a soil sterilizing liquid and change the type of compost that is used.

6.7.1.6. Bronze Butterfly

i) Damage: This butterfly has an insatiable appetite for geraniums and its caterpillars cause the damage. The butterfly eggs are laid on the flower buds and the caterpillars then burrow into them, eating there way into the stem.

ii) Management: Spray with a strong insecticide every two weeks to keep this pest under control.

6.7.1.7. Bud Worm, *Heliothis (Helicoverpa) virescens*

i) Damage: The adult bud worm moth has light green wings with a series of lighter colour bands that span a distance of 2.5-3.75 mm. Adults lay eggs singly on leaves of plants during early evening. Eggs are laid on the flower buds of geranium. The larval bud worms are striped and quite variable in colour from dark brown to red or even green depending in part on the plant they are feeding. Young larvae tunnel primarily through small buds and stems while larger caterpillars chew holes in flower buds, petals and leaves. Larvae take one month to eat their way to maturity. It completes two generations in a year. The damage by the small larvae as they tunnel through plant stems results in the introduction of various disease organisms into the plant and results in stems that break readily.

The budworm over winters as pupa in earthen cells several inches beneath the soil surface. In areas where winter temperatures dip below 20°F, pupa do not successfully survive the winter. Adults normally emerge in late spring.

ii) Management : Hand picking of the caterpillars is strongly recommended. A braconid wasp *(Cardiochiles nigriceps)* and a sarcophagid fly parasitize caterpillars. *Polistes* wasps also destroy many larvae.

Insecticides which have shown some degree of success are Orthene, Sevin, the pyrethroids Talstar and Scimitar and the biological control treatment with *Bacillus thuringiensis*. Insecticidal treatments should be repeated often to disrupt the life cycle. Alternating insecticides used in these control efforts will help to reduce the chances of this pest developing additional insecticidal resistance.

6.7.1.8. White Ants (Termites), *Microtermes obesi, Odontotermes obesus*

i) Damage: The adults are brownish dirty white in colour and damage root and stem through out the year.

ii) Management: Application of chlorpyriphos at 600-800 ml/ha with irrigation water gives effective control of termites.

6.8. CHAMOMILE

Chamomile (*Matricaria recutita*), an important medicinal and aromatic herb, is widely cultivated for its flowers which are used as medicinal tea (which acts as mild sedative and digestive and provides relief in cough and cold) in Europe and the USA. Their extracts are being used in antiflogistic creams and ointments. The oil of chamomile is utilized in alcoholic and non-alcoholic beverages, ice creams, candies, baked foods as well as in high-grade perfumery throughout the world. It is traditionally employed in manufacturing of pain-relieving balms. It is used in scenting of shampoos, pomades and face creams. It is grown in small area in North-Central parts of Uttar Pradesh and Himachal Pradesh.

6.8.1. Insect Pests

6.8.1.1. Aphid, *Aphis faba*

i) Damage: The nymphs and adults suck cell sap from the inflorescence of plants.

ii) Management: Spraying dimethoate at 2 ml/litre of water is effective for the management of aphids.

6.8.1.2. Lygied Bug, *Antographa chryson*

i) Damage: Lygied bug is one of the serious pests of chamomile. The pest causes defoliation of the plant. Both nymphs and adults cause damage by sucking cell sap from the plants. The adults are 6-7 mm long, black with white patches on the wing, which extend slightly beyond the abdomen. They are mostly prevalent during February and March.

ii) Management: Spraying 0.05% malathion or dimethoate at 2 ml/litre of water gives effective control of *A. chryson*.

6.8.1.3. *Nysius minor*

i) Damage: The pest is responsible for flower shedding.

ii) Management: Spraying carbaryl 50 WP at 3 g/litre of water is effective for the management of *N. minor*.

6.9. CITRONELLA

Citronella (*Cymbopogon winterianus*) is a large, perennial, stemless, aromatic grass. Its oil from leaves has a strong characteristic lemonic odour extensively used in scenting soaps, cosmetics, deodorants and mosquito repellent products. The oil is used as starting material to produce pure citronellol, geraniol and related upstream high-value aroma chemicals used in perfumery industry. It is cultivated in 6,000-10,000 ha land in high rainfall tracts of Assam, Arunachal Pradesh, Meghalaya, Nagaland and Manipur. It is also cultivated on small scale in Karnataka, coastal Andhra Pradesh and Orissa.

6.9.1. Insect Pests

6.9.1.1. Aphid

i) Damage: The adults which are yellowish green in colour damages leaves from vertical sides during January-February.

ii) Management : Spraying dimethoate at 0.3-0.7 kg a.i./ha gives effective control of aphids.

6.9.1.2. Army Worm, *Mythimna separata*

i) Damage: The larvae which are green in colour cause defoliation during March-April.

ii) Management : Spraying malathion or quinalphos each at 0.05% gives effective control of army worm.

6.9.1.3. Grass Hopper, *Macrosiphum muscanthii*

i) Damage: Adult and nymphs with stout body, grayish in colour cause defoliation during August.

ii) Management : Spraying malathion at 0.05% gives effective control of grass hopper.

6.9.1.4. Shoot Borer, *Chilo infuscatellus*

i) Damage: Larvae are dirty white in colour with fine longitudinal strip on the body, bore the stem of young crop and causes dead heart during January.

ii) Management: Spraying chlorpyriphos at 600-800 ml/ha with irrigation is effective in controlling shoot borer.

6.9.1.5. Termite, *Microtermes obesi*

i) Damage: Adult workers are whitish in colour and destroy roots of plants throughout the year.

ii) Management: Spraying chlorpyriphos at 600-800 ml/ha with irrigation is effective in controlling termites.

6.9.1.6. Thrips, *Anaphothrips sudanensis, Chloethrips holorphnus*

i) Damage: Nymphs and adults of *A. sudanensis* which are yellowish brown in colour and *C. holorphnus* are completely brown in colour bore into the stem of young crop and cause dead heart.

ii) Management: Spraying 0.05% dimethoate or phosphamidon at 0.3-0.7 kg a.i./ha is effective in controlling thrips.

6.9.1.7. White Grub, *Holotrichia consanguinea*

i) Damage: Grub/larva is dirty brown in colour with dark brown colour head which feed on roots during July.

ii) Management: Spraying chlorpyriphos at 600-800 ml/ha with irrigation is effective in controlling white grubs.

6.10. PALMAROSA

Palmarosa (*Cymbopogon martinni* var. *motia*), also called as Rusa, Rusha and Rosha, is grown in Andhra Pradesh, Madhya Pradesh, Maharashtra, Karnataka, Tamil Nadu and Uttar Pradesh. The major trading centres include Amravati, Akola and Elichpur in Maharashtra; Khandwa and Betul in Madhya Pradesh; Zaheerabad and Mehboobnagar in Andhra Pradesh. Oil of palmarosa is used for fine perfumery and valued for its geraniol content. Oil imparts rose like aroma to wide variety of soaps and tobacco products. Oil contains 70-80% geraniol.

6.10.1. Insect Pests

6.10.1.1. White Grub, *Holotrichea consanguinea*

i) Damage: White grubs attack on roots and leaves.

ii) Management: Spraying chloropyriphos 20 EC at 600- 800 ml/ha with irrigation gives effective control of white grubs.

6.10.1.2. Aphid, *Aphis gossypii*

i) Damage: Nymphs and adults suck the cell sap from the inflorescence of plants.

ii) Management: Spraying dimethoate or folidol each at 0.1% is effective in managing aphids.

6.10.1.3. Termite, *Microtermes* sp.

i) Damage: Newly planted woody seedlings are more vulnerable and the seedlings die after stem from the base is damaged.

ii) Management: Application of chloropyriphos 20 EC at 600- 800 ml/ha with irrigation is effective for the management of termites.

6.10.1.4. Thrips, *Haplothrips* sp.

i) Damage: Nymphs and adults suck the sap from young shoot tips and leaves, also feed on floral parts, thus affecting seed set.

ii) Management: Spraying dimethoate or folidol each at 0.1% gives effective control of thrips.

Chapter 7

TUBER CROPS

Tuber crops form an important part of the diet in many areas of the tropics and subtropics. They are most important after cereals and grain legumes. They are mostly grown for the carbohydrates stored in underground storage organs viz., roots or rhizomes. In addition, the leaves or leaf stalks of some of the tuber crops are rich in protein, vitamins and minerals and are used for either human consumption or cattle feed. Tuber crops are a major source of food and calories in many tropical countries. The annual tropical root crop production (sweet potato and cassava) is in the range of 8.921 million metric tonnes in 0.367 million ha with productivity of 20.1 tonnes/ha (Table 7.1). Root and tuber crops are predominantly grown in the upland or rainfed areas, mainly by resource poor small and marginal farmers. The major tuber crops include cassava (tapioca), sweet potato, colocasia, amorphophyllus, yams and coleus. Cassava and sweet potato are grown to a larger extent in southern and eastern India. Their cultivation has spread to Maharashtra, Gujarat and North-Eastern states as well. The Salem belt in Tamil Nadu and the Samalkot belt in Andhra Pradesh are known for cassava as an industrial crop. Yams and Colocasia are popular in Andhra Pradesh, Tamil Nadu, West Bengal, Uttar Pradesh and Orissa states.

Table 7.1. Area, production and productivity of tuber crops in India (2005-06)

Tuber crop	Area ('000 ha)	Production ('000 tonnes)	Productivity (tonnes/ha)
Sweet potato	122.8	1066.5	8.9
Cassava (Tapioca)	244.6	7854.9	31.4
Total	367.4	8921.4	20.1

(Source: Indian Horticulture Database 2006)

Tuber crops, especially cassava can be successfully substituted as alternative source of feed for livestock in both tropical and subtropical countries. Cassava has also been used as a substrate for the production of ethyl alcohol. The tropical root crops, in general, have a great potential in meeting basic food and energy needs of the developing world. Tuber crops are very efficient in solar energy transfer, having a clear superiority over cereals in biological efficiency and are comparatively free from severe pests and diseases. Cassava and sweet potatoes account for about 30% of the total production of root crops from developing countries. The potential of root crops as nutritionally rich sources of β-carotene, anti-oxidants, dietary fibre and minerals like calcium have begun to be recognized as a result of the multifarious research programmes worldwide.

7.1. YAM

Yam (*Dioscorea* spp.) belongs to the family Dioscoreaceae, which is widespread throughout the tropics and subtropics. Normally yams are consumed as boiled, baked or fried vegetables. They contain 18-20% starch with mucilaginous substance. Yam also contains alkaloids, tannins and steroids which have pharmaceutical value. *Dioscorea alata* (greater yam), *D. rotundata* (white or white Guinea yam), *D. cayenensis* (yellow Guinea yam) and *D. esculenta* (lesser yam) produce edible tubers. *D. alata* was first cultivated in S.E. Asia and is widely grown in Asia and the Pacific, is less favoured in Africa, but much grown in the Caribbean. *D. rotundata* is the most important species in W. Africa, where it is originated. *D. cayènensis* is a native of W. Africa, where it is extensively cultivated. *D. esculenta* originated in Indo-China and is extensively grown in S.E. Asia and elsewhere in the tropics.

In India, greater yam and lesser yam are under cultivation. Madhya Pradesh, North-Eastern states, West Bengal, Bihar, Orissa, Uttar Pradesh, Kerala, Tamil Nadu, Gujarat and Maharashtra are major yam-growing states.

7.1.1. Insect Pests

7.1.1.1. Scale, *Aspidiella hartii*

i) Distribution: It is widespread in distribution and present in Old World Tropics and Pacific.

ii) Symptoms: Adult and nymphal scale insects (whitish yellow in colour) are present on the tubers and sometimes on the aerial portions of the plant which do not thrive. Stored tubers may be thickly covered by scales and shrivel. The problem is perpetuated by the use of infested tubers as planting material.

iii) Management

a) Cultural Methods: Scale insects on tubers may be controlled by using clean planting material. Yams should not be planted after ginger, as this crop also supports the pest.

b) Chemical Methods: Dipping the tubers in 0.3% aldrin or in a mixture of 5.6 ml diazinon 20% EC and 425 ml white oil in 13.5 litres of water before planting has been recommended. Fumigation of tubers with carbon disulphide has also been suggested.

7.1.1.2. Yam Beetle, *Heteroligus meles*

i) Damage: The yam beetle is a pest that starts as a curled grub that changes into a beetle. Yam beetles may spoil the tubers by eating and making holes into them. Then the tuber rots.

ii) Management

- If the yam beetle is infesting the tubers, wash the planting pieces in lead arsenate and Bordeaux mixture. This will stop the grubs from eating into the tubers and spoiling them. Follow crop rotation with non-host crop.
- Dusting of holes or soil surface with aldrin, endosulfan or grammalin gives effective control of beetles.

7.1.1.3. Chrysomelid Beetle, *Lilioceris livida*

i) Damage: Larvae and adults feed on the young leaves and shoots. These larvae cover themselves with their own excreta.

ii) Life Cycle: Pink elongate eggs 2 mm long are laid on the underside of the leaves in groups of about six. As they develop they darken to reddish brown. The larvae are thick and fleshy. Adults are basically reddish-brown to black, with light yellow to brown elytra on which are dark brown to black spots.

iii) Management : Removal of larvae by hand or by spraying with 0.2% carbaryl or 0.075% dichlorvos or 0.075% endosulfan dust gives effective control.

7.1.1.4. Mealy Bugs, *Planococcus citri*

i) Damage: The mealy bug feeds on the roots and/or tubers of yams. Adult female mealy bugs are small, segmented, whitish-grey, wingless insects covered in a mealy wax secretion. The mealy bugs are usually noticed when the yams are harvested, and can multiply often to considerable numbers causing shriveling on stored yams.

ii) Life Cycle: The female adults lay eggs from which nymphs emerge and develop through several immature stages. Any male nymphs present develop wing buds and eventually fully developed wings.

iii) Management: The pest can be controlled by using a clean planting stock. The white oil-diazinon mixture may be sprayed to control mealy bugs.

7.1.2. Vertebrate Pests

i) Damage: Wherever cassava, yams, taro or sweet potatoes are grown, they shall be attacked and eaten by rodents. The damage may be seen immediately when the roots are eaten at ground level or may not be apparent until the crop is harvested if rats have burrowed and eaten away the roots from beneath. Losses of the order of 5-15% in West and Central Africa and up to 50-60% in the Cayman Islands in the West Indies were reported. Rat attacks on sweet potatoes begin when the crop is about 100 days old. Rodent attack may continue after harvest in farm and storage. The species responsible for damage in storage include *Bandicota bengalensis* in Indian sub-continent, *Mastomys natalensis* and *Rattus exulans* in Far East and South Pacific region.

ii) Management: There are three main methods to control rodents in the field. Trapping has been employed since the earliest times, and much skill has been developed in its use. Poisoning and environmental alteration have the biggest contribution to make to increasing crop yields.

7.2. SWEET POTATO

Sweet potato (*Ipomea batatas*) is an economically important member of the family Convolvulaceae and probably originated in Central America, where it has long been cultivated. It is now grown extensively throughout the tropics and subtropics. It is one of the life sustaining crops rich in carbohydrates that stand between man and starvation. Sweet potato ranks fourth and sixth on the list of dry matter production per hectare and edible energy production per hectare, respectively. It is also used for making starch, syrup and alcohol. It occupies a prime place in terms of calories production/unit area and time. It is grown as a starchy food crop throughout the tropical, sub-tropical and frost-free temperate climatic zones in the world. It is being cultivated in 0.129 million hectares producing 1.148 million tonnes of sweet potatoes with an average yield of 8.87 tonnes per hectare (Table 7.2). India ranks sixth in area but the productivity is very low (8.9 t/ha), lower than the world productivity level.

Table 7.2. Area, production and productivity of leading sweet potato growing states in India (2005-06)

State	Area (in '000 ha)	Production (in '000 MT)	Productivity (in MT/ha)
Orissa	47.1	394.3	8.4
Uttar Pradesh	20.0	240.8	12.1
West Bengal	25.8	208.0	8.1
Maharashtra	5.0	70.0	14.0
Karnataka	3.6	42.1	11.6
Chhattisgarh	3.0	38.2	12.6
Tamil Nadu	1.4	29.6	21.1
Assam	8.5	29.5	3.5
Madhya Pradesh	4.7	28.2	6.0
Meghalaya	5.4	18.9	3.5
Others	4.9	49.1	10.0
Total	129.4	1148.6	8.87

(Source: Indian Horticulture Database 2006)

7.2.1. Insect Pests

7.2.1.1. Leaf Folder, *Brachemia* spp.

i) Distribution: *B. convolvuli* is distributed in Canary Island, Mauritius, Comoro Island, Southeast Asia and Jawa. *B. macroscopa* has been recorded in India.

ii) Symptoms: Cream and black larvae spin webs close to the main veins of the leaves, on the underside, while older lager larvae spin webs across the upper surface of the leaf and draw the two sides together. The leaves eventually dry up, die and drop off.

iii) Biology: *Brachmia convolvuli* eggs are laid on the lower surface of leaves and hatch after 3 days. There are 5 larval instars. The mature larva is 11-12 mm long and cream coloured with dorsal black bands on segments 8-10 of the abdomen; the head, prothoracic shield and all of abdominal segment 5 are black. The larvae pupate for 7 days in the folded leaf. Adults of *B. convolvuli* are 4 mm long with grayish brown fore wings, marked with white hind wings fringed with grey. The entire life cycle, egg to adult takes 24 days.

iv) Management: Sprays of 0.2% carbaryl or 0.025% fenitrothion are recommended in India for the control of *B. convolvuli*.

7.2.1.2. Horn Worm, *Agrius convolvuli*

i) Distribution: Horn worm is distributed in most of Africa, Bangladesh, Burma, South China, Southern Europe, India, Indonesia, Iran, Malaysia, New Zealand, Pacific Islands and Papua New Guinea.

ii) Symptoms: Leaves are eaten by horn worms.

iii) Biology: Spherical eggs (1 mm in diameter) are laid singly on the host plant, usually on the upper leaf surface. The larvae, which concentrate on the young shoots are variable in colour from green to brown and distinctively patterned. There are 5 larval instars and larvae mature after 3-4 weeks and reach a length of about 95 mm. Pupation takes place in the soil and may take 3 weeks or several months depending upon the climate. The adult hawk moths (wing span 80-120 mm) are grey with black lines on the wings; the abdomen is marked laterally with pink bands. They feed upon nectar in flowers.

iv) Management

a) Cultural Methods: Ploughing the land between crops to expose the pupae reduces infestation. Hand picking of the larvae may be quite effective in small areas.

b) Chemical Methods: A 0.2% solution of carbaryl in a high volume spray, 0.05% endosulfan or trichlorphon (7 days waiting period before harvest) at 100 g a.i./100 litres of water should be applied when the larvae are first seen. Early instars can be controlled more easily than fifth instars. Particular care should be taken to spray the young leaves.

7.2.1.3. Bihar Hairy Caterpillar, *Diacrisia obliqua*

i) Distribution: Bihar hairy caterpillar is distributed in South and East Asia.

ii) Symptoms: Numerous larvae covered in yellow hairs skeletonize the leaves. The entire vines may be defoliated.

iii) Biology: Eggs are laid on the leaves in rows or clumps and after 4-9 days the larvae emerge. Mature larvae are 30-40 mm long with 7 orange bands across the body on which grow tufts of yellow hairs. At first, the larvae feed in groups but when they are about a week-old, they disperse to new feeding sites. The larval period lasts 28-34 days. Pupation takes place in the soil in reddish brown cocoons and may take 51 days.

iv) Management: Endosulfan and fenitrothion both at 0.05% are recommended for the management of the pest. Diazinon or dichlorvos or carbaryl at 0.2% are also suggested for the control of larvae.

7.2.1.4. Leaf Miners, *Bedellia* spp.

i) Distribution: *Bedellia ipomoeae* from Fiji, *B. somnulentella* from Papua New Guinea and Queensland, and a number of *Bedellia* spp. from Africa have been reported attacking sweet potato.

ii) Symptoms: Narrow mines widening to blotches appear on the leaves. The plants have a withered appearance and growth is reduced.

iii) Biology: Eggs are laid on the leaf. The emerging larva is distinctly segmented with a rather pointed head and abdomen. Larvae may be red or yellowish grey in colour. After making several mines, they pupate on a web of silken threads among the leaves. When moving outside the mines, the larvae travel by a looping movement. The life cycle can be completed within 4 weeks. Adult *Bedellia* spp. are small (3 mm long with a wing span of 8 mm) tan to grey moths.

iv) Management: Generally *Bedellia* spp. are controlled by predators, but these may be killed by foliar sprays of insecticide. A serious breakdown can cut down the effective leaf surface and lead to reduced yields of tubers. Carbaryl at 1 kg a.i./ha, chlorfenvinphos at 0.6 to 0.9 kg/ha, diazinon at 55 ml/ha, dimethoate at 140 ml/ha, endosulfan at 70 ml/ha, malathion at 140 g/ha or trichlorphon at 100 g/ha have been recommended for the management of the pest.

7.2.1.5. Tortoise Beetles, *Aspidomorpha* spp., *Conchyloctaenia punctata*

i) Distribution: *C. punctata* from South Africa; *A. adheares, A. miliaris, A. quadriradiata* and *A. socia* from Papua New Guinea; *A. miliaris* from India and *A. dissentanea* from Angola, Cameroon and Guinea have been reported attacking sweet potato.

ii) Symptoms: Adults and larvae feed on the leaves making large round holes and eventually skeletonizing the leaves.

iii) Biology: The Eggs are laid on the underside of the leaves in batches cemented to the leaf. The larvae are flattened, spiny and usually carry excreta and the skins of earlier instars on anal projections. Adult tortoise beetles (Sub-family Cassidinae) are broadly oval in shape and hence the common name. They are bright, sometimes patterned or metallic in colour.

iv) Management

a) Cultural Methods: Weeds of the family Covolvulaceae should be controlled as tortoise beetles can breed on them.

b) Chemical Methods: Carbaryl (0.2%) at 1000 litres/ha is recommended in India and Papua New Guinea, while endosulfan at 500-750 g a.i./ha elsewhere.

7.2.1.6. Weevil, *Cylas* spp.

i) Distribution: *Cylas formicarius* from Pan Tropical, *C. punctcollis* from Africa (countries south of Sahara) and *C. brunneus* from West Africa have been recorded on sweet potato.

ii) Symptoms; Tubers and vines are tunneled by small white, legless larvae. The tunnels may be partially filled with frass. Ant-like weevils, either having glassy blue black elytra, reddish brown legs and thorax and a black head (*C. formicarius*) or entirely shiny black (*C. punctcollis* or *C. brunneus*) may be found on the leaves or in the tunnels. Yield, storage life and plant vigour are reduced.

iii) Biology: The total length of development is 4-5 weeks. Eggs are laid in holes in the vines or tubers and hatching takes place within a week if the temperature is between 20 and 30°C. The larvae tunnel in the vines and tubers for about 14 days. When fully developed (fourth instar) the larvae are 3 mm in diameter. Pupation takes place inside the plant and lasts about one week.

iv) Management

a) Cultural Methods: Sweet potatoes should be planted only in fields that have had no *Cylas* infestation within the last 12 months and preferably more than 1 km away from any infested land, and only *Cylas*-free planting material should be used. Earthing-up of the plants, particularly those varieties with a tendency to push out of the ground reduces attack. After harvest, the field should be cleared of trash, which should be burnt or fed to livestock.

Crop rotation (paddy-sweet potato-cowpea) can minimize weevil damage. Boemeryl acetate, a kairamone present in the periderm of sweet potato (Coleus) can attract both male and female weevils which offers immense scope for the management of weevils.

b) Chemical Methods: Some of the chemicals recommended for the management of weevils are presented in Table 7.3.

Table 7.3. Recommendations for chemical control of weevils

Insecticide	Dose/Conc.	Application method
Carbaryl	0.1%	3 times at 4 weekly intervals.
Carbofuran	1.4 kg/ha	Soil application around vine cutting at planting.
Carbophenothion	0.035%	-----
Chlorfenvinphos dust	600-900 g/ha	30 day waiting period before harvest.
Chlorfenvinphos EC	240 g/ha	-----
Diazinon	0.3%	Dip for planting material or foliar spray.
Dieldrin	1 kg/ha	4 times at weekly intervals beginning 40 days after planting. Five weeks waiting period.
Disulfoton	1.8 kg/ha	4 weeks after planting.
Endosulfan EC	2.25 kg/ha	Application at planting time.
Endosulfan granules	2.25 kg/ha	Soil application at planting.
Fenitrothion	0.05%	-----
Fenthion	0.1%	3 times at 4 weeks intervals
Malathion EC	28.5 ml	Dip for cuttings or foliar spray.
Mephosfolan	1.4 kg/ha	Soil application around cutting at planting.
Methomyl	1 kg/ha	Four times at 2 week intervals beginning 40 days after planting.
Methyl-demeton	0.05%	Spray at beginning of 2nd, 3rd and 4th month of growth.
Monocrotophos	0.05%	4 times at monthly intervals.
Phorate	1.4-2.0 kg/ha	Soil application around cutting at planting and 4 weeks after planting.
Phosphamidon	0.05%	Spray at beginning of 2nd, 3rd and 4th month of growth.

c) Host Resistance: Deep rooting varieties with long necks between the tubers and the stems being less susceptible because of adult weevil cannot burrow downwards more than 1 cm. Early maturing varieties escape serious damage.

7.2.1.7. Striped Weevil, *Alcidodes* spp.

i) Distribution: *Alcidodes dentipes* from tropical Africa, *A. erroneous* from Kenya and *A. fabricii* from India have been recorded attacking sweet potato.

ii) Symptoms: *Alcidodes dentipes* and *A. erroneous* attack both stems and tubers. Adults girdle the stems causing wilting, while larvae bore into stems and tubers making tunnels slightly larger than those of *Cylas* larvae and causing galls to form on the stems. *A. fabricii* has been recorded breeding inside the seed capsules of sweet potato.

iii) Biology: The life cycle of the weevils is similar to that of *Cylas*. *Alcidodes dentipes* is about 14 mm long and brown with 4 white streaks along its elytra. *A. erroneous* is similar in size but brown-black in colour with an irregular white patch on each elytrum. *A. fabricii* is reddish brown with four longitudinal white stripes on the elytra. The eggs are laid at the base of the fruit; the larvae bore into the fruit and feed on the seeds. Pupation takes place within the fruit.

iv) Management : The methods outlined for *Cylas* should control *Alcidodes* spp.

7.2.1.8. Stem or Vine Borer, *Omphisa anastomosalis* (Lepidoptera: Pyraustidae)

i) Distribution: Generally widespread in Southeast Asia; China, India, Japan and Sri Lanka.

ii) Symptoms: Stems wilt and die and black-headed yellowish-white larvae, marked with brown tubercles can be found boring in the stems. Frass can be found close to the attacked stems. These larvae may also bore into the tubers.

iii) Biology: The flattened yellow, oval eggs are laid in groups of 2 or 3, usually on the stem and occasionally on the leaves. The leaves emerge after an incubation of 5-8 days and eventually reach 30 mm in length. They tunnel in the stem for 28-50 days and pupate in a thin cocoon within the tunnel, close to the tubers. Adults emerge after about 14 days. The moth is 15 mm long with a wing span of 33 mm, with a reddish brown body and reddish brown markings on the wings, which are white.

iv) Management: Crop hygiene and crop rotation reduce infestation.

7.3. CASSAVA (TAPIOCA)

Cassava (*Manihot esculenta*), popularly known as tapioca, is a member of the family Euphorbiaceae, which has long been cultivated in Central America and Brazil and was later carried to Africa (where it is now an important subsistence crop), Asia and the Pacific Islands. An economic crop can be obtained even under marginal conditions or on exhausted soils. The crop requires very little attention after planting. It has draught tolerance (mainly due to in-built mechanism to shed or drop the leaves under adverse moisture conditions) and produces more calories/unit area. The tubers contain 25-30% starch, 1.5% protein and 2.1% minerals. It is being cultivated in 0.252 million hectares producing 7.925 million tonnes of cassava with an average yield of 31.36 tonnes per hectare (Table 7.4). The Salem belt of Tamil Nadu and Samalkot belt of Andhra Pradesh are known for cassava as an industrial crop.

Table 7.4. Area, production and productivity of leading cassava growing states in India (2005-06)

State	Area (in '000 ha)	Production (in '000 MT)	Productivity (in MT/ha)
Tamil Nadu	127.1	4857.4	38.2
Kerala	95.0	2585.1	27.2
Andhra Pradesh	19.0	388.7	20.5
Meghalaya	4.2	23.1	5.5
Pondicherry	0.7	19.5	27.9
Nagaland	1.3	15.9	12.2
Assam	2.9	14.0	4.8
Karnataka	1.1	12.0	10.7
Others	1.4	9.2	6.6
Total	252.7	7924.9	31.26

(Source: Indian Horticulture Database 2006)

7.3.1. Insect Pests

7.3.1.1. White Fly, *Aleeurodicus dispersus*

i) Damage: White flies suck the sap and devitalize the plant (Fig. 7.1).

Fig.7.1. Whiteflies on leaves of cassava.

ii) Management: The field release of endoparasitoid, *Encarsia* sp. gave 69.53% parasitization during August and 76.22% during September. The predators like *Cybocephalus* sp., *Cryptolaemus montrouzieri* and *Mallada astur* can also be effectively used for the biocontrol of spiraling white fly.

7.3.1.2. Stem Muscle Scale, *Aonidomytilus albus*

i) Distribution: The pest is distributed in West and East Africa, Argentina, Brazil, India, Madagascar, Mexico, Taiwan, West Indies and USA.

ii) Symptoms: Grey, elongate adult scales 2.5-3.0 mm long and pale nymphs, infests the lower stems of the plant (Fig. 7.2). Infestation may become severe in cassava cuttings kept for planting; the consequent drying up reduces germination success.

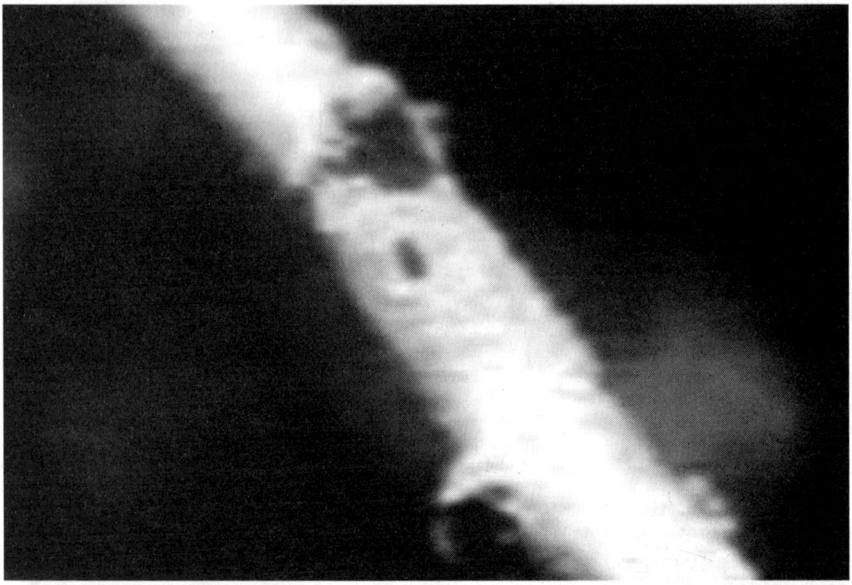

Fig. 7.2. Scale infestation on stem of cassava.

iii) Biology: About an hour after hatching the nymphs begin to feed, during the next two weeks they secrete a cottony wax cover beneath which molting and then secretion of a new scale takes place. After another two weeks a second molt occurs. Males subsequently enter the pupal phase and females mature. Male pupation takes less than one week to complete. The entire life cycle takes 5-6 weeks.

iv) Management

a) Cultural Methods: Clean healthy cuttings should be used for planting.

b) Chemical Methods: Spraying or dipping the cuttings in 0.1% demeton-s-methyl, 0.5% parathion or 0.1% malathion is recommended.

7.3.1.3. White Peach Scale, *Pseudaulacaspis pentagona*

i) Distribution: It is cosmopolitan in distribution.

ii) Symptoms: Circular papery female scales (2 mm in diameter) and long (0.75 mm) male scales can be seen on the stem and leaf bases. Weak or old plants are particularly susceptible.

iii) Biology: The nymphs emerge from eggs beneath the female scale and begin to feed. They secrete cottony threads but these disappear when a thin skin is secreted over the nymph. At the first molt, the insects become disc-like. Cast skins at each molt are added to the scale. Males molt 3 times and then pupate. Females mate after the second molt and begin egg laying.

iv) Management: Spraying of oil emulsion or 0.05% parathion will control this scale.

7.3.1.4. Black Scale, *Parasaissetia nigra*

i) Distribution: It is distributed in most warm countries.

ii) Symptoms: Colonies of dark brown to black domed soft scales 2-4 mm long can be seen on the stems. *Saissetia coffeae, S. miranda* and *Ceroplastes cirripediformis* have also been reported on cassava stems and are similar in appearance to black scale, i.e. hemispherical and waxy in appearance.

iii) Biology: The scale is the female insect. Beneath the scale she lays eggs which hatch into crawlers that migrate over the plant before settling down to feed. Although females do not move around they retain their legs. Female crawlers molt twice before becoming adult and males, which have two wings, molt 3 times.

iv) Management: Cassava cuttings must not be allowed to develop heavy infestations of scale insects since they will produce poor plants when planted or do not germinate. Healthy plants will not suffer heavy yield loss due to scale insects. Kerosene emulsion and parathion have been recommended for soft scale control.

7.3.1.5. Mealy Bugs, *Phenacoccus gossypii, P. manihoti, Ferrisia virgata*

i) Symptoms: Leaves and shoots become distorted and yellow due to mealy bugs feeding on them. Adult female mealy bugs are segmented soft-bodied insects without wings, about 2.5 mm long and covered in white, waxy meal.

ii) Biology: Generally the female lays groups of 200-400 eggs, which are protected by waxy filaments. In a few hours, nymphs very much like females in appearance but much smaller, emerge. The nymphs pass through several immature stages, if males are present, then they pupate, emerge and mate with the females. The life cycle, egg to adult, may be completed within 6 weeks.

iii) Management

a) Cultural Methods: Clean cuttings should always be used for planting.

b) Chemical Methods: Infected cuttings should be dipped or dusted with aldrin, BHC or DDT. Mephosfolan and parathion sprays or dusts may be applied to the growing point if it is seriously affected by mealy bugs and should control the pest.

7.3.2. Mite Pests

7.3.2.1. Red Spider Mite, *Tetranychus cinnabarinus*

i) Distribution: The pest is distributed throughout the semitropics and in protected cultivation in temperate zones.

ii) Symptoms: Lower leaves develop yellow dots along the veins, these dots become reddish and if the dry season is prolonged, the infestation spreads to higher leaves (Fig. 7.3). Leaves dry up and fall and some plants die. The mites responsible are brownish-red with dark lateral internal markings, produce webs.

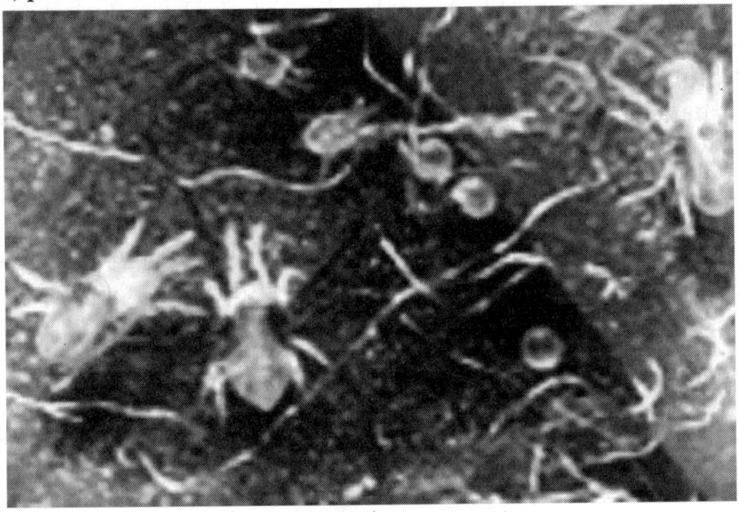

Fig. 7.3. Red spider mite infestation on leaves of cassava.

iii) Biology: Reproduction takes place during dry weather; the optimum temperature for development is 32°C. Single eggs are laid on the under surface of the leaves or on the webbing. Development of egg to adult takes about 17 days and there may be 19 broods per year.

iv) Management: Spraying of 0.04% binapacryl, 0.05% chlorobenzilate, 0.03% dimethoate or parathion, 0.01% lime sulphur, 0.16% mancozeb, 0.001% naled or 0.07% phosalone is recommended for the management of spider mites.

7.4. COLOCASIA

Colocasia (*Colocasia esculenta*), also known as cocoyam, taro, dasheen and eddoe, is a member of the family Araceae and occurs wild in S.E. Asia. It is cultivated throughout the humid tropics and is of great importance in the Pacific Islands. It is mostly a staple food or subsistence crop grown commercially in some countries.

7.4.1. Insect Pests

7.4.1.1. Aphid, *Aphis gossypii*

i) Distribution: The pest is cosmopolitan in distribution, but absent from the cooler pats of Asia and Canada.

ii) Symptoms: Nymphs, winged and wingless adult aphids are feeding on leaves, causing specking, leaf curl and a generally sickly appearance. A sooty mold may develop on the honey dew secreted by the aphids in addition ants are attracted to the honey dew.

iii) Biology: Adults may give birth to two nymphs/day. Winged adults (all female) have a black head with two green bands. The abdomen is green with black cornicles. Wingless adults (also all female) are yellowish green to black.

iv) Management: Avoid planting colocasia close to alternate hosts such as melon, cucumber or other cucurbits or cotton. The following chemicals have been recommended for the management of aphids (Table 7.5).

Table 7.5. Recommended chemicals for the control of aphids

Insecticide	Dose	Application method
Carbaryl WP	1 kg	In 450 litres (high volume) or 112-168 litres (low volume) of water.
Diazinon EC	11 ml	In 13.5 litres of water, 14 days minimum waiting period.
Dimethoate EC	4.4 ml	In 13.5 litres of water, 14 days minimum waiting period.
Malathion EC	14.2 ml	In 13.6 litres of water, 7 days minimum waiting period.

7.4.1.2. White Fly, *Bemisia tabaci*

i) Distribution: The pest is recorded in most parts of the tropics and subtropics.

ii) Symptoms: Adults, nymphs and pupae are found on the lower leaf surface. When infestation is heavy, the leaves turn yellow, wilt and die. The growth of the colocasia may be stunted.

iii) Biology: Yellow elliptical eggs, attached by filaments to the leaf, are laid on the under surface of the leaves. The first instar nymphs emerge and feed. After the first molt, they are fixed to one place on the leaf as yellow elliptical scale-like insects. The third molt gives rise to the puparium which is less flat than the previous nymphal stages and the red eyes of the adult can be seen through the integument. Egg to adult development takes 2-3 weeks. The adult has a yellow body, white waxy broad tipped wings and is 1 mm long with a wing span of about 2 mm.

iv) Management: The following chemicals have been recommended for the control of white flies (Table 7.6).

Table 7.6. Chemical control of white fly on colocasia

Insecticide	Dose (a.i./ 13.6 litres of water)	Application
Carbaryl WP	34 g	7 days minimum waiting period before harvest.
Malathion WP	14 g	7 days minimum waiting period before harvest.
Trichlorphon WP	23 g	7 days minimum waiting period before harvest.

Chapter 8

PLANTATION CROPS

Plantation crops serve a variety of human needs such as food, oil, industrial raw materials, beverages and confectionary items. They generate huge employment opportunities directly or indirectly to several million people in their production, processing, marketing and international trade sectors. Plantation crops are a large group of crops. These are grown over an area of 3.282 million ha with production of 11.262 million tonnes in India (Bijay Kumar, 2007) (Tables 8.1 and 8.2). The plantation crops contributed 6.09% of total horticulture production, in 2006-2007. India is the largest producer and consumer of areca nut, cashew nut, tea, third largest producer of coconut, fourth largest producer and consumer of rubber and sixth largest producer of coffee in the world. While the major plantation crops include coconut, areca nut, oil palm, cashew, coffee, tea and rubber, minor plantation crops are cocoa and betel vine. They play an important role in export as well as domestic requirement, employment generation and poverty alleviation particularly in rural sector. The annual income generated is more than Rs. 5,52,590 million and export earning is approximately Rs. 93,350 millions.

Table 8.1. Area, production and productivity of plantation crops in India (2005-06)

Plantation crops	Area ('000 ha)	Production ('000 MT)	Productivity (MT/ha)	% Share of India production
Coconut	2028.9	10190.1	5.0	90.5
Areca nut	381.1	483.1	1.3	4.3
Cocoa	29.5	10.2	0.3	0.1
Cashew	843.4	579.4	0.7	5.1
Total	3282.9	11262.8	3.4	

Table 8.2. State-wise area, production and productivity of plantation crops in India (2005-06)

State	Area in '000 ha	Production in '000 MT	Productivity in MT/ha
Kerala	1096.6	4545.1	4.1
Tamil Nadu	497.7	3415.0	6.8
Karnataka	735.1	1095.9	1.5
Andhra Pradesh	286.1	706.6	2.4
Maharashtra	180.0	376.1	2.0
Orissa	170.8	267.0	1.5
West Bengal	44.3	261.9	5.9
Assam	104.3	217.8	2.0
Goa	81.9	115.8	1.4
Gujarat	20.4	99.1	4.8
Andaman & Nicobar	30.0	64.7	2.1
Lakshadweep	2.7	36.5	13.5
Meghalaya	17.9	21.9	1.2
Pondicherry	5.2	20.3	3.9
Tripura	6.7	11.7	1.7
Mizoram	2.0	5.3	2.6
Nagaland	1.1	2.0	1.8
Total	3282.9	11262.8	3.4

There is a good scope for switching over the production of plantation crops viz., coffee, tea, coconut and areca nut to organic farming, as it is possible to realize higher returns from the unit quantity exported when grown organically. The plantation crops are generally grown in ecologically fragile hilly tracts, adopting organic farming methods would protect environment and also prevent contamination thereby producing products of high nutritional quality. Further, the plantation crops are amenable for organic farming as they produce huge amount of waste biomass for recycling, which can meet major portion of nutrients required.

Export of Plantation Produce

India is the largest producer in the world of cashew nuts, coconuts and tea. India is largest producer, importer, processor, exporter and second largest consumer of cashew. Indian exports during 2006-2007 were US$ 866 million in tea and coffee (Table 8.3).

Table 8.3. Export of plantation crops

Commodity	Apr-Mar 2006	Apr-Mar 2007	% Growth	% Share
Tea	390.92	432.00	10.51	0.34
Coffee	358.83	434.77	21.16	0.34
Cashew	585.76	553.41	-5.52	0.44

8.1. COFFEE

Coffee (*Coffea arabica, C. canephora*) is an important commercial plantation crop mostly grown on the hilly slopes of the traditional coffee tracts in South India (Karnataka, Kerala, Tamil Nadu and Andhra Pradesh) and non-traditional Orissa, Maharashtra, Madhya Pradesh and North-Eastern Council States. The coffee industry is well established and flourishing enterprise. The major coffee plantations in India are in the Southern States with Karnataka being the major coffee producing state. Recently some coffee plantations have come up in the North Eastern States of India as well. Besides being one of the major foreign exchange earner to a tune of around Rs. 2000 million annually, coffee industry supports more than four lakh people to earn their livelihood. The two main species of coffee in cultivation are *Coffea Arabica* (Arabica) and *Coffea canephora* (Robusta). *C. arabica* produces finer seeds which are more valued and grows best at higher elevations (1000 and 1500 m MSL). *C. canephora* is sturdier and better adapted to the warm climate prevailing in the lower elevations (500 and 1000 m MSL). Robusta, in general, is more tolerant to many diseases and pests including nematodes, as compared to Arabica. Coffee in India is grown under a canopy of shade trees of various species along with other plantation crops like black pepper, cardamom, banana, cocoa, etc. The present area under coffee in India is above 100,000 hectares, nearly three-fourth of which is under the fine quality Arabica species with most of the remaining area under Robusta.

India produced 290,000 tonnes of coffee accounting for approximately 4% of the global production. Over 80% of the coffee produced in India is exported. Although over 50 varieties are produced, the popular ones are the Indian Robusta which is used for blending and Arabica. There is a strong demand in the world market for production of high value, high quality, especially organic coffee. Today, the area under certified organic coffee is 2500 ha with an estimated production of 1500 MT of which 1000 MT is being exported currently.

8.1.1. Insect Pests

8.1.1.1. White Stem Borer, *Xylotrechus quadripes*

i) Damage: Arabica coffee is most preferred. Infested plants show ridges round the stem. It causes yellowing and wilting of plants. The adult beetle makes tunnel and emerges outside (Fig. 8.1). Grubs feed on the wood of the stem and thick primaries and tunnel in all directions, sometimes right down to the tap root. Tunnels are packed with chalk-like excreta.

ii) Management

- Maintenance of optimum shade.
- Regular tracing and destruction of infested plants.

- Scrubbing of main trunk and thick primaries with coir gloves to remove loose scaly bark.
- Repeated application of neem oil (2-5%) at short intervals of 1-2 weeks.
- Use of pheromone traps for monitoring incidence.
- Application of 10% lime on stem.
- Mass release of the parasitoid, *Apensia* sp. help to manage the white stem borer.
- Swab the stem with thick emulsion of lindane or carbaryl in April-May and October-December. This is also the time when adult beetles emerge, take to flight, mate and oviposit.

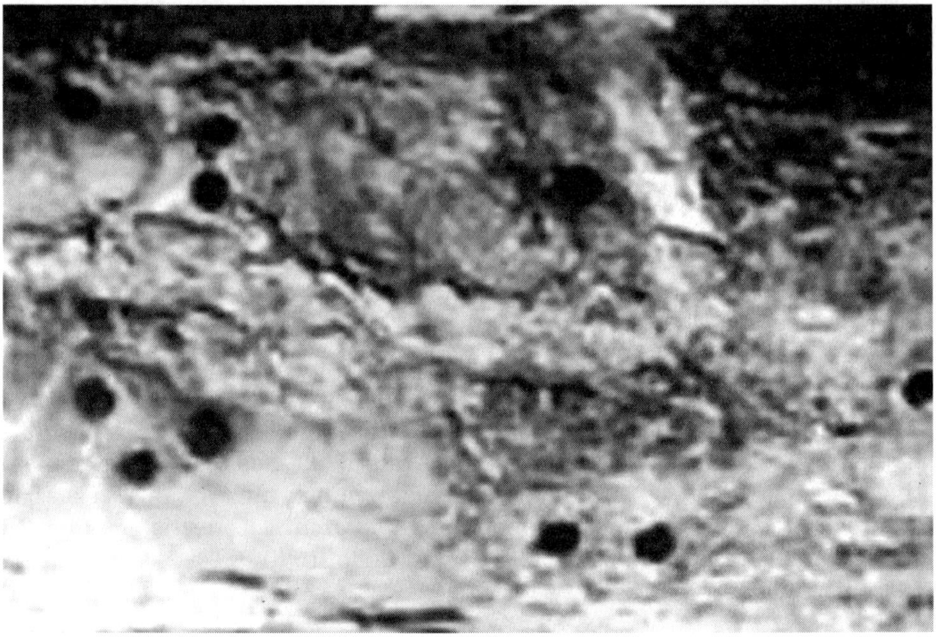

Fig. 8.1. White stem borer infestation on coffee.

8.1.1.2. Berry Borer, *Hypothenemus hampei*

Coffee berry borer, the most serious pest of coffee, is believed to be native to Northeast and Central Africa. It has spread to most coffee growing countries (58) in all the continents throughout the tropics except Papua New Guinea, Hawaii and Solomon Islands. In Asia, it is found in Brunei, Cambodia, India, Indonesia, Laos, Malaysia, Philippines, Saudi Arabia, Sri Lanka, Thailand and Vietnam.

The pest entered India in 1990 when it was reported from Gudalur in the Nilgiris. The pest probably got accidentally introduced through coffee seeds brought by refugees from Sri Lanka or through illegally imported coffee seeds. It has now spread into many coffee growing areas of Tamil Nadu

(Gudalur, Palani Hills, Bodinayakanur and Kilkotagiri), Kerala (Wynad and Idukki) and Karnataka (Kodagu, Hassan, Chickmagalur and Sakleshpur). It has spread to more than 1.2 million hectares accounting for about 35% of the total coffee growing area of the country.

i) Damage: *Hypothenemus hampei* is known to feed and reproduce only in the seeds of coffee. The pest attacks both Arabica and Robusta types and is also found on other species of coffee. The female enters the coffee berry by cutting an irregular hole, generally at the tip of the berry (Fig. 8.2). Berries in all stages of development are prone to attack. Breeding occurs within berries with hardening or hard endosperm. The damage is mainly due to feeding inside the berries in the field. In general, only one of the two beans in a berry is damaged and hence maximum loss could be 50% of infested berries. Damage from 5-90% of the berries has been reported. Berries of the previous season left on the plant or fallen on the ground serve as pest reservoirs. This serves as a source for the next season crop.

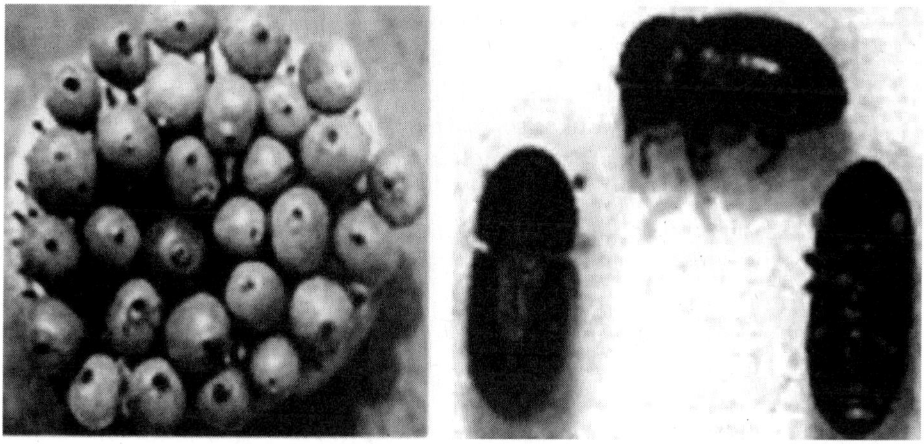

Fig. 8.2. Berry borer infestation on coffee.

ii) Biology: The adult female after entering the berry tunnels into the bean, excavating irregular galleries while feeding and laying bean shaped eggs. The brownish black female adult measures 1.5-1.9 mm. The males are smaller and generally incapable of flight. The population is usually female dominated (10: 1). The females can lay from 30-70 eggs during its life. The eggs hatch in 4-9 days and the larvae feed and make extensive tunnels. The larval period lasts for 10-26 days. The total development from egg to adult lasts for 25-35 days. On emergence, the female remains inside the berry for a few days, during which period mating takes place between siblings. The females then leave the berry to infest fresh berries. Adults are long lived, a female can live for 282 days with an average of 156 days. Females can survive up to 2 months in buried beans and mobile females can survive by feeding on immature berries. Three to four generations a year have been reported.

iii) Ecology: The optimum temperature and RH requirements for reproduction and development were found to be around 25-26°C and 90-95%, respectively. An elevation of 500-1000 metres has been found to be optimum for infestation. Intermittent rain favours emergence and development. The incidence was found to be more under dense levels of shade as compared to open or less dense shade.

iv) Management

a) Cultural Methods: Cultural and phytosanitary measures that aim at disrupting the breeding are especially useful in places like India where there is a distinct gap between harvest and availability of suitable fruits in the new crop. Timely and clean harvest (removing fallen berries) helps in reducing the infestation.

Ungleaned estates recorded an average infestation of 13% as compared to 1.5% in gleaned estate. The use of picking mats while harvesting also helps in reducing the gleanings (fallen berries on the ground) to an extent of 73% and in reducing the infestation to the extent of 72% in the next crop as compared to areas harvested without picking mats.

Coffee dried to 11.5% moisture helps in killing all stages as also dipping infested fruits in boiling water for 2 min or cold water for 2 days.

The beetles from migrating drying yards can be trapped by using oil coated polythene sheets and also in the berries on plants surrounding the drying yard by using them as trap plants, which can be clean harvested and destroyed.

Thinning of shade trees and training and pruning of unwieldy bushes to avoid excessive shade help in reducing the infestation.

b) Chemical Methods: Spraying 0.07% endosulfan was found to be the most effective among all insecticides tested. It is important to spray during the period when the females are entering the berry i.e. during September-October. Spot application in infested patches also proved useful to prevent the spread of infestation. Spraying of chlorpyriphos, quinalphos, methomyl and malathion is also effective.

Mass trapping using chemical attractants (1: 1 mixture of ethanol and methanol is best) could be used in the management of the berry borer. A trap density of about 70 traps/ha was found optimal. Large number of beetles can be trapped during the harvest and post-harvest periods.

c) Biological Methods: Among indigenous natural enemies, *Beauveria bassiana* was the most abundant and caused epizootics under high humidity conditions. High volume sprays gave better results than low volume sprays. The fungus is most effective at a temperature of 23-28°C and RH of more than 90%. A spore load of 1 x 10^{10} per tree has been found effective. Trials have shown that treatment during September with 10^{8} conidia/ml of spray solution resulted in 51-60% mortality of the berry borer.

Among parasitoids, only 4 are considered promising. They are *Prorops nasuta, Cephalonomia stephanoderis, Phymastichus coffea* and *Heterospilus coffeicola,* which are all native of Africa. Of these 4 promising parasitoids, three have been introduced into India. *Prorops nasuta* and *Cephalonomia stephanoderis* were imported from Mexico in 1995 and *Phymastichus coffea* and *P. nasuta* from Columbia in 1999. The parasitoid, *C. stephanoderis* has established in several areas of Kodagu district, Karnataka; Wynad district, Kerala and Lower Palanis, Tamil Nadu.

d) Integrated Methods: The pest could be effectively tackled by adopting the following measures:

- Maintenance of optimum shade and good drainage.
- Thorough and clean harvesting of the crop in time.
- Use gunny bags/picking mats/polythene sheets while harvesting.
- Collecting the gleanings and dipping the infested berries in boiling water for 2 minutes or in cold water for 48 hours for killing all the stages of the borer inside the berry.
- Spreading the polythene sheet smeared with castor oil over the harvested coffee heap during the day for trapping all the beetles.
- Maintenance of trap plants around drying yard followed by clean harvesting from the trap plants and treating the berries in boiling/cold water.
- Timely spraying of entomopathogenic fungus, *Beauvaria bassiana.*
- Mass trapping of beetles using lures.
- Fumigating the storage material and transport structures effectively check the pest.

8.1.1.3. Shot Hole Borer, *Xylosandrus compactus*

i) Damage: The shot hole borer is a serious pest of robusta coffee. It attacks tertiary twigs, young primaries and suckers. The infestation starts by mid-June and increases by September. The adult female bores a hole the size of a pinhead on the lower side of a twig in the internodal region. Withered or dried branches with shot holes can be observed on infested plant. Leaves at point of attack fall prematurely. Terminal leaves wilt, droop and dry up. Extensive tunneling within the branches is noticed. The tunnel which extends 38 mm to both sides of the hole is lined with the ambrosia fungus (*Monilia* sp.) on which the larvae feed (Fig. 8.3).

Fig. 8.3. Shot hole borer infestation on coffee.

ii) Management: The incidence on Robusta coffee could be minimized by providing optimum thin shade, good drainage, regular pruning and burning of affected twigs and suckers, especially during April-May and September-December.

Swab the stem with thick emulsion of lindane or carbaryl in April-May and October-December. This is also the time when adult beetles emerge, take to flight, mate and oviposit.

8.1.1.4. Green and Red Scales, *Coccus viridis, Sassietia coffeae, S. hemisphaericum*

i) Damage: Sporadic cosmopolitan; the bugs infest succulent shoot regions (Fig. 8.4). All stages of the pest occur crowded on the underside of the leaves. The upper side of the leaves get blackened owing to the growth of a sooty mould on honey dew secreted by these bugs. Young plants are affected more commonly. Green scales sucks sap from tender parts, congregating down on under surface of leaves close to mid rib and veins. Infested leaves curl up and tender twigs droop. Sooty mould develops on honey dew secreted by scales. Other symptoms include curled or distorted young leaves covered with a black fungus and dropping of tender berries.

Fig. 8.4. Green scale infestation on coffee.

ii) Management: Prune the affected parts and spray 0.03% dimethoate, diazinon, quinalphos, dicrotophos or monocrotophos.

Control ants by dusting ekalux 1.5% or malathion 5% around the bushes and shade trees. Destroy ant nests. Spray for the shoots and drench the roots of grown up plants with any one of the following: ekalux 25EC, folithion 1000, lebaycid 1000 in 200 litres of water or freshly prepared neutral Bordeaux mixture along with 1 kg of fish oil rosin soap with 100 ml of sandovit or vettoplant. Spray quinalphos 2.5 ml/l of water.

8.1.1.5. Cockchafer Beetle, *Phyllophaga (Holotrichia) conferta*

i) Damage: Grubs (Fig. 8.5) damage roots, especially in nurseries. More serious in areas full of weeds and grasses. The pest is active during the monsoon (July to September-November). They cut and eat feeder roots from May to October-November.

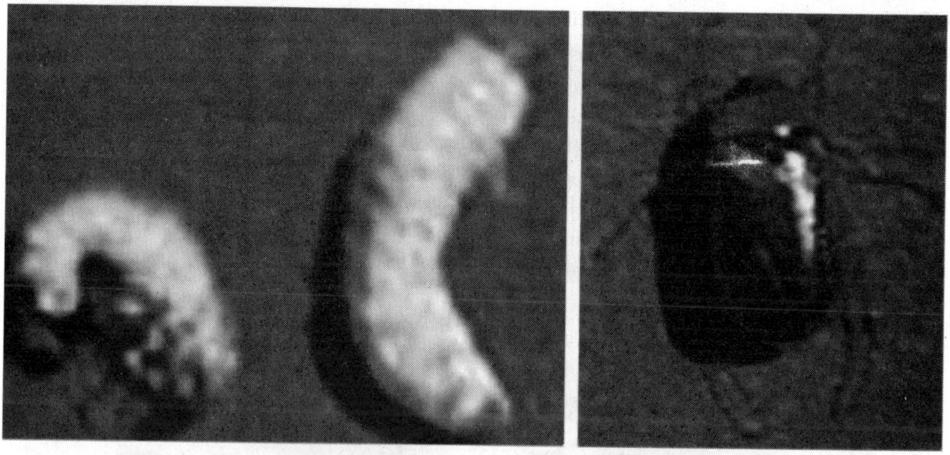

Fig. 8.5. Cock chafer beetle larvae and adult.

ii) Management

- Build up good shade.
- Drench soil around the base of the plants with lindane 20 EC at 750 ml in 200 litres of water.
- Apply 5g of phorate10G or aldicarb10G at the time of planting.
- Soil treatment with 5% aldrin or dieldrin.

8.1.1.6. Thrips, *Scirtothrips bispinosus, Heliothrips haemorrhoidalis*

i) Damage: Thrips are tiny fragile insects that lacerate and feed on leaves. Affected leaves become distorted, crinkled and mottled. Severe in hot weather, though found throughout the year. Females have saw-like ovipositor. Eggs are laid inside the leaf tissue. Larvae and adults grasp and scrape the epidermis and suck the sap, resulting in local scars, and distortion of young leaves.

ii) Management

- Maintain good shade, spray for green bug (except lannate) only in exposed fields till such time that shade trees cover up.
- Remove weeds and grasses that harbour thrips.
- Spray with 0.03% phosphamidon, dimethoate, methyl demeton or diazinon.

8.1.1.7. Shoot and Root Mealy Bugs, *Planococcus lilacinus, P. citri, P. perissopneum*

i) Damage: The pest (Fig. 8.6) causes yellowing, wilting and defoliation leading to stunting of plants. The other symptoms of damage include development of sooty mould, death of the feeder roots, and berry drop. Ants help to distribute the pest. The pest is active from February to June.

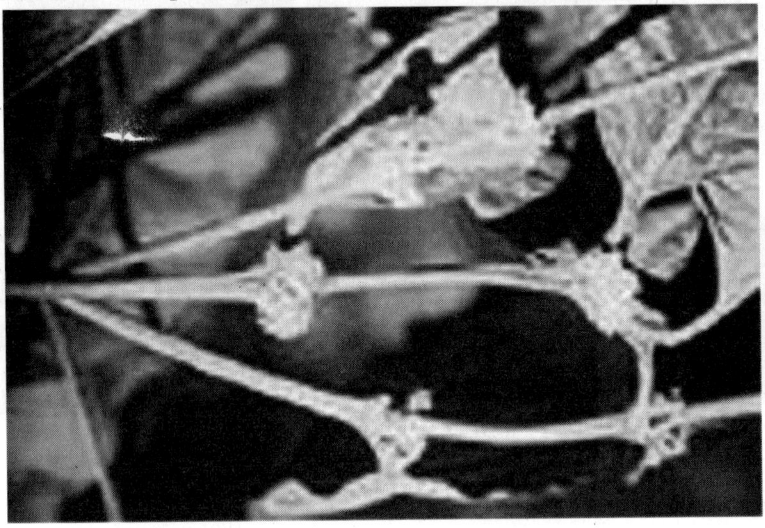

Fig. 8.6. Mealy bug infestation on coffee.

Leaves become chlorotic, flower buds abort and berries become small. Sooty mould develops on honey dew secreted by the bugs. Leaves of infested plants become black.

ii) Management: Control ants by dusting ekalux 1.5% or malathion 5% around the bushes and shade trees. Destroy ant nests. Remove mealy bug weeds. Treat with phorate or Disunity 5 G. Spray the shoots and drench the roots of grown up plants with any one of the following: ekalux 25EC, folithion 1000, lebaycid 1000 in 200 litres of water or freshly prepared neutral Bordeaux mixture along with 1 kg of fish oil rosin soap with 100 ml of sandovit or vettoplant.

Spray any systemic insecticide at twice the recommended concentration. After a month, release predatory beetle, *Cryptolaemus montrouzieri* at 10 beetles/plant.

8.1.1.8. Hairy Caterpillar

i) Damage: Appear generally on shade trees and later come down to coffee. Occur in congregations.

ii) Management

- Collect and kill caterpillars wherever possible.
- Burn pupae from January to May.

- Install light traps for destroying moths.
- For killing caterpillars, spray ekalux 25 EC at 350 ml or thiodan 35EC at 280ml along with 1 kg of fish oil rosin soap.

8.1.1.9. Snail, *Ariophanta solata*

i) Damage: The pest occurs from June-October on dadap and coffee. It feeds on bark and leaves causing damage to foliage.

ii) Management: Hand-picking and clean weeding help to reduce the incidence of the pest. Broadcast snailkil (metaldehyde product) at 37.5 kg/ha.

8.1.2. Vertebrate Pests

8.1.2.1. Birds

i) Damage: Birds viz. small green barbet and red whiskered Babul (*Pycnonotus jocosus*) cause economic damage to coffee. These birds feed only on ripened coffee berries. They punctured the pericarp and siphoned in the sweet contents and dropped the husk and seeds on the ground. Monkeys also cause economic damage to coffee by feeding on sweet, succulent and palatable stem tissue of young plants.

ii) Management: Timely scaring of birds and monkeys using trained dogs effectively prevented the loss.

Table 8.4. Management of minor pests of coffee

Pest	Nature of damage	Control measures
Red borer, *Zeuzera coffeae*	The adult moth emerges in Apr-May. Larvae damage the stem of young plants and thick primaries during Sept-Nov. Affected branches show wilting and reveal one or more holes through which excrement of larvae hangs out in the form of pellets.	Destroy caterpillars, Prune and burn infested branches generally below pest infestation level.
Coffee leaf-miner, *Agromyza coffeae*	Has gained importance as a major pest recently; larvae mine leaves.	Prune and destroy affected leaves; spray 0.1% BHC + 0.1% DDT.
Coffee grasshopper	A minor pest; defoliates the plants.	Dust 5-10% BHC.
Cutworm, *Agrotis segetum*	A sporadic pest; feeds on leaves.	Dust 5-10% BHC.
Coffee bean weevil, *Araecerus fasciculatus*	The weevil infests the berries in the field and dry beans in storage.	Dust 5-10% BHC.
Mites: *Oligonychus coffeae, Brevipalpus obovatus, B. phoenicis*	Mites feed in large colonies on the undersurface of the leaves; attacked leaves dry up in patches.	Dust sulphur or spray 0.05% wettable sulphur.

8.2. TEA

India is the largest producer and consumer of tea (*Camellia sinensis*) in the world. During 2006-2007, India produced approximately 950,000 tonnes of tea of which around 218,000 tonnes was exported. The total area under tea is 5,07,200 ha. It accounts for 20% of the total area under tea in the world, 28% of world production, 22% of global tea consumption and 15% of the total global tea exports. Tea accounts for approximately £1 billion in turnover in India. Tea is grown mainly in three major regions in India-Assam, Darjeeling in North East and the Nilgiris region in Southern India. The largest producers of tea are Assam (50.7%), West Bengal (22.1%), Tamil Nadu (15.9%) and Kerala (8.3%).

Other areas where tea is grown to a small extent are Karnataka, Tripura, Himachal Pradesh, Uttarakhand, Arunachal Pradesh, Manipur, Sikkim, Nagaland, Meghalaya, Mizoram, Bihar and Orissa. The Northern States of India account for around 75% of the tea grown in India and the Southern States accounting for around 25% of the production. The tea industry provides direct employment for one million workers of which sizeable number are women.

More than half of the one million hectares of tea cultivation is in India and Sri Lanka, on areas ranging from 225,000 to 350,000 hectares. Between them, these two major producers account for more than three-quarters of the world's export trade of 600,000 metric tonnes of this commodity. This apparently disproportionate share in the export trade is because in these two countries, domestic consumption is far outstripped by total production.

With the current low price trends which are highly unremunerative to the growers, that most of the growers have shifted the cultivation practices to organic farming system. Today, more than 10,000 ha have been certified under organic tea cultivation.

8.2.1. Insect Pests

8.2.1.1. Shot-Hole Borer, *Xyleborus fornicatus*

i) Damage: The shot hole borer is a major pest of tea. Its infestation is characterized by small pinholes leading into the tunnel which are lined by a fungus (*Monilia* sp.). The infested branches get discoloured around shot holes and the leaves on the distal portion turn yellow, droop and drop. Peak infestation of the pest occurs during October-December.

ii) Management: The incidence could be minimized by removing affected branches, avoiding pruning during dry weather followed by spraying of *Beauvaria bassiana* at 1.5 kg/ha during May-October when humidity levels are high and also by placing cut stems of a jungle plant, *Montonoa bipinnatifida* in infested plantation during 3rd and 4th year to attract shot hole borer beetles (Fig. 8.7). Cut stems (22 to 28 days old) measuring 90 cm long and 25 to 30 mm thickness attracted large number of adult beetles. The number of beetles attracted to the stem trap was more during June to October with a peak in July. On an average of four cut stems were required for an area of 100 sq. m. and approximately 400 cut stems are needed per hectare. The number of beetles attracted to the stems declined after 15 to 20 days indicating the need for replacement of cut stems.

Fig. 8.7. Cut stems of *Montanoa bipinnatifida* to attract shot hole borer beetles.

An indigenous strain of *Beauveria bassiana* (available as a wettable powder 'Bio-Power' formulation) gives effective control of shot hole borer (Fig. 8.8).

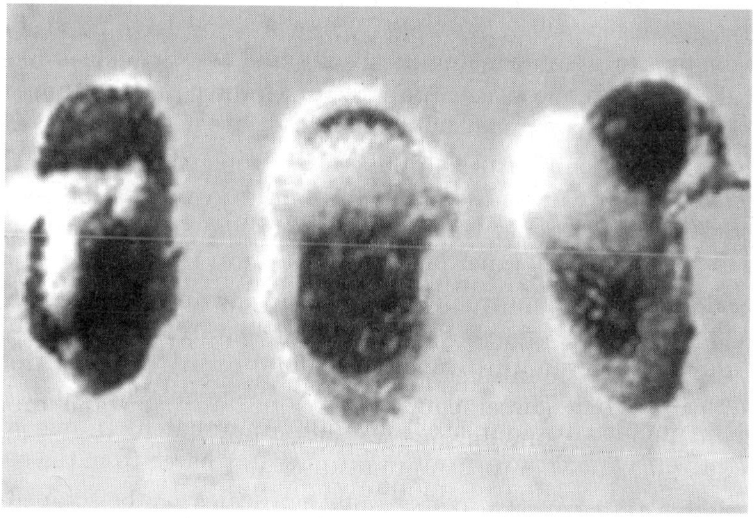

Fig. 8.8. *Beauveria bassiana* infected adults of shot hole borer.

8.2.1.2. Mosquito Bug, *Helopeltis theivora*

i) Damage: Tea mosquito bug (Fig. 8.9) laid large number of eggs on the broken ends of plucked shoots. Leaves curl up, become gradually deformed and remain small.

Fig. 8.9. Mosquito bug infestation on tea leaf.

ii) Management: Intensive removal of stalks during plucking is recommended to reduce the incidence of this pest. Weeds offer excellent hiding places and serve as alternate hosts for *Helopeltis*. Growth of weeds and wild host plants in and around tea fields may be controlled and this will help to reduce the incidence of pest.

8.2.1.3. Scales

i) Damage: The most common insect pests of tea are scales. Scale insects feed on plants by piercing plant tissue and sucking sap. Scales do not look like typical insects. They are small, immobile and have no visible legs. They vary in appearance depending on species and sex. Some look like small fish scales attached to the plant. As a result of their unusual appearance, populations can reach damaging levels before they are noticed.

On tea, scales usually attach to leaves but some species also attach to stems. Their feeding weakens the plant. With a heavy infestation, symptoms include yellowing of the upper leaf surface, fewer and smaller blossoms, leaf drop, twig dieback and sometimes death.

Tea scale has an oblong shape with a ridge down the centre parallel to the sides. It is a small scale with the female about 1/20 inch. The male is about two-thirds the size of the female. The females vary in colour from dark brown or grey to nearly black. Males are white. The female lays 10 to 16 eggs, which remain protected under her body until they hatch. In one to three weeks, bright yellow immature forms called crawlers hatch from the eggs.

ii) Management : With a light infestation, scales can be scraped off the plant and discarded. If only a few leaves are infested, hand picking and destruction of infested leaves is very effective. The best time to spray with a refined horticultural oil or insecticides (acephate, malathion, cyfluthrin, carbaryl) is in spring, after the plants have finished blooming and the danger of cold weather has passed. This will kill many adults, crawlers and eggs by smothering them. Spray twice at 10 days interval when temperatures are between 40 and 85° F.

Most insecticides are effective only against the crawlers. In addition, using an insecticide against scales can result in the death of naturally occurring enemies of scales. As such, insecticides should be avoided unless required.

Monitor the crawler emergence with sticky cards, tape wrapped around a branch, or by putting an infested shoot or leaf into a bag and watching for crawler movement. Crawler activity often coincides with the flush of new plant growth in the spring. However, some scale species may have overlapping generations with an extended crawler emergence period, such as along the coast.

8.2.1.4. Phassus Borer, *Sahydrassus malabaricus*

i) Damage: Phassus borer causes long cylindrical tunnels in thick branches. Entry hole is plugged with chewed wood and silk

ii) Management : Spraying of endosulfan or quinalphos gives effective control of the pest.

8.2.1.5. Thrips, *Scirtothrips bispinosus*

i) Damage: Leaves curl and show parallel lines of feeding marks on either side of the midrib

ii) Management : Spraying of chlorpyriphos or dimethoate or phosalone is effective for the management of thrips.

8.2.2. Mite Pests

8.2.2.1. Red Spider Mite (RSM), *Oligonychus coffeae*

i) Damage: Red spider mite (Fig. 8.10) has emerged as an important pest of tea in the last few years. Studies on its population ecology showed that the incidence of RSM was high during January to May and low during June to December and reached a peak in March/April. Crop loss studies revealed that the mite can cause more than 18 % loss in crop when the infestation is severe. Leaves turn ruddy bronze due to feeding.

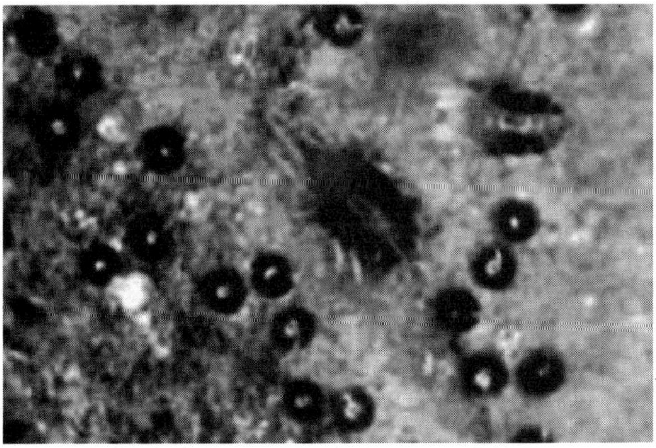

Fig. 8.10. Red spider mite infestation on tea.

ii) Management: An entomopathogenic fungus, *Paecilomyces fumosoroseus* (available as a wettable powder 'Mycomite' formulation) gives effective control of red spider mites (Fig. 8.11).

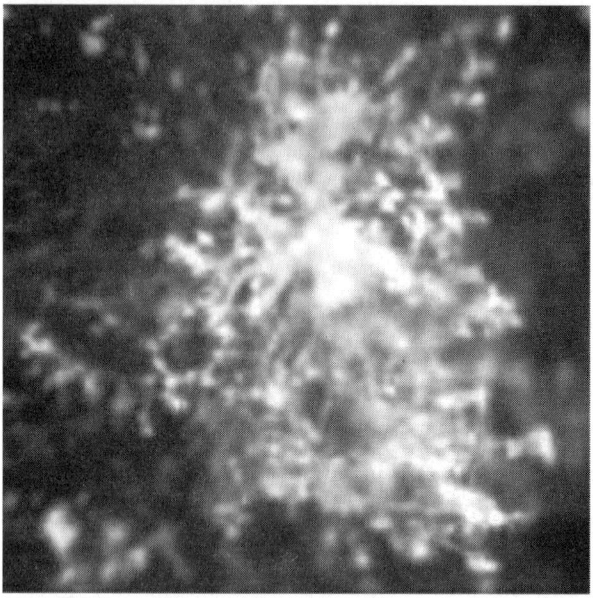

Fig. 8.11. Biological control of red spider mites using *Paecilomyces fumosoroseus.*

Use of hydrocarbon oil formulations of sulphur is effective against red spider mites. Recently, spray oil from paraffinic base has been found effective against red spider mites. Since this oil does not leave any residues in tea, it is incorporated into the mite control programme in tea.

IPM

- Populations of leaf folding caterpillars such as flush worms, leaf roller and tea tortrix can be reduced by their removal while plucking.
- Bushes may be pruned to reduce the damage of *Helopeltis* and shot hole borer.
- Unshaded areas are more prone to attack by thrips and mites. Proper shade management will help to prevent excessive build up of thrips and mites.
- Growth of weeds like *Mikania cordata, Bidens bifernata, Emillia sp., Polygonum chinense, Lantana camera* and wild host plants near tea fields may be controlled to reduce mosquito bug population.
- Application of higher levels of K (N: K_2O=1: 2) in the pruned year helps to reduce shot-hole borer incidence.
- *Amblyseius herbicolus* and *Euseius ovalis* are 2 phytoseid mites feeding on the eryophid mites-*Acaphylla theae* and *Calacarus carinatus.* The stigmatid, *Agistemus fleschneri* is an important

predator of eggs and nymphs of *Oligonychus coffeae* in North Eastern India.

- Coccinellids - *Cryptogonus bimaculatus, Jauravia quadrinotata, Menochilus sexmaculatus* and *Stethorus gilvifrons* prey upon mites.
- *Chrysoperla carnea* released at 500-1000 adults/ha feeds on thrips and mosquito bugs.
- Anthocorids – *Anthocorus* and *Orius* spp. and predatory thrips – *Aeolothrips intermedius* and *Mymarothrips garuda* are important predators of thrips.
- Scale insects and mealy bugs are attacked by parasitoids – *Coccophagus cowperi* and *Encyrtis infelix* heavily parasitize *Saissetia coffeae.*
- The larval parasitoid, *Apanteles aristaeus* is efficient against flush worms.
- The leaf roller, *Caloptilia theivora* is heavily parasitized by the Eulopid, *Sympiesis dolichogaster* (20 to 83%).
- The tea tortrix, *Homona coffearia* is severely affected by Ichneumonid larval parasitoid, *Phytodietus spinepes.*
- The eggs of mosquito bug are parasitized by Mymarid, *Erythmelus helopeltidis* (52-83%). The Neuropteran, *Chrysoperla carnea* (at 500-1000 adults/ha) feeds on mosquito bugs in the field and gives effective control in North Eastern India.
- *Bt* is effective for the management of looper caterpillars, cut worms, flush worms and other lepidopterous pests. *Beauvaria bassiana* is effective against shot-hole borer.
- Entomopathogenic fungi – *Verticillium lecanii, Paecilomyces fumosoroseus* and *Hirsutella thompsoni* were found effective against pink, purple and red spider mites. *Beauvaria bassiana* at 1 kg/ha is recommended in North Eastern India against mosquito bugs.
- Azadirachtin is effective against pink, and purple mites and caterpillars.

Table 8.5. Maximum residue limit (MRL) of certain pesticides on tea

Name of the Chemical	Pesticide residues MRL (mg/kg)	Name of the Chemical	Pesticide residues MRL (mg/kg)
Acephate	0.1	Fenazaquin	-
Aldrin/Dieldrin	0.02*	Fenitrothion	0.5
Bitertanol	0.1	Glyphosate	0.1
Cartap hydrochloride	0.1	Gluphosinate ammonium	-
Carbofuran	0.2	Hexaconazole	0.05
Carbosulfan	0.1	Heptachlor	0.02*

Chlorpyrifos	0.1	Lambdacyhalothrin	1
Chlordane	0.02*	Malathion	0.5
Cypermethrin	0.5	Monocrotophos	0.1*
Diflubenzuron	0.05	Oxydemeton methyl	0.05*
DDT	0.2	Paraquat	0.1
Deltamethrin	5	Propargite	5
Diazinon	0.05	Propiconazole	0.1
Dichlorvos	0.1	Profenofos	0.1*
Dicofol	20	Phosalone	0.1
Dimethoate	0.05	Copper Oxychloride	Exempted
Endosulfan	30	Surplus Formulations	Exempted
Ethion	2	Quinalphos	0.1
Fenvalerate	0.05*	Tridemorph	20
Fenpropathrin	0.02(G)	2, 4-D	0.1

* Indicates lower limit of analytical determination

8.3. COCONUT

The coconut palm (*Cocos nucifera*), originated in South East Asia, is now cultivated widely throughout the tropics of the world. The largest producers of coconut are the Philippines, Indonesia, India and Sri Lanka. In all the coconut producing countries, it is an important source of export earning despite being a regular constituent of the food. Coconut is an important small holders' plantation crop. The total area under coconut cultivation in India is 2.029 million ha with an annual production of 10.190 million tonnes and the average productivity of 5.0 tonnes/ha is the highest in the world (Bijay Kumar, 2007) (Table 8.6). It provides proteins, fats and some vitamins. Coconut contributes 700 billion rupees to the GDP of the country. Coir and coir products exported from India earned Rs. 2260 million during the above period. The contribution of the crop to the total edible oil pool in India is around 6 per cent. It is supporting the livelihood of 10 million people across the country. The four Southern States – Kerala, Tamil Nadu, Karnataka and Andhra Pradesh together accounts for 90% of the total area and production. Coconut is predominantly grown under rainfed conditions in Kerala and parts of coastal Karnataka, Maharashtra and Tamil Nadu. In rest of the country, it is mainly grown under irrigated conditions.

Table 8.6. State-wise area, production and productivity of coconut in India (2005-06)

State	Area in '000 ha	Production in '000 MT	Productivity in MT/ha
Kerala	897.8	4352.3	4.847
Tamil Nadu	370.5	3348.5	9.037

Karnataka	467.5	832.5	1.780
Andhra Pradesh	104.0	613.7	5.900
Maharashtra	18.0	188.1	10.450
Orissa	50.8	189.0	3.720
West Bengal	24.9	222.6	8.939
Assam	19.1	141.0	7.382
Goa	25.3	86.2	3.407
Gujarat	16.4	95.1	5.798
Andaman & Nicobar	25.6	59.9	2.339
Lakshadweep	2.7	36.5	13.518
Pondicherry	2.1	19.2	9.142
Tripura	3.3	4.8	1.454
Nagaland	0.9	0.8	0.888
Total	2028.9	10190.1	5.0

8.3.1. Insect Pests

8.3.1.1. Rhinoceros Beetle, *Oryctus rhinoceros*

i) Damage: Rhinoceros beetle feeds on unopened fronds and spathes of coconut. Characteristic fan like geometric cuts in the newly emerged frond are observed (Fig. 8.12). Infestation on spathe often destroys the infloresc-ence and prevents the production of nuts.

Fig. 8.12. Rhinoceros beetle damage on coconut and adult beetle.

ii) Management: Releasing of 10-15 Baculovirus infected beetles/ha has been found to be effective in checking the population of the rhinoceros beetle. These releases initiate a self perpetuating disease in the beetles through copulation, contamination through excreta, etc. The virus can infect all stages of the pest and the infection is passed down to the progeny also. However, re-release of the pathogen is needed after 3-5 years.

The green muscardine fungus, *Metarrhizium anisopliae* is another biocontrol agent which can infect all stages of rhinoceros beetle. The fungus can be easily multiplied in coconut water, tapioca chips, etc. On inoculating the fungus in decaying organic matter, it thrives well and cause a good degree of mycosis of the grubs resulting in reduction in black beetle population.

Clerodendron infortunatum, a commonly found weed is a promising botanical. Grubs or black beetles become deformed and sterile when fed on feed containing this plant. The leaves, shoot, flowers and root can be incorporated in the composting material and the larvae that hatch out in such pits shall be affected by the contaminated feed.

Integrated pest management includes hooking out of beetle, placement of perforated polythene sachets containing phorate in leaf axil and adoption of biological control measures using baculovirus.

- Treat manure pits with carbaryl/chlorpyriphos dust to kill the grubs.
- Soak castor cake at 1 kg in 5 litres of water and keep in mud pots in the garden to attract adults. Collect the adults and allow them to swim in a tray containing *Oryctes* baculovirus and release them back in the field.
- Hook out and kill the adults. Fill the crown with the mixture of chlorpyriphos dust and fine sand (1:1 by volume).
- A commercial slow release formulation of pheromone to attract adult beetles is also available.

8.3.1.2. Red Palm Weevil, *Rhyncophorus ferrugenius*

i) Damage: Red palm weevil grubs feed on the inner tissues of trees. Infected palms show holes on trunks and a viscous brown fluid oozes out. Chewed up fibre is extruded through the holes. Longitudinal splitting of leaf base and wilting of inner leaves can be observed (Fig. 8.13). Gnawing sound produced by grubs feeding inside will be audible. Symptoms become clear at an advanced stage when the crown topples. Eventually the tree dies.

Fig. 8.13. Red palm weevil damage on coconut and adult beetle.

ii) Management

- Injection of 1% endosulfan into the tunnels in the stem, trapping of adult weevils using baits and field sanitation are effective in controlling the pest.
- Pyrecon-E (derivative from Pyrethrum plant) has been a time tested effective chemical for the control of red palm weevil.
- Fill the crown with the mixture of chlorpyriphos dust and fine sand (1:1 by volume).
- Set up attractant traps (mud pots) containing 2.5 litres of toddy + 5 ml acetic acid + 5 g yeast.
- Scoop out the frass from the hole. Then insert one aluminium phosphide tablet inside the tunneled trunk. Plug the hole with mixture of wet clay and cow dung.
- A commercial slow release formulation of pheromone to attract adult weevils is also available known as Ferrolure (Fig. 8.14).

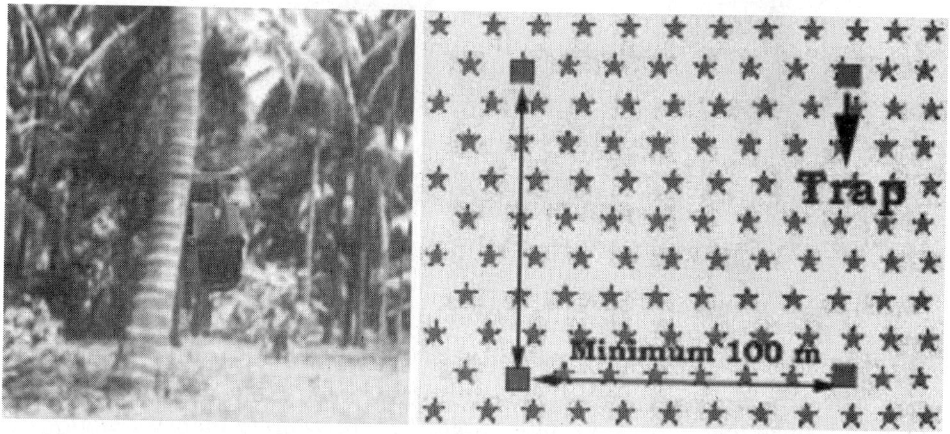

Fig. 8.14. Ferrolure pheromone for attracting adult red palm weevil of coconut.

Fig. 8.15. Black headed caterpillar damage on coconut.

8.3.1.3. Black Headed Leaf Eating Caterpillar, *Opisina arenosella.*

i) Damage: Leaf eating caterpillar eats away the green portion of the leaves giving a blighted appearance (Fig. 8.15). Caterpillar feeds on green matter from the lower leaf surface, remaining in galleries made of silken web and frass. Palm present a burnt up appearance.

ii) Management: Neem oil/soap sprays gave effective control of this pest.

Root Feeding Technique: Expose some roots by excavating the soil at the base of the palm (No need to trace out the tip of the root). Select a matured dark brown pencil thick root near the base and cut without damaging the cut end of the root and insert it in a small (15 x 10 cm) polyethylene cover with monocrotophos at 10 ml/palm diluted with an equal quantity of water. Adjust the cover so that the cut end of the root is completely immersed in the solution. The solution will be absorbed by the palm within 24 hours. The pest will be controlled within 7 to 10 days after treatment (Fig. 8.16).

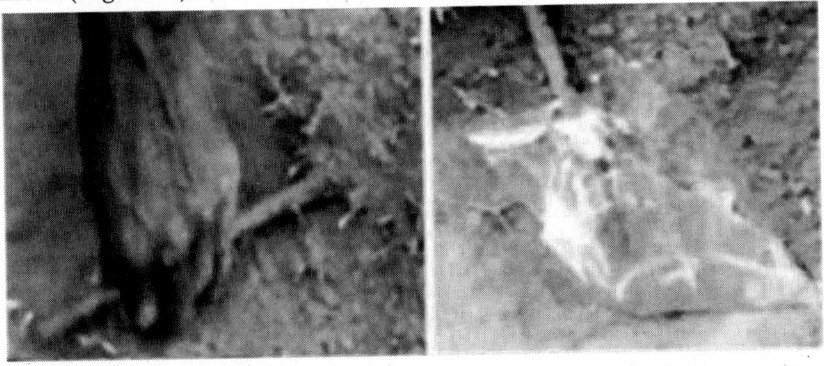

Fig. 8.16. Root feeding of monocrotophos to control black headed caterpillar.

Effective biological control measures have been evolved using parasitoids at different stages. An exotic parasitoid, *Trichogramma embryophagum* has been found to effectively parasitize the eggs in silken galleries. *Apanteles taragamae* is an important endoparasitoid of 1st and 2nd instar caterpillars. The frass of the young larvae acts as a kairomone and attracts the females for parasitism. This parasitoid holds good potential for pest suppression. *Goniozus nephantidis* is quite hardy and gregarious. It parasitizes medium and fully developed larvae. *Bracon hebetor* parasitizes medium sized or fully developed larvae in the coastal areas. A single parasitoid parasitized 22.6±7.18 late instar caterpillars during its life period. The elasmid parasitoid, *Elasmus nephantidis* is the only true prepupal parasitoid of *O. arenosella*. It is monophagous and a gregarious ectoparasitoid. It is well adopted to thrive during summer season, which is the peak period of activity of the pest. The bethylid parasitoid, *Brachymeria nosatoi* is found to possess the major attributes of an efficient biocontrol agent of the pest. It is a sturdy parasitoid prevalent in nature almost throughout the year. Its natural incidence is quite high (up to 38.8%), even during summer season, which is incidentally peak period of pest abundance. There has always been a preponderance of females, which have greater searching ability and locates and parasitizes host pupae remaining inside the cocoons in silken galleries. The 5th and 6th instar caterpillars were most readily accepted by the parasitoid. The pupal parasitoid, *Xanthopimpla punctata* is an outstanding example of a polyphagous species of parasitoid which turned out to be a better killer of the pest than even other host specific and gregarious species of parasitoids.

The optimum doses of release of indigenous parasitoids for effective suppression of *O. arenosella* were 20.5% for *G. nephantidis,* 49.4% for *E. nephantidis* and 31.9% for *B. nosatoi*, when the parasitoids were released individually and 40.4% when a combined release was made. The release at above doses effected 81 and 83% reduction in population density of the pest in case of individual releases of *G. nephantidis* and *E. nephantidis*, respectively. Release of the parasitoids in the early stage of pest build up was proved absolutely necessary for effective suppression of the pest.

8.3.1.4. Coreid Bug, *Paradasynus rostratus*

i) Damage: Coreid bug (nut crinkler) has become a serious problem in many parts of Kerala. It causes damage to the buttons and tender nuts. Nuts are shed in severe cases of infestation and also deform the nuts with characteristic crevices on the husk just below the perianth (Fig. 8.17). There will also be gummy exudation from such crevices. Mostly such nuts will be barren.

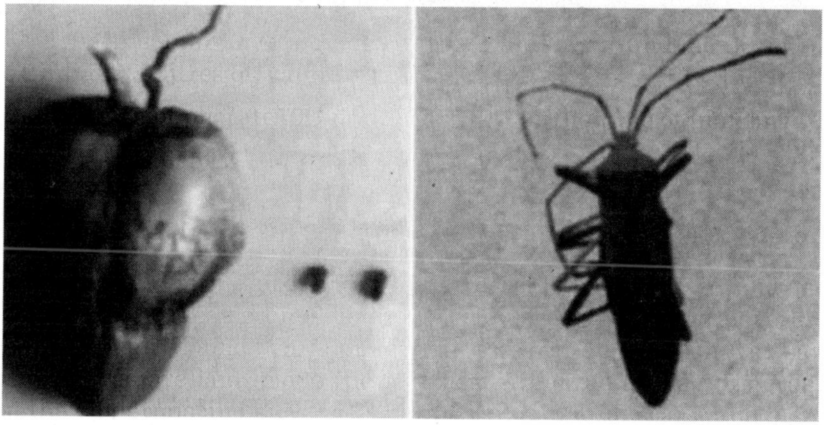

Fig. 8.17. Coreid bug damage on coconut and adult bug.

ii) Management: Spraying 0.1% carbaryl or endosulfan on the unopened spathes and bunches (except the newly opened inflorescence) will control the pest.

8.3.1.5. Lace Bug, *Stepanitis typicus*

i) Damage: Lace bug sucks sap from coconut foliage and acts as a vector in transmission of phytoplasma from root wilt affected palms to healthy palms.

ii) Management: This can be controlled by spraying endosulfan or monocrotophos both at 0.01%.

8.3.1.6. Termites, *Odontotermes obesus*

i) Damage: It is estimated that nearly 20% of the coconut seedlings are damaged by termites particularly in laterite soil. Seedlings are attacked through the nut or at the base of the young shoot and older palms through trunk to crown. Affected seedlings wither and die.

ii) Management: Cover the germinating nuts with a layer of sand. Drenching the nursery with 0.05% chlorpyriphos twice at 20-25 days interval and swabbing the affected trunk with the same chemical are affective. 5% carbaryl dust should be incorporated into the soil.

8.3.1.7. Scale, *Aspidiotus destructor*

i) Damage: The scales are severe during hot summer. In severe cases, leaflet turn yellowish, wither and dry. Scales may also congregate on green nuts and suck the sap (Fig. 8.18).

Fig. 8.18. Scale infestation on leaf and nut of coconut.

ii) Management: Spraying with 0.03% diazinon, dimethoate, phosphamindon or monocrotophos is effective for the management of scales.

8.3.1.8. Slug Caterpillars, *Latoia lepida, Macroplectra nararia, Contheyla rotunda*

i) Damage: Caterpillars are fleshy, slug like and have a series of tufts of spines highly irritating to touch. They pupate in hard shell like grayish cocoon in trunk. Caterpillars initially feed gregariously on the leaves and then disperse (Fig. 8.19).

ii) Management : Spray carbaryl at 2 g/litre of water when the larvae are in congregation. Root feeding also controls the pest.

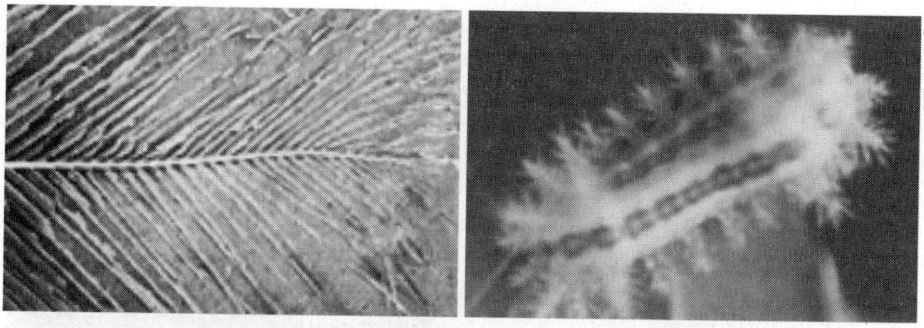

Fig. 8.19. Slug caterpillar damage on leaf of coconut.

8.3.1.9. White Grub, *Leucopholis coneophora*

i) Damage: Grubs (Fig. 8.20) are found in the roots. Leaves are yellow and sick pale. Severely infested palms exhibit button shedding and tapering of crown.

Fig. 8.20. Grubs and adult of white grub on coconut.

ii) Management : Incorporate in the top soil (up to 15 cm) phorate 10 G at 100g/palm prior to the on set of monsoon and also after the cessation of monsoon (May-June, September-October). Provide copious irrigation.

8.3.2. Mite Pests

8.3.2.1. Eriophyid Mite, *Aceria guerreronis*

i) Damage: Recently coconut gardens in some parts of Kerala, Tamil Nadu, Andhra Pradesh and Karnataka were seriously affected by a non insect pest called eriophyid mite. These mites infest by sucking sap from the soft tissues of buttons. In the initial stages, symptoms are seen as triangular patches close to perianth. Later, because of the continuous desapping by various stages of mites present beneath the inner bracts of perianth, brown coloured patches are formed (Fig. 8.21). As the nuts grow in size, the injured patches become warts and then develop into longitudinal splits on the surface of nuts. The liquid oozing from these patches dries and as a result dried decayed matter is noticed. The damage thus caused affects the quality of husk and dehusking becomes difficult.

ii) Management: Neem/garlic oil/soap emulsion has emerged as one of the potent measures for the control of mites. The fungus, *Hirsutella* *thompsonii* has been found to be an effective microbial agent for annihilating this pest.

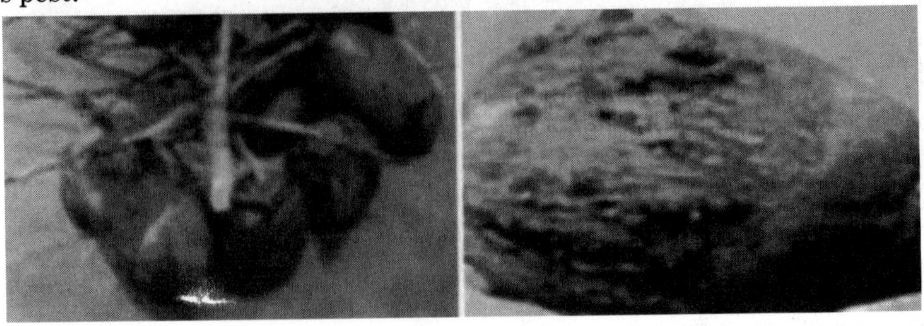

Fig. 8.21. Eryophid mite damage on coconuts.

Dissolve 1 kg biocatch in 200 litres of water and keep it overnight. Filter the solution using a muslin cloth. To the filtrate add 500 ml of Nimbecidine. Spray the resultant solution on the frond and newly developing nuts at 15 days interval.

8.3.3. Vertebrate Pests

i) Damage: Monkey damaged the maximum number of nuts (150-200/15 days in March 1997), rodents mainly *Bandicota bengalensis* damaged 15-20 nuts/15 days and wild boar the least, 0-4 nuts/15 days in Dakshina Kannada district of Karnataka.

ii) Management: Trunk banding (metal), burrow baiting, clean basin cultivation, crown cleaning, lacing base of grown-up tree trunks with waste oils or covering base of seedlings with thorny bamboo or a mat of dry sticks, timely harvests and pick-up of fallen nuts (against porcupine, rat and squirrel) have proved effective.

a) Integrated Method: Clean cultivation, crown cleaning, trunk banding, burrow baiting, Ibex fencing with at least 1 m of open space on either side, and covering base of seedling (up to 2-3 years) with tightly packed cover of thorny bamboo sticks or porcelain pipes wherever coconut palms were regularly prone to rodents damage saved 46% nuts over control against rodents, monkey and wild boar depredations. These practices to some extent afforded protection against raids by bison and elephants also.

8.4. ARECA NUT

Areca nut (*Areca catechu*), otherwise known as the betel nut, has its origin in the humid regions of Asia and Malay Islands. India continues to dominate the world in area, production and productivity of areca nut and has achieved self sufficiency in areca nut production. It is an important cash crop of India. India is the largest producer of areca nut in the world covering an area of 3.811 lakh ha with annual production of 4.831 lakh tonnes (Bijay Kumar, 2007) (Table 8.7). Most of the production is domestically consumed. It is masticatory and the ripe fruits are sometimes used as an antihelminthic and astringent in Europe. It has a toning effect and stimulates the nerves. The extract after boiling the nuts contains tannins which are used for tanning leather. Areca nut industry forms the economic back bone of nearly 6 million people in India and for many of them it is the sole means of livelihood.

Table 8.7. State-wise area, production and productivity of areca nut in India (2005-06)

State	Area in '000 ha	Production in '000 MT	Productivity in MT/ha
Kerala	108.6	119.3	1.098
Tamil Nadu	4.8	10.3	2.140
Karnataka	161.6	215.7	1.334
Andhra Pradesh	0.3	0.2	0.666
Maharashtra	2.0	5.0	2.500
West Bengal	9.4	29.3	3.128
Assam	71.2	66.8	0.938
Goa	1.6	2.6	1.625
Andaman & Nicobar	4.4	4.8	1.090
Meghalaya	11.5	15.5	1.356
Pondicherry	0.1	0.1	1.000
Tripura	3.4	6.9	2.029
Mizoram	2.0	5.3	2.650
Nagaland	0.2	1.3	6.500
Total	381.1	483.1	1.026

8.4.1. Insect Pests

8.4.1.1. Spindle Bug, *Carvalhoia arecae*

i) Damage: This is a serious pest of young areca plantations of Kerala, Karnataka and parts of Tamil Nadu. This pest multiplies rapidly with close of monsoon. Adult bugs are brightly coloured red and black. They inhabit the inner most leaf axils, usually below the spindle. The light violet brown nymphs are greenish yellow with the border of body. They suck sap from tender spindle resulting in loss of vigour and consequent death of plants. The feeding injury is caused on the lamina and petiole. The affected leaves show dry brown patches (Fig. 8.22). Severe leaf damage causes stunting of palms.

ii) Life cycle: Eggs are laid singly between the leaflets of the spindle. The eggs hatch in 9 days. There are five nymphal stages and the life cycle is completed in 15-24 days.

iii) Management: Filling the inner-most leaf axils around the spindle with phorate 10% granules (10 g/palm) at 3 months interval is effective. Conveniently this can be placed in sachets of 2 g (2 per tree).

Fig. 8.22. Spindle bug on areca nut.

Spray crowns with 0.1% carbaryl. The spray should reach the leaf axils. Repeat spraying after 30-35 days if pest incidence continues. Placement of 2 g phorate 10G sachets on the top most two leaf axils prevents the pest attack.

8.4.1.2. Root Grub, *Leucopholis burmeisteri, L. lepidophora* (Coleoptera: Melolonthidae)

i) Damage: Root grubs or white grubs (Fig. 8.23) occur in low lying and clayey soils where the water table is high. They are voracious feeders on areca roots and are found in the areca tracts of Kerala and Karnataka. Adult beetles (cockchafers) emerge during May-June after few days of pre-monsoon showers i.e., after 8-10 days of showers, between 6.30 to 7.30 PM. These beetles lay eggs in soil mostly up to 10 cm depth. Eggs hatch out in about three weeks. The early instar grubs feed on the roots of grasses and other humus. The grub period with three instars is completed in 7 to 8 months. The pupation is in soil in cocoons of mud. This period lasts about one month. The adult beetle is chestnut brown in colour. The second and third instar grubs of these beetles feed on tender and mature roots of the palm, in cases of severe infestation; the bole of the palm is also eaten up. They feed on roots of grasses and intercrops like banana, cocoa, tapioca, yams etc.

Fig. 8.23. Grub and adults of root grubs on areca nut.

In areca seedlings, the feeding on roots results in dropping and drying of leaves. Affected seedlings come off easily since the entire root system is usually eaten up. Palms with few years of infestation show a sickly appearance, with yellowing of leaves, tapering of stem, and reduction in yield. The palms may topple in case of severe loss of root system.

ii) Management : Keeping the garden clean, well-drained and free of weeds reduces its attack. Loosen soil around the base of palms to a depth of 10-15 cm and drench with 0.04% chlorpyriphos suspension twice, once in May just before the onset of southwest monsoon and again in September-October towards the close of the monsoon. Repeat application for 2 or 3 years consecutively to secure a complete eradication of the pest.

Root grubs can also be controlled by soil application of phorate around the palms. Apply phorate at 15 g/palm twice a year during May and September. Collect the adult beetles in the evening hours for 8-10 days after pre monsoon showers and destroy them.

8.4.1.3. Inflorescence Caterpillar, *Tirathaba mundella* (Lepidoptera: Pyralidae)

i) Damage: This is reported from Karnataka and Kerala. The adult moth lays eggs in the mechanically damaged portions of the spadices and the emerging caterpillars bore inside the spadices. The egg period lasts for five days and the larval period for about 26 days with five instars. Pupal period lasts for 9-11 days. The caterpillars féed on the inflorescences especially the tender female flowers and rachille and web together the rachille into a wet mass with silken threads formed by the caterpillars and take shelter in it (Fig. 8.24). Mature caterpillars can damage newly opened inflorescences also. In severe cases of incidence, these bore into the tender buttons and tender nuts as well.

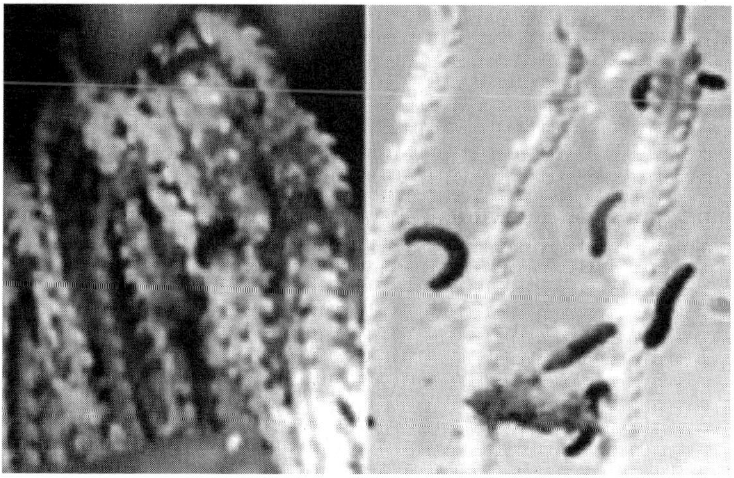

Fig. 8.24. Inflorescence caterpillars on areca nut.

As a result of webbing and feeding, the inflorescence is unable to exert the natural pressure on the spathe necessary for its opening and thus spathe opening is delayed. Yellowing of spadices, presence of small holes with frass and drying patches on the spathe are the external symptoms of attack.

ii) Management : Force open the inflorescence out of the enclosing spathe and spray 0.05% monocrotophos. Control slugs, which predispose inflorescence to the attack of caterpillar, by using bait of metaldehyde.

8.4.1.4. Pentatomid Bug, *Halymorpha marmorea* (Hemiptera: Pentatomidae)

i) Damage: This bug causes tender nut drop in areca. Incidence of this bug is seen from March/April to July/August. The later instar nymphs and adult bugs pierce the tender nuts of different maturity and suck the kernel sap. As a result of this feeding, the kernel dries up and the tender nuts drop. Characteristic pin prick black marks are seen at the point of feeding sites, which lead internally to the kernel. The adults are bronze coloured with brown spots and in young stages they are black with white spot on the leg. When tender nuts are not available, the insect migrates to other hosts like cowpea and bitter gourd.

ii) Management : The crops should be closely watched and the bug when noticed should be mechanically removed and destroyed. One round of spray with endosulfan (0.05%) on the bunches is effective.

Many coleopteran predators, chiefly coccinellids of areca nut mites, are used. They are *Aspects indicus, Cyleophalus semipictus, Stethrus keralicus* and *Stethorus parcepunctatus.*

8.4.1.5. Scale Insect, *Aonidiella orientalis* (Homoptera: Diaspididae)

i) Damage: Scale insects colonize the leaves, spathes, leaf sheaths and bunch and suck sap from the tissues. Continuous feeding on nuts results in pre-mature yellowing of nuts and in severe infestation, the kernel may not develop and may turn black and shrivel up. The scale insects are present through out the year, but are more serious during October to February.

ii) Management: Spraying 0.04% diazinon, dichlorvos (DDVP), phosphamidon, monocrotophos or quinalphos gives effective control of the pest.

8.4.2. Mite Pests

8.4.2.1. White Mite, *Oligonychus indicus* (Acari: Tetranychidae)

i) Damage: Adults and nymphs of this spider mite colonize the lower surface of leaves (Fig. 8.25). The colony is found below a white web on the leaves. Life cycle is of short duration. The total duration of the immature stages varies from 6.5 to 9.0 days. They suck sap from lower surface of leaf causing them to turn yellow or bronzed in appearance.

ii) Management : Removal of heavily infested and dried leaves and burning help in eradicating the source of infection. Dusting sulphur or spraying 0.05% wettable sulphur is effective for the management of mites. Spray dicofol 2ml/litre of water to the undersurface of infested leaves. Repeat the spray after a fortnight if there is a recurrence of the pest.

Fig. 8.25. White mite damage on areca nut leaves.

8.4.2.2. Palm or Red Mite, *Raoiella indica* (Acari: Tenuipalpidae)

i) Damage: Nymphs and adults of this mite are seen in large numbers on the lower surface of leaves. In severe cases of infestation, they may be seen on the leaf stalks and on the spindles. The life cycle lasts for 12.9 days. This mite also attacks palms like coconut, date and ornamentals. Population of these mites start building up after the monsoon rains and reaches its maximum in April-May. Neglected and poorly irrigated gardens and nurseries, particularly those in exposed conditions are prone to severe infestation. Colonies of these mites start declining with the onset of rains in June.

ii) Management : Removal of heavily infested and dried leaves and burning helps in eradicating the source of infection. Orange coloured mites can be controlled by spraying bunches with dimethoate at 0.05%. Dusting sulphur or spraying 0.05% wettable sulphur also gives effective control of the pest. Spraying under surfaces of leaves and crown with dicofol (2 ml/litre) or dimethoate (1.5 ml/litre) is effective.

8.4.3. Vertebrate Pests

i) Damage: Monkeys, rats, wild boars and bats damage nuts. Monkey damage nuts by irregular peeling off fruit epicarp, partially or completely. The unriped fruits are plucked-off and dropped. Bat damage was restricted to ripened or over ripened fruits leaving the fruit surface damaged. Wild boars feed only on fallen ripened nuts. They chewed the whole nut leaving the crushed nuts. Nuts damaged by monkeys are collected and sold by farmers as second quality nuts at 20% reduced rates than healthy nuts.

ii) Management: Growers protected the crop by gun patrolling, Ibex fencing, cracker bursting, scaring and by lighting and repellent odours like waste tire burning, bonfire, etc. Damage to areca seedlings from wild boar and rat could be avoided by using locally available porcelain pipes. Monkey damage can be reduced by disfavouring the habitat like debranching shade trees, keeping dog for chasing monkeys, etc.

a) Integrated Method: Clean cultivation, crown cleaning, trunk banding, burrow baiting, Ibex fencing with at least 1 m of open space on either side, and covering base of seedling (up to 2-3 years) with tightly packed cover of thorny bamboo sticks or porcelain pipes wherever areca nut palms were regularly prone to rodents damage saved 46% nuts compared to control against rodents, monkey and wild boar depredations. These practices to some extent afforded protection against raids by bison and elephants also.

8.5. BETEL VINE

Betel vine (*Piper betel*) is a perennial, dioecious, evergreen creeper grown in India. It is a chewing stimulant of brain, lungs and heart. The leaves contain Vitamins B and C. It is an important cash crop in Andhra Pradesh, Assam, Bihar, Karnataka, Kerala, Madhya Pradesh, Maharashtra, Orissa, Tamil Nadu, Tripura, Uttar Pradesh and West Bengal with an annual turnover of about Rs.700 crores. Its leaves are exported to Pakistan, Bangladesh, Indonesia, Malaysia, Burma and Thailand.

8.5.1. Insect Pests

8.5.1.1. Mirid Bug, *Disphincuts politus, D. maesayum*

i) Damage: The mirid bug (Fig 8.26) de-saps the tender leaves and shoots. The nymphs and adults damage the leaves by puncturing and sucking the juice causing the leaves to shrivel, fade and dry up. Adult bugs are slender, active and reddish brown; both nymphs and adults suck the sap from tender leaves which develop black spots. The market value of the leaves is reduced. The pest is active from June to October. Warm and humid climate is favourable for development of the pest.

Fig. 8.26. Mirid bug on betel vine.

ii) Management: Spraying 0.1% malathion or 0.002% pyrethrins is effective. In serious infestations, apply 1.5% fish oil soap against the mirid bugs. The leaves should be harvested only after 15 days of insecticide application. The treated leaves should be consumed/marketed after thorough washing in water. Spraying 4% Neem Seed Kernel Extract also gives effective control of the pest.

8.5.1.2. Scale Insects, *Lepidosaphes* sp., *Pseudococcus* sp.; Mealy Bugs, *Ferrisia virgata*

i) Damage: Scale insects and mealy bugs occur on the stems. Adults and nymphs infest tender shoots and suck the sap and reduce the vitality of the vines. The leaves are covered with white masses of mealy bugs which sucks the sap from leaves and shoots.

ii) Management: Pruning and destroying the affected parts helps in reducing the incidence of the pests. Spraying 0.025% quinalphos or 0.5% malathion is effective against the pests. Against scale insects, restrict insecticide application on the stem only. The leaves should be harvested only after 15 days of insecticide application. The treated leaves should be consumed/ marketed after thorough washing in water.

8.6. OIL PALM

Oil palm (*Elaeis guineensis*) is the highest oil-yielding plant (4-6 tonnes oil/ha) among perennial oil-yielding crops, producing palm oil and palm kernel oil. These are used for culinary as well as industrial purposes. It occupies an area of 44,413 ha in Andhra Pradesh (80% area), Karnataka, Assam, Gujarat, Goa, Kerala, Maharashtra, Orissa, Tamil Nadu, Tripura and West Bengal. Average productivity in oil palm plantations is now 4-5 tonnes/ha which compares favourably with other countries.

8.6.1. Insect Pests

8.6.1.1. Rhinoceros Beetle, *Oryctes rhinoceros*

i) Damage: *Oryctes rhinoceros* is a severe pest of oil palm. The pest causes severe damage to emerging fronds and spindle. The adult beetle feeds on the softer tissue of the rachis, resulting in snapping off of the fronds and spears at the feeding sites.

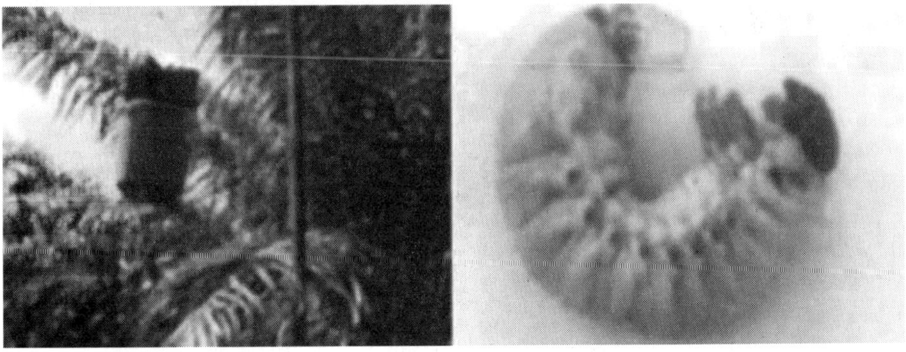

Fig. 8.27. Pheromone trap for rhinoceros beetle and virus infected larva.

ii) Management : A combination of cultural, chemical and biological control is now being used in an integrated manner to successfully manage outbreaks of the pest. Field sanitation and elimination of breeding sites are essential components of the pest management operation. This pest can be suppressed by using pheromone traps and the virus *Baculovirus oryctes* (Fig. 8.27).

8.6.1.2. Bag Worms, *Cremastophysche pendula*, *Mahasena corbetti*, *Metisa plana* and Nettle Caterpillar, *Setora nitens*

i) Damage: Bag worms and nettle caterpillars are opportunistic pests that can devastate large areas of oil palm if not detected and treated early. Early detection is achieved by carrying out regular inspection and counting of pest numbers in the field.

ii) Management : Action is taken to control the pests by use of selective insecticide application, e.g., trunk injection or spot spraying with pyrethroid or *Bacillus thuringiensis* formulations once when an economic threshold level of pest population is exceeded.

More proactively, methods of direct or indirect biological control are utilized. Direct biocontrol involves the use of non-occluded spherical and granulosis viruses and *Cordyceps* entomopathogenic fungi for control of nettle caterpillars. Predatory pentatomid bugs such as *Cantheconidea furcellata* and *Platynopus melacanthus* are also mass bred for release into fields.

Indirect biocontrol is achieved by the cultivation of nectariferous plants like *Euphorbia heterophylla, Cassia cobanensis, Antigonon leptopus* and *Turnera subulata* in and around infested fields. These plants encourage populations of parasitoids and predators of bagworms and nettle caterpillars by providing them with nectar and shelter. The increased populations of natural enemies would in turn exert biocontrol on the pests.

8.6.1.3. Red Palm Weevil, *Rhynchophorus* spp.

i) Damage: Red palm weevil (Fig. 8.28) is a major pest of oil palm in India. These weevils lay their eggs at the cut end of petioles or other wounds. The emerging larvae tunnel into the crown and feed on the growing tissues. Palms infested by red palm weevil start wilting and leaves show gradual increase in chlorosis and fracture when strong wind blows.

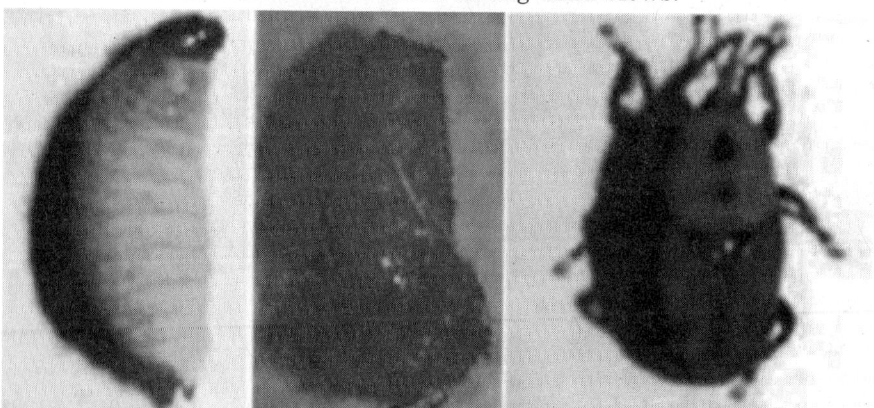

Fig. 8.28. Larva, pupa and adult of red palm weevil.

ii) Management : If detected early, treatment of affected palm with 0.2% solution of endosulfan or 0.1% carbaryl would save the palms.

8.6.2. Vertebrate Pests

8.6.2.1. Rats

i) Damage: Rats (Fig. 8.29) are implicated in heavy losses of seedlings. Rats, if uncontrolled, can bring about at least 5% loss of crop in oil palm. These pests were traditionally controlled with poison baits with potential undesirable secondary effects.

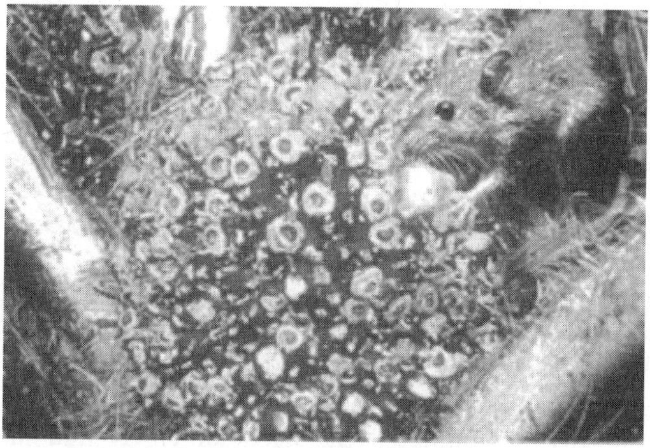

Fig. 8.29. Rat damage on oil palm fruit bunches.

ii) Management : For biocontrol of rats the use of the barn owl *Tyto alba* has been implemented. By setting up owl boxes at an intensity of 1 box per 10 hectares of oil palm in a regular grid, success in biocontrol was achieved (Fig. 8.30). Owls have been recorded to suppress fresh fruit damage by rats to below the economic threshold of 5% for the past 12 years without having to use rodenticide baits.

Fig. 8.30. Biological control of rats by barn owl.

8.6.2.2. Birds

i) Damage: Many birds such as the forest crow, the house crow and the common Indian myna cause severe damage to oil palm fruit bunches. Parakeets, crows and common myna are the major pests in Karnataka. The roseringed parakeet cause severe damage to oil palm when fruit bunches are borne above 3 m, i.e. when the palm is over 5 year old. These birds feed on the mesocarp of the oil palm fruits.

ii) Management: The damage can be minimized by scaring the birds and covering the ripe bunch with wire net, 150 days after fruit set. Wire mesh covered bunches recorded 10.9% damage, plastic bag covered bunches 9.2%, while the k.k. bunch cover recorded the least fruit damage of 1.6% as compared to 48.95% damage in control bunches (Chakravarthy *et al.*, 1998). Covering the bunches with used bags proved effective at Hyderabad, Andhra Pradesh.

a) Integrated Management: Covering ripening bunches with gunny bags or thatch of farm wastes or paddy straw, timely harvests and scaring away vertebrates by odours, watch and ward saved 62% fruits from birds, rodents and wild boar compared to control.

There are many examples of how managing species other than the crop, rather than attempting to eliminate them, has benefits in oil palm. For example:

- *Elaeidobius kamerunicus*, a weevil introduced into South East Asia in the 1970's eliminated the need for costly and inefficient hand pollinating, resulting in a sharp increase in oil yields.
- Owls are encouraged for rodent control.
- Leaving less competitive species of weeds provides a habitat for predators of insect pests and helps to prevent soil erosion.

8.7. CACAO

Cacao (*Theobroma cacao*) is an important commercial crop grown the world over. It is native of Amazon Valley of South America. The total area under cacao cultivation in India is 0.295 lakh ha with an annual production of 0.102 lakh tonnes with an average productivity of 0.345 tonnes/ha (Bijay Kumar, 2007) (Table 8.8). In India, it is being cultivated in Kerala, Karnataka, Andhra Pradesh and Tamil Nadu. Kerala account for about 76% of the area and 78% of the total production in the country.

Table 8.8. State-wise area, production and productivity of cacao in India (2005-06)

State	Area in '000 ha	Production in '000 MT	Productivity in MT/ha
Kerala	10.2	6.5	0.637
Tamil Nadu	1.4	0.2	0.142
Karnataka	6.0	2.8	0.445

Andhra Pradesh	11.8	0.7	0.059
Total	29.5	10.2	0.345

8.7.1. Insect Pests

8.7.1.1. Mealy Bug, *Planococcus lilacinus, P. citri*

i) Damage: This cosmopolitan group of insects with exceptional powers of dispersal is often seen colonizing cacao in most of the cacao growing tracts of South East Asia like India, Sri Lanka and Papua New Guinea. Though few species of mealy bugs colonize cacao trees, two species of *Planococcus* and one species of *Planococcoides* are the most important and most abundant of cacao insects. They are the cacao mealy bug, *P. lilacinus*, the citrus mealy bug, *P. citri* and the West African mealy bug, *Planococcoides njalensis*. In India, the cacao mealy bug, *Planococcus lilacinus* is reported as a serious pest causing damage to cacao and is present in all cacao tracts of the country. It occurs consistently on the plant, and is present throughout the year. This colonizes the tender parts of the plant such as the growing tips of the shoots, the terminal buds, the flower cushions, the young cherelles and mature pods (Fig. 8.31).

Fig. 8.31. Mealy bug infestation on cacao.

Mealy bug poses major problem for cacao. During summer months mealy bugs cause damage to spindlo leaves, spathes and bunches. The infested portions turn yellow and finally dry up. Mealy bugs suck the sap from developing pods and tender shoots. Retarded growth and excessive branching at undesired height are the initial symptoms of damage. Later, it causes abortion and wilting of cherelles.

ii) Management: The mealy bugs can be controlled with two rounds of spray of 0.1% fenthion or monocrotophos. Spot application of monocrotophos at 125 ml in 100 litres of water is effective against mealy bugs. Spraying 0.03% diazinon, dimethoate, or phosphamindon is also effective.

8.7.1.2. Aphids, *Toxoptera aurantii*

i) Damage: Two species of aphids occur on cacao. The abundant and cosmopolitan one is the citrus black aphid, *Toxoptera aurantii.* The Adults and nymphs of aphids colonize the flower buds, flowers, flush leaves, tender stems and very small cherelles. Leaf crinkling and shedding of flowers are the outcome of infestation.

ii) Management : A number of natural enemies (syrphids, cecidomyiids, coccinellids, hemerobiids) feed on the aphids and reduce the population. In severe cases of infestation, spraying with dimethoate at 1.5 ml/litre of water is effective for the management of aphids.

8.7.1.3. Ash Weevil, *Myllocerus viridanus, M. maculosus (Curculionidae)*

i) Damage: Ash weevils are recorded in the Nilgiris and Yercadu regions of Tamil Nadu as early as in 1967. *M. viridanus* occur as a major insect in all cacao growing tracts of Kerala and Tamil Nadu. Adults are seen in large numbers on the underside of leaves and feed on the interveinal tissues leaving the veins intact. The flaccid tender flushes are not preferred for feeding. Population peak is observed in July-September. Infestation is relatively severe on young plants and quite often, the entire foliage of such plants is skeletonized leading to growth retardation. The incidence of ash weevils is more serious in coconut-cacao plantations.

ii) Management : Spraying of quinalphos or chlorpyriphos at 2 ml/litre of water is effective against weevils. Drenching the soil with carbaryl (Sevin 50% WP) at the rate of 20g in 10 litres of water is effective in controlling the pest.

Fig. 8.32. Stem borer infestation on cacao.

8.7.1.4. Stem Borer, *Zeuzera coffeae*

i) Damage: The red borer of coffee, *Zeuzera coffeae* (Cossidae) is a pest of cacao in many countries. Caterpillars of this leopard moth bore into young branches and make unramified hollow tunnels inside the stem. The symptoms of attack are a round hole on the stem (Fig. 8.32), drying up of the upper portions above the hole, and excreta and chewed up fibres strewn out on the ground. If the main stems of young plants are attacked, the plants die. When the branches are attacked, only the branches dry up and simple pruning will save the trees.

ii) Management : Prune and destroy the affected branches. Smear the affected portion with an emulsion containing carbaryl wettable powder.

8.7.1.5. Leaf Feeding Caterpillars, *Lymantria* spp., *Euproctis* spp., *Dasychira* spp.

i) Damage: Cacao lymantriids, are the most commonly encountered leaf feeding insects of cacao in India. The caterpillars of this moth feed on the tender leaves and on the pericarp of cherelles and unripe pods. Three genera of lymantriids, viz., *Lymantria, Euproctis* and *Dasychira* are so far reported on cacao. The caterpillars of these moths feed on the tender leaves, and on the surface tissues of cherelles and green pods.

L. ampla is commonly and abundantly observed leaf-feeding insect of cacao in India. These are seen more in the field after the monsoon rains. The caterpillars cause severe damage on leaves in young plants. The early instars of this moth feed on leaves or the surface tissues of growing pods during day and night, but later instar caterpillars are nocturnal in habit. During day time they hide on fallen leaves at the base of the tree or the basal surface of the main stem.

Caterpillars of *Euproctis* spp. are also defoliators of cacao. These caterpillars feed on tender leaves, surface of cherelles and green pods. The sporadic high population of *Euproctis* spp. can cause severe damage of tender leaves. The period of attack is seen from June-July.

ii) Management: If the damage is very severe, spraying could be given by mixing 16ml of dimethoate in 10 litres of water.

The indigenous natural enemies of this lymantriid include the insect pathogen of the mature larvae, *Paecilomyces fumosoroseus*; a pupal parasite, and a braconid larval parasite, *Apanteles* sp.

8.7.2. Vertebrate Pests

8.7.2.1. Rats, *Rattus* spp., *Bandicota* spp.

i) Damage: Rodents are the major problem in cacao interplanted with coconut in all the countries since they have the ability to exploit new food sources. Rats and squirrels are the main groups of rodents that damage cacao in all the cacao cultivating tracts of the world. These damage the pods and the attack follows a fairly constant pattern. Ripe pods are usually chosen and a large hole is bitten through the pod husk (Fig. 8.33). The beans are then extracted and after the sweet mucilage is eaten, these are discarded. Many species of rats damage both young and mature pods. Rat damage is reported from most of the tropical countries including the islands. The most important rat species observed is the black rat *Rattus rattus* which is very serious pest of coconuts in all coconut growing areas. The black rat is recorded in India as a major pest occurring in areca-cacao and coconut-cacao mixed gardens. Black rats damage cacao grown as monoculture also since under South Indian conditions, the vicinity of preferred nesting hosts like coconut is assured.

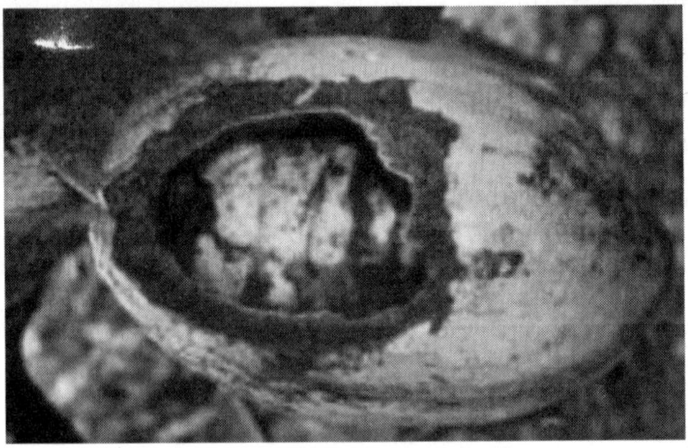

Fig. 8.33. Rat damage to cacao pod.

Rats inhabit the crown of nearby coconut palms. They descend during night and damage the cacao pods.

ii) Management : Only the cooperative efforts of plantation owners/ small farm holdings can achieve management of rodent population. Otherwise the local trials done by individual farm owners have no impact on the population levels of rodents and the damage they inflict on cacao and other crops. This is more pronounced in the case of rats. Baiting and trapping methods are available. The rats can be controlled by keeping 10g bromadiolone (0.005%) wax cakes on the branches of cacao trees twice at an interval of 10-12 days. Squirrels are best controlled by trapping with wooden or wire mesh single catch live trap with ripe coconut kernel as the bait. The success is more if trapping is carried out during the lean periods of the crop (September-November) and when the alternate foods such as paddy, cashew apples and jack fruit are not available. The traps are to be inspected daily and the trapped animals are to be killed to minimize the chances of communication with other animals. Timely harvest of the pods will help in increasing the efficiency of poison baiting as well as trapping.

8.7.2.2. Striped Squirrel, *Funambulus palmarum*

i) Damage: Two species of squirrels are reported as pests of cacao in India. They are the Western Ghats squirrel, *Funambulus tristriatus*; the South Indian palm squirrel, *F. palmarum*. Of these the Western Ghats squirrel is the most serious pest of cacao in India. Squirrels usually make oval holes on the central or terminal portion of the pods for feeding (Fig.

8.34). Squirrels are diurnal in habit and damage ripe pods. Striped squirrel gnaws the bronzing pods and extracts beans along with mucilaginous pulp.

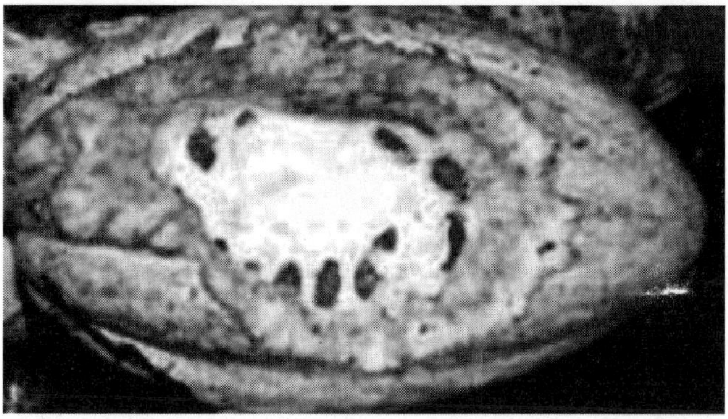

Fig. 8.34. Squirrel damage on cacao pod.

ii) Management: Cover pods with punched polybags of 150 gauge smeared with bitumen kerosene mixture.

8.7.2.3. Palm Civet, *Paradoxurus hermaphroditus*

i) Damage: The palm civet, *Paradoxurus hermaphroditus* (Viverridae) also known as toddy cat, is a nocturnal tree climber. It is present in many cacao-growing tracts of India and Malaysia. This carnivore damages the pods by biting and breaking the husk of cacao pods. The damage caused by this to cacao is reported from South India. The palm civet bite and break the husk of cacao pods. The pieces of broken chunks are 2.0 to 3.0 cm in diameter. There is no distinct pattern for the damage. The terminal half or one side of the pod is broken. While feeding, the beans are swallowed and as such no trace of beans is visible directly under the tree. Piles of defecated beans are seen scattered around the cacao plantations.

ii) Management : Palm civets could be easily controlled by poison baiting with 0.5g of carbofuran granules using ripe bananas as baits. Two such banana fruits have to be tied to the trunk/jorquettes of five to six cacao trees per hectare.

8.7.2.4. Other Vertebrate Pests

i) Damage: Jungle cat (*Felis chaus*), monkey (*Macaca radiata*), jackal (*Canis aureus*), bison (*Bibos gaurus*), stag (*Cervus canis*) and birds constitute other vertebrate pest complex. Jungle cat damage commenced during pod ripening stage (October-November). Monkey, jackal and birds were found damaging pods during ripening stage. Jungle cat is the most important pest incurring over 20% loss in pod yields in hill region of Karnataka and loss due to all vertebrates exceed 35% (Chakravarthy, 1993).

ii) Management: Thyagaraj *et al.* (1996) conducted a baiting trial with carbofuran 3G and warfarin 0.05% simultaneously with wooden snap trapping (15-18 traps/ha) at 3 locations and banana fruits of cv. Puttabale served as a poison carrier. The number of healthy pods/plant increased significantly in the treated plots compared to untreated plots. In Northern Karnataka, wet jaggery and dry sea fish are used as poison carriers.

Integrated Management: Cacao plantations were effectively protected from jungle cat, rodents, jackal, monkey and stag damage by baiting carbofuran 3G in local banana fruits or wet jaggery, intercropping with areca or coffee, timely harvest, clean cultivation and snap trapping at peak production period (August/September and December/January) (Thyagaraj *et al.*, 1996).

8.8. RUBBER

India is the fourth largest producer of natural rubber with a share of 8.8% in world production in 2006. The two major states with rubber plantations are Kerala and Tamil Nadu, with Kerala alone accounting for over 80% of the area under plantation. India produced an estimated 852,000 tonnes of natural rubber during 2006-2007 and exported 56,545 tonnes while it imported 89,699 tonnes. In 2006 the yield per hectare in India at 1879 kg, was amongst the highest in the world.

8.8.1. Insect Pests

8.8.1.1. Bark Feeding Caterpillar, *Aetherastis circulata, Ptochoryctis rosaria*

The pest is distributed in Nagercoil, Nedumangad, Punalur, and Thrissur localities.

i) Damage: The caterpillars build galleries with fecal matter and silk all over the trunk region and branches of trees. Generally larvae feed on dead bark (Fig. 8.35) and occasionally on live bark causing exudation of latex. Deep scars are found at the regions of feeding.

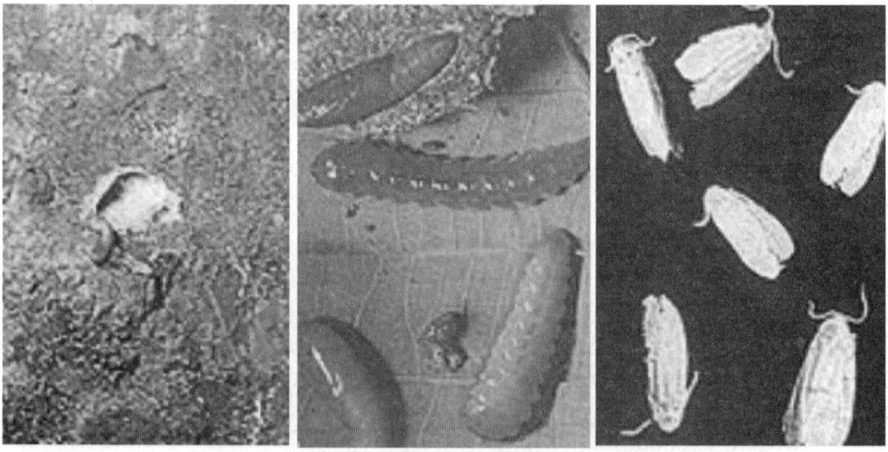

Fig. 8.35. Bark eating caterpillar on rubber.

ii) Management: Remove the mud galleries along with the larvae. Mix 100 g carbaryl in a litre of water. Dip a paint brush in this solution and swab around the affected portion.

When the infestation is severe, apply carbaryl 5% dust at the rate of 10 kg per hectare or fenval 0.4% dust at the rate of 7 kg per hectare with a power duster. Spraying the trunk with fenvalerate 0.02% is also effective.

8.8.1.2. Cockchafer Grub, *Holotrichia serrata, H . rufoflava, Anomala varians*

The pest is more prevalent near forest areas and in loose soils.

i) Damage: Cockchafer grub (Fig. 8.36) feeds on the roots of seedlings in the nursery and young plants. The affected plants droop and fall over.

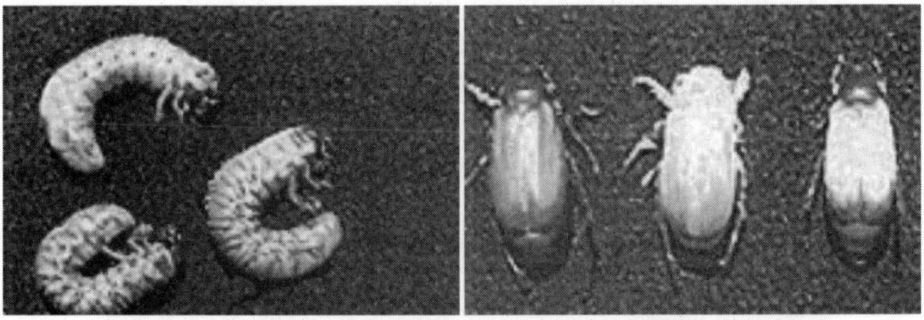

Fig. 8.36. Cockchafer grubs and adults.

ii) Management: Incorporate 25 kg per hectare of phorate 10G in the soil at the time of bed preparation.

8.8.1.3. Mealy Bug, *Ferrisia virgata*

The pest is seen mostly in nurseries in all rubber growing areas.

i) Damage: Mealy bugs are soft bodied small insects with white mealy outer covering. Occurrence and damage are similar to scale insects. Occur on tender shoot (Fig. 8.37), suck the sap and devitalize the plant. Association with sooty mould is also observed.

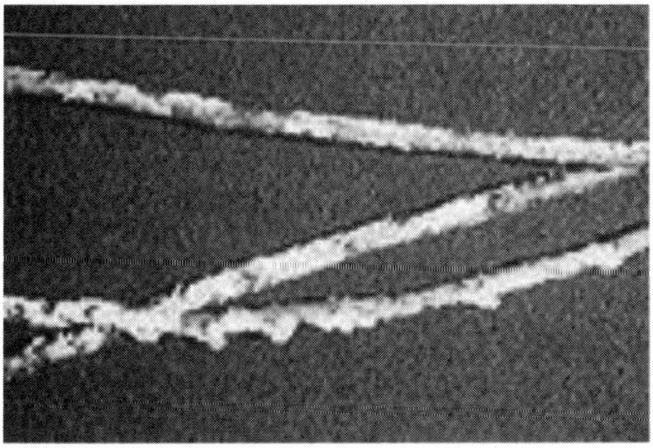

Fig. 8.37. Mealy bug infestation on rubber.

ii) Management: In the normal course, its own parasite will keep this pest under check. But, when the population increases beyond a level, then spraying with monocrotophos at 2 ml/litre of water is recommended. Spraying with organophosphorus insecticides like malathion 0.1% or quinalphos 0.075% also gives effective control of the pest.

8.8.1.4. Scale Insect, *Saissetia nigra*

The pest is seen generally in young plantations and nurseries in almost all rubber areas.

i) Damage: Scale insects are small insects with an outer black, dome-shaped covering. Occur on leaflet petioles and tender shoot portions and suck the sap (Fig. 8.38). Severely affected portions dry up and die. Ants and sooty mould are associated with this insect.

Fig. 8.38. Scale infestation on rubber stem.

ii) Management: Natural enemies like insect parasites and entomopathogenic fungi keep this pest in check. When severe infestation is noted, spray organophosphorus insecticides like malathion at 0.05% concentration.

8.8.1.5. Slugs and Snails, *Mariaellae dussumieri, Crytozona (Xestina) bistrialis*

The pest is distributed randomly in rubber areas in Kanjirappally, Thrissur and Calicut.

i) Damage: They feed on latex by lacerating the tender leaves and buds. Growth of affected buds is arrested and side shoots develop giving a bunchy appearance. Slugs drink latex from the tapping cut and collecting cup also (Fig. 8.39). Nursery seedlings are more susceptible.

ii) Management: Broadcast 2.5% metaldehyde bait pellets (snailkill 2.5) at the base of the infested plants or seedling beds or paint the base of the stem with metaldehyde. To repel slugs and snails, brush 10% Bordeaux paste around the stem above the bud union to a length of 30 cm. Repeat the application after 30-45 days if the attack continues. Spraying metaldehyde 0.05% or 0.1% of one day old suspension in the infected area is more effective than broadcasting.

Fig. 8.39. Snail infestation on rubber.

8.8.1.6. Termite (White Ant), *Odontotermes obesus*

The pest is distributed in dry regions of Central Kerala (Thrissur and Palakkad) and non-traditional areas like Dapchari in Orissa.

i) Damage: Feeds on the dead bark of trees and young plants (Fig. 8.40). Builds covered passageways of soil on the tapping panel and collection cup. Sometimes young plants dry up due to attack.

Fig. 8.40. Termite infestation on rubber.

ii) Management: Drench the soil at the base of affected plants with chlorpyriphos 0.1% solution. When mulch is present, spray the mulch also.

8.8.2. Mite Pests

8.8.2.1. Mite, *Hemitarsonemus dorsalis*

i) Damage: Mites are responsible for crinkling and shedding of leaves.

ii) Management: Spraying sulphur 50WP at 2g/litre of water gives effective control of mites.

8.9. CASHEW NUT

India is the largest producer, importer, processor, exporter and second largest consumer of cashew. In cashew nut, area has increased from 0.176 million hectares in 1961 to 0.843 million hectares in 2005-06. The production in cashew has gone up from 0.079 million tonnes in 1961 to 0.579 million tonnes in 2005-06 (Bijay Kumar, 2007) (Table 8.9). India exported cashew kernels worth US$ 553 million during 2006-07. Export of cashew is rising at 27% per annum. These export earnings are exceeded only by coffee and rice among agri-exports.

Table 8.9. State-wise area, production and productivity of cashew nut in India (2005-06)

State	Area in '000 ha	Production in '000 MT	Productivity in MT/ha
Kerala	80.0	67.0	0.837
Tamil Nadu	121.0	56.0	0.479
Karnataka	100.0	45.0	0.450
Andhra Pradesh	170.0	92.0	0.541
Maharashtra	160.0	183.0	1.143
Orissa	120.0	78.0	0.650
West Bengal	10.0	10.0	1.000
Assam	14.0	10.0	0.642
Goa	55.0	27.0	0.490
Gujarat	4.0	4.0	1.000
Meghalaya	6.4	6.4	1.000
Pondicherry	3.0	1.0	0.333
Total	843.4	579.4	0.7

8.9.1. Insect Pests

8.9.1.1. Tea Mosquito Bug, *Helopeltis antonnii, H. schoutedeni, H. anacardii*

i) Damage: Tea mosquito bugs, also known as *Helopeltis* bugs or mirid bugs, are the most important and serious pests of cashew. These bugs are slender, delicate insects, about 7-10 mm long with long legs and antennae, the antenna being nearly twice as long as the body. The females are red and the males are brown to yellowish red. They lay eggs inserted into the soft tissue near the tips of flowering or vegetative shoots. Nymphs (immature bugs) are yellowish in colour. Both adults and nymphs feed on young leaves, young vegetative and flowering shoots, and developing fruits.

Feeding on the stalks of the tender shoots causes elongated green lesions, sometimes accompanied by exudation of gum. Severely damaged shoots show die-back due to the effect of bug saliva in combination with fungi, which enter the plant tissue through the feeding lesions (Fig. 8.41); the subsequent development of numerous auxiliary buds causes a bunched terminal growth known as 'witches' broom'. In case of serious infestations, the trees may appear as if scorched by fire.

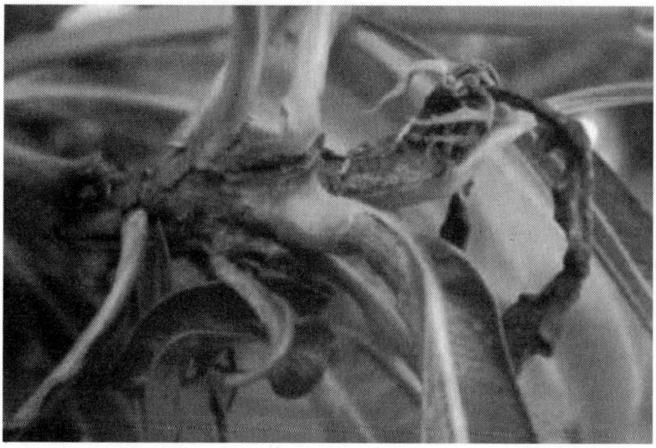

Fig. 8.41. Typical damage of tea mosquito bug on a young shoot of cashew by *Helopeltis* bugs. Note necrosis on the shoot (Courtesy: A. M. Varela, ICIPE).

Attacked leaves are deformed and show angular lesions, particularly along the veins, which may drop off, so that the leaves appear as if attacked by biting insects. Bug feeding on developing apples and nuts causes brown sunken spots. The growth of trees is seriously retarded and fruit formation of attacked flowering shoots is reduced. The production loss from the tea mosquito bug is estimated to be about 30 per cent.

ii) Management

- Monitor the crop regularly. *Helopeltis* attack occurs very suddenly and great vigilance is very important to control this pest, particularly during the rainy season or when water is available leading to flushing (production of young shoots) when *Helopeltis* populations normally build up.
- Conserve natural enemies. Weaver ants build nests on cashew trees providing good protection against this and other bug pests (Fig. 8.42).

Fig. 8.42. Weaver ants foraging on cashew apples and nuts (Courtesy: A. M. Varela, ICIPE).

- Do not interplant cashew with crops that are host for *Helopeltis* bugs, such as cotton, tea, sweet potato, guava and mango.
- The pest can be easily controlled by spraying 0.05% endosulfan (15 ml in 10 litres of water). Three sprays should be given for complete control of the pest. The first spray of endosulfan should be given in October-November when the new flushes emerge, second of systemic insecticides like monocrotophos and dimethoate in December-January during emergence of inflorescence, and followed by third spray of endosulfan in February at the time of fruit set.

8.9.1.2. Coconut Bug, *Pseudotheraptus wayi*

i) Damage: Adult bugs are reddish brown in colour and 12 to 14 mm long. Nymphs are red brown to green brown in colour and have long antennae. Bug feeding causes necrotic bruise-like depressions; a hard lump develops, which can be easily removed when the fruit is peeled. The bug sucks on the developing fruits causing pock marks. The kernels are also affected showing spots, which lower their market value.

ii) Management

- Conserve natural enemies. Weaver ants nest on cashew trees deterring and feeding on coconut bugs.

8.9.1.3. Cashew Weevil, *Mecocorynus loripes*

i) Damage: The cashew weevil is a large weevil (Fig. 8.43), about 20 mm long, and of a knobbed appearance. It is dark grey-brown in colour. The female weevil lays eggs singly in small holes in the bark of the trunk or branches. The larvae are legless grubs, whitish in colour with a brown head. They bore through the bark and move downwards tunneling under the bark while feeding on the sapwood. Brown-black gummy frass is seen on the trunk and main branches. Heavily attacked trees become ringed by damaged sapwood and eventually die. Neglected plantations are likely to be severely attacked. Fully-grown larvae pupate in a chamber about 2 cm below the bark.

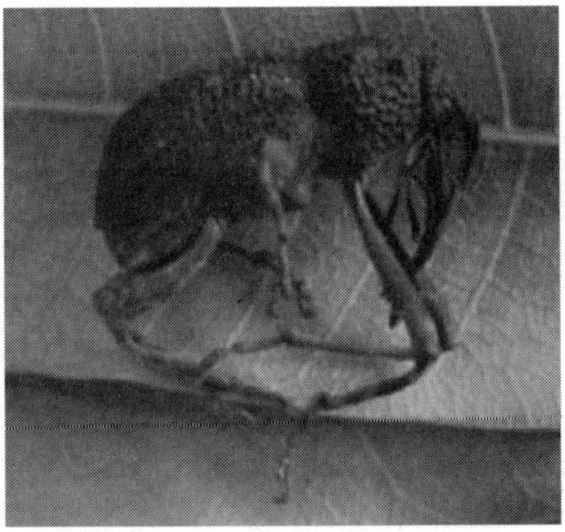

Fig. 8.43. Adult cashew weevil (*Mecocorynus loripes*). Real size: about 20 mm long (Courtesy: A.M. Varela, ICIPE).

ii) Management

- Cut away bark from damaged areas of lightly infested trees and kill the larvae and pupae underneath. Repeat this every month for a further six months if required
- Destroy severely infested trees. First collect and destroy all adult weevils; then fell the tree and remove the bark to expose all larval galleries; kill all larvae and pupae and burn the tree.

8.9.1.4. Red-Banded Thrips, *Selenothrips rubrocinctus*

i) Damage: Adults of the red banded thrips are dark brown or blackish. Nymphs are pale yellow with a broad transverse red band on the dorsal side of the abdomen (Fig. 8.44). Thrips attack older leaves, flowers and shoots. Attacked leaves drop off leaving bare shoots with few young leaves at the tip. Infestation of flowers causes poor fruit formation. Locally limited infestations may cause considerable damage.

Fig. 8.44. Immature stage of the red-banded thrips. Note typical transverse red band on the abdomen (Courtesy: A. M. Varela, ICIPE).

ii) Management

- Conserve natural enemies. Anthocorid bugs are important in natural control of thrips

8.9.1.5. Mealy Bugs, *Pseudococcus longispinus*

i) Damage: The long-tailed mealy bug (*Pseudococcus longispinus*) attacks shoots, inflorescences, apples and nuts. Affected parts appear completely white. Trees infested during the flowering stage fail to produce fruits, whereas those infested at the nut swelling stage produce discoloured nuts, which result in a lower grade. However, cutting tests showed no difference in kernel outturn between clean and discoloured nuts. Mealy bugs have been a problem for cashew growers in Tanzania.

ii) Management

- Conserve natural enemies. Mealy bugs are usually controlled by a wide range of natural enemies. However, use of pesticides may kill these natural enemies leading to mealy bug outbreaks.

8.9.1.6. Stem Girdler, *Paranaleptes reticulata*

i) Damage: Adults are long horn beetles, with a body length of 25-35 mm and with antennae longer that the body. The head and the thorax are dark brown; the wing cases are orange with large black blotches giving them a reticulate appearance. Adult beetles girdle branches from 3-8 mm in diameter leaving a V-section cut; only a narrow, central pillar round the pith zone is left, which eventually breaks off. Female beetles lay elongated eggs in transverse slits made in the bark of the girdled branch at points above the girdle. Larvae are yellow in colour, 45 mm in length when fully grown. They mine in dead wood of the girdled branches. Pupation takes place in the dead wood. The life cycle takes one year. Neglected plantations may be severely damaged.

ii) Management: Once a year (in November or December) all girdled branches should be collected and burned. Only the dead or dying part of the branch above the girdle needs to be collected.

8.9.1.7. Stem and Root Borer, *Placaederus ferrugineus*

i) Damage: This is a serious pest of cashew and is capable of killing the tree outright. The adult is medium sized, reddish brown longicorn beetle, the head and thorax of which are dark brown or almost black. The female lays eggs into the crevices of loose bark in the trunk or the exposed portion of the roots of cashew trees. The grubs that hatch out bore into the bark and feed on the soft tissues (Fig. 8.45). The attack could be identified by presence of small holes in collar region, gummosis, extrusion of chewed up fibres and excreta and discolouration of bark. If no control measures are taken up, the leaves start yellowing followed by drying of branches and finally resulting in the death of the tree. Cashew stem and root borer is responsible for killing 5-10 percent of the productive trees.

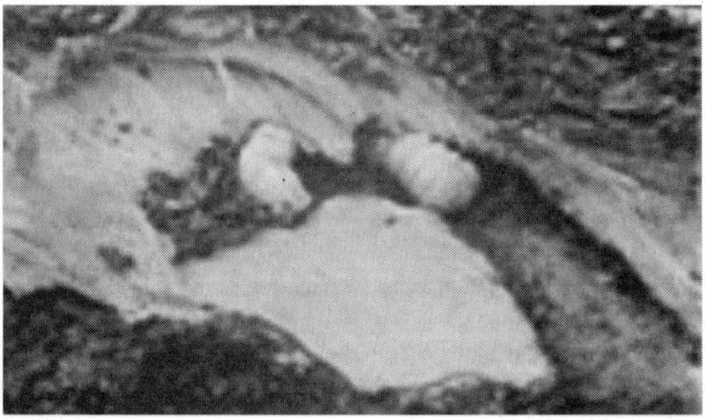

Fig. 8.45. Stem borer on cashew nut.

ii) Management: It is very difficult to control the pest in the advanced stage. Therefore, control measures should be taken up at early stage of infestation. For effective control of the pest, swab the trunk of the tree with coal tar and kerosene (1: 2) up to one metre height from the ground. Treat the soil with 500 g BHC 10% to a radius of 45 cm. The trees showing the severe infestation should be removed along with roots and burnt. Heaping the organic waste around the trunk should be avoided.

The grubs are removed by peeling the bark with chisel and hammer and destroyed. The peeled portion of the bark is smeared with Bordeaux paste and the trunk and exposed roots should be swabbed with carbaryl emulsion at 10 g/l of water. Drench the soil around the trunk of the tree with carbaryl at 4 g/l of water. Keep the orchard clean.

8.9.1.8. Fruit and Nut Borer, *Thylocoptila paurosema*

i) Damage: It is a pinkish dark actively moving caterpillar which bores into the tender apples and nuts. The entrance of the hole is plugged with fecal matter. Nuts are damaged in case of severe infestation. The caterpillars attack the fruits at all stages and cause the shriveling and premature fall of nuts.

ii) Management: This pest can be effectively controlled by spraying 0.05% endosulfan.

8.9.1.9. Twig Borer, *Niphonoclea albata*

i) Damage: This insect pest is common during the dry season. The adult beetle girdles the small branches causing them to dry up or break and drop to the ground. Its creamish larvae bore into the pith of the branches. As they feed, they move downward until they pupate. All affected twigs and small branches may eventually die.

ii) Management : Remove or collect all affected twigs as well as dried twigs on the ground. Dispose them properly by burning before applying chemical sprays with malathion, carbaryl and/or methyl parathion. Repeat application at 7-10 days interval when necessary.

8.9.1.10. Leaf Miner, *Acrocercops syngramma*

i) Damage: Young plants in the nursery and in the orchard are more affected by these pests. Caterpillars of this silvery grey moth mine through the tender leaves, thus, severely damaging them.

ii) Management: Spray 0.05% phosphamidon as soon as infestation is detected on new leaves.

8.9.1.11. Saw-Toothed Grain Beetle, *Cryzaephillus surinamensis*

i) Damage: This pest is known to attack the nuts during storage.

ii) Management: Nuts should be thoroughly dried and placed in air tight containers. Surface treatment is recommended. For finished products, fumigation is recommended.

8.9.1.12. Termites, *Neotermes gardneri* and *Trinevitermes biformis*

i) Damage: Termites attack the roots and the trunk of cashew trees. They burrow on the bark of roots and branches especially of old trees (Fig. 8.46). They build their soil mounds or nest on dead parts of the tree trunks and feeds on tree bark. They feed on roots that are remaining underground. Trees succumb when the population is high.

Fig. 8.46. Termite infestation on cashew nut tree.

ii) Management: Soil mounds must be destroyed and the tree trunks should be swabbed with coal tar. The queen should be located and killed either mechanically or by spraying with 2% Chlordane. Chlordane should not be applied on living parts of the tree because of its long residual effect. Cistin powder could be applied to any part of the tree infested with termites.

8.9.1.13. Bark Eating Caterpillar, *Indarbela tetraonis*

i) Damage: Caterpillars bore short distances into the trunk or branches, usually at the forks and takes refuge during the day. During nights, larvae comes out and feeds on bark of the tree moving in the rope like web gallery of wood, silk and excreta (Fig. 8.47). Small trees may succumb and large trees fail to put forth a new flush when the infestation is severe.

Fig. 8.47. Bark eating caterpillar infestation on cashew nut.

ii) Management : Remove the frass and swab the trunk with carbaryl 50% WP emulsion at 10 g/litre of water.

8.9.1.14. Leaf and Blossom Webber, *Lamida moncusalis*

i) Damage: Leaf and blossom webber occur during emergence of new flushes and panicle. Caterpillars web the shoot and inflorescence together.

They remain inside and feed. Apples and nuts are also covered with webs and their surfaces are scraped (Fig. 8.48). Nuts crack and fail to develop. The shoot in the webbed portion dries up.

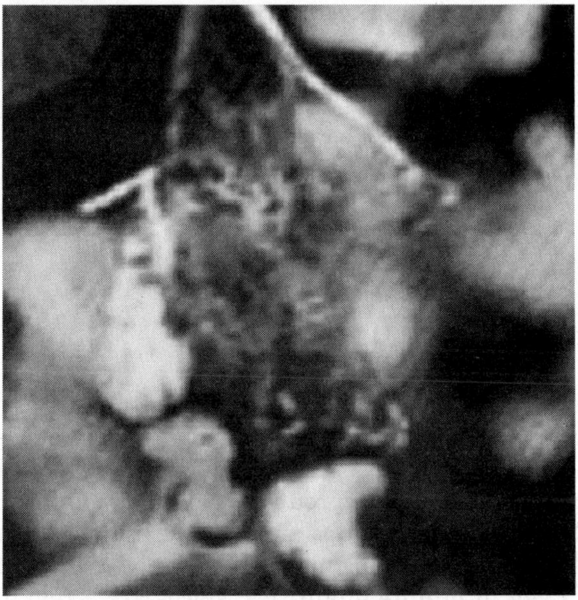

Fig. 8.48. Blossom webber infestation on cashew nut.

ii) Management : Spraying of endosulfan or chlorpyriphos at 2 ml/litre of water gives effective management of the pest.

8.9.2. Vertebrate Pests

i) Damage: Birds, deer, squirrel, monkey, wild boar, porcupine damaged cashew in hill region of Karnataka. Seedlings are damaged by rat, wild boar and porcupine. The total loss due to vertebrate pests in cashew at Mudigere, Chikmagalur, in Karnataka was 17% of apple and 21% of nut during 1991-92. A troupe of 8 monkeys on an average, caused 24% nut losses (Chakravarthy, 1993).

Chapter 9

SPICE CROPS

From time immemorial, India is regarded by the rest of the world as the "magic land of spices". Spices are natural plant products used to improve flavour, aroma, taste and colour of food products, beverages, liquors, pharmaceutical, cosmetic and perfumery products. No other country in the world has such a diverse variety of spice crops as in India. Indian spices are also noted for their excellent aroma, flavour and pungency not so easily matched by any other country. Even in minute quantities, spices are a real delight to all our sensory organs, making the food dishes more palatable. Usage of spices in food processing industries is increasing quite fast in all the countries. Hence, the world demand for spices in recent years is continuously on the rise.

India is the largest producer, consumer and exporter of spices in the world. India is the largest producer in the world of ginger, turmeric and black pepper. The annual production of spices is more than 5 million tonnes valued at US$ 1,500 million. India's share of the world spices trade is estimated at 45 to 50% by volume and 25% by value. The present annual production of spices in the country is 5.107 million tonnes from over 2.857 million hectares (Bijay Kumar, 2007) (Table 9.1). The spices contributed 2.76% of total horticulture production in 2005-2006. About 8.5% of India's export earnings from agricultural and allied products come from spices. Indian spices flavour foods in over 134 countries. The important spice crops include black pepper, cardamom, ginger, garlic, turmeric and chilli.

Table 9.1. State-wise area, production and productivity of major spices in India (2005-06)

State	Area in '000 ha	Production in '000 MT	Productivity in MT/ha
Andhra Pradesh	326.9	920.5	2.8
Tamil Nadu	130.0	784.0	6.0
Karnataka	258.8	619.8	2.4
Gujarat	426.1	505.6	1.2
Madhya Pradesh	292.3	345.7	1.2

Rajasthan	347.9	297.1	0.9
Uttar Pradesh	91.0	283.8	3.1
West Bengal	108.0	226.7	2.1
Orissa	145.6	195.3	1.3
Kerala	428.1	191.8	0.4
Chhattisgarh	28.9	168.2	5.8
Maharashtra	112.5	106.0	0.9
Others	162.1	463.2	2.9
Total	2857.7	5107.7	1.8

Since spices form a part of traditional/ethnic medicines, the demand for organically produced spices is growing. Export of organic spices from India has started in right earnest. The country at present exports around 50 tonnes of different varieties of organic spices. Export will get a significant boost in the coming years as more and more farmers switch to organic methods.

Export of Spice Produce

India is the largest producer, consumer and exporter of spices and spice products in the world and produces more than 50 spices. Spice export figures for the year 1996-97 and 2006-07 shows the phenomenal growth of the Indian spice industry. The Indian spice export for the year 1996-97 was 2,25,000 metric tonnes which was worth $350 million; in the year 2006-07 the spice exports touched the figure of 3,75,000 MT which was worth US$ 800 million. Thus, these figures show the exponential growth of the Indian spice industry in a very short time. Spice exports from India have touched US $ one billion mark during 2007-08. The major destination of Indian spice exports is USA followed by European Union, Sri Lanka, Japan and Middle East Countries.

9.1. BLACK PEPPER

Black pepper (*Piper nigrum*), known as the "King of spices", has its origin in the South-Western hills of India. It is a branching and climbing perennial shrub belonging to the family Piperaceae and is grown in the hot and humid parts of the world. Pepper is loved for its spicy factor, piperine. It is used in various culinary seasonings, as a preservative for meat and fish, in medicine and perfumery (pepper oil). In India, it is a common practice to use live trees as standards for climbing pepper. Pepper is the most important spice produced in the country and accounts for 50-60% of the total export earnings. It is being cultivated in 0.223 million hectares producing 0.070 million tonnes of black pepper with an average yield of 1.3 tonnes per hectare.

9.1.1. Insect Pests

9.1.1.1. Pollu Beetle, *Longitarsus nigripennis*

i) Damage: Adult beetles feed and damage tender leaves and spikes. Grubs bore into and feed on internal tissues. The infested spikes turn black and decay (Fig. 9.1).

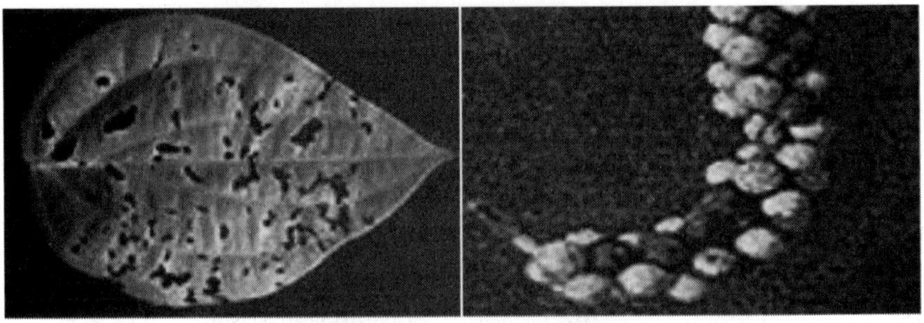

Fig. 9.1. Pollu beetle damage on black pepper leaf and spike.

ii) Management : Regulation of shade in the plantation reduces the population of the pest in the field. The pests can be managed by spraying endosulfan or quinalphos each at 0.05% during June-July and September-October. Neemgold (neem-based insecticide) at 0.6%, neem oil at 0.4% during August, September and October are also found effective for the management of the pest. The underside of leaves (where adults are generally seen) and spikes are to be sprayed thoroughly. An integrated strategy involving regulation of shade and spraying of Neem Gold at 0.6% at 21 day intervals during June-October was found effective against pollu beetle.

9.1.1.2. Top Shoot Borer, *Cydia hemidoxa*

i) Damage: Larvae bore into tender terminal shoots and feed on internal tissues, resulting in blackening and decay of affected shoots.

ii) Management : Spraying monocrotophos or endosulfan each at 0.05% on tender terminal shoots and repeating at monthly intervals during July-October protects the emerging new shoots against the pest.

9.1.1.3. Leaf Gall Thrips, *Liothrips karnyi*

i) Damage: Leaf margins curl downwards and inwards resulting in formation of marginal leaf galls. Infested leaves become crinkled and malformed (Fig. 9.2).

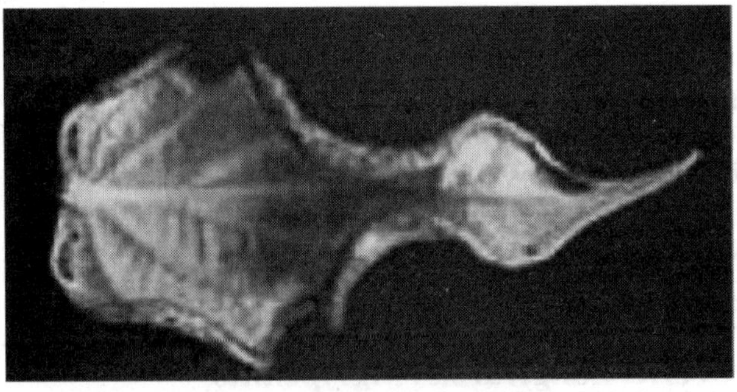

Fig. 9.2. Leaf gall thrips damage on black pepper.

ii) Management: Spraying monocrotophos or dimethoate each at 0.05% during emergence of new flushes in the field or cuttings in the nursery gives effective control of the pest.

9.1.1.4. Scale Insects, *Lepidosaphes piperis, Aspidiotus destructor, Lecanium* sp.

i) Damage: Scale insects are sedentary and appear as encrustations on stem, leaves and berries (Fig. 9.3). Insects feed on plant sap and cause yellowing and wilting of infested portions.

Fig. 9.3. Scale infestation on black pepper leaf.

ii) Management: Clipping off and destroying severely infested branches reduce the pest incidence. Spraying monocrotophos or dimethoate each at 0.05% during early stages of pest infestation and repeating the spray after 21 days is effective in controlling the pest. Scales are controlled by shade management during May-June and application of neem oil (0.6%) during October-November. The pest can be managed by spraying tobacco decoction. Spraying of neem-garlic suspension at 2% was found effective. *Chilocorus nigrita* is a potential natural enemy against scale insects.

9.1.1.5. Leaf Feeding Caterpillar, *Synegia sp.*

i) Damage: The pest damages the leaves and spikes of younger vines.

ii) Management: Spraying 0.05% quinalphos gives effective control of leaf feeding caterpillar.

9.1.1.6. Mealy Bug, Gall Midge and Aphids

i) Damage: Mealy bugs, gall midges and aphids infest tender shoots especially in nurseries.

ii) Management: Spraying 0.05% monocrotophos if the infestation is severe gives effective control of these pests. Mealy bug infestation on roots can be controlled by drenching the soil with 0.075% chlorpyriphos.

9.2. CARDAMOM

Cardamom (*Elettaria cardamomum*), also known as the "Queen of Spices" enjoys a unique position in the world market has its origin in the evergreen rain forests of Western Ghats of South India. India enjoyed near monopoly in the field till recently. Guatemala has emerged as a competitor to

India. India and Guatemala are the main producers and exporters of cardamom. Other cardamom growing countries are Tanzania, Sri Lanka, El Salvador, Vietnam, Laos, Kampuchea and Papua New Guinea. The area under cultivation in India during 2004-05 was 73,237 ha, with an annual production of 11,415 metric tonnes. The export earning during 2006-07 was Rs. 274.7 million by exporting 650 million tonnes of small cardamom. Kerala, Karnataka, and Tamil Nadu produce cardamom in India. While Kerala accounts for 60 per cent of the production, Karnataka contributes 30 per cent, and Tamil Nadu 10 per cent. In Karnataka, cardamom is cultivated mostly in Kodagu, Hassan, and Chikmagalur districts. The plants last for 15 years. They grow in moist, elevated, and shady forest loams. They require a fairly distributed annual rainfall ranging from 2,500 mm to 5,000 mm and are found within an altitude ranging between 600 m and 1,200 m above the mean sea level. It is used for flavouring various food preparations, confectionery, beverages, liquors and medicines.

9.2.1. Insect Pests

9.2.1.1. Rhizome Weevil, *Prodioctes haematicus*

i) Damage: This is a serious pest in the secondary nursery especially where seedlings are raised continuously year after year. The grubs feed on the rhizome and basal portion of the stem. This results in drying of leaves and breaking of stem at the base.

ii) Management: Drenching the nursery beds with chlorpyriphos at 0.04% can control the pest.

9.2.1.2. Shoot Fly, *Formosina flavipes*

i) Damage: The pest is observed in the nursery during January to May. Dead heart or decay of the central spindle is the external symptom.

ii) Management: Spraying of 0.025% quinalphos or application of phorate granules at 1 g a.i./m^2 is recommended for control of the pest.

9.2.1.3. Thrips, *Sciothrips cardamomi*

i) Damage: This insect is a serious pest of cardamom. Thrips are pronounced during dry months of December-May. Injury is caused by thrips feeding in young leaf sheaths and basal ends of unopened flower bracts. Nymphs and adults feed with their rasping-sucking mouth parts, which lacerate young tissues and suck the juices that ooze from ruptured cells. First the injured area develops a silvery sheen because of air occupying the emptied cell cavities. Due to oxidation, this area becomes a mixture of white, yellow, and brown splotches and streaks. Infestation on the panicle and flower buds results in stunted growth of panicles, shedding of flower buds and warty growth on the surviving capsules (Fig. 9.4). The infested capsules are light in weight, inferior in quality and fetch very low price in the market.

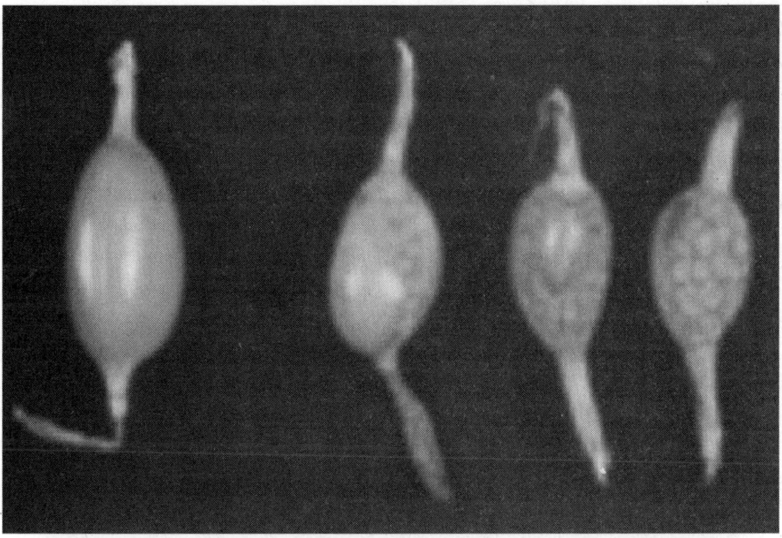

Fig. 9.4. Thrips damage on capsules of cardamom.

The cardamom thrips are difficult to find, since feeding is confined to young tissues in dark, confined areas such as the sheaths of young leaves, the base of unopened flower bracts, and possibly the spindle of top shoots. The nymphal stages are sluggish in movement. Adults are also sluggish and rarely take to flight. *Sciothrips cardamomi* is suspected to be a vector of mosaic disease that affect cardamom in India.

ii) Biology: The life cycle (egg to adult) is completed in 25 to 30 days. Reproduction is mostly sexual, though virgin females are able to lay viable eggs during the summer. The white kidney shaped eggs are inserted in leaf sheaths or flower bracts by the sharp ovipositor of the adult. Eggs hatch in about a week. The transparent larvae are approximately 1/25 inch in length. The larval stage consists of three larval stages separated by molts in the course of 15 to 21 days. Pupae resemble the larvae except for the presence of wing pads (incompletely developed wings). There are 2 pupal stages, the prepupa and pupa, neither of which are feeding stages. Pupae become adults in 10 to 15 days. The head and abdomen of this thrips are dark grey brown and the thorax and legs are pale yellowish brown. Females measure approximately 1/4 inch in length; males are slightly smaller.

iii) Management: General cultural practices involving sanitation, such as the removal of alternative hosts and plant debris within and outside the field and regulation of shade helps in reducing the pest infestation. The removal of dried shoot peels at the base of the stem resulted in a reduction of thrips populations by one-third.

Chemical control is difficult because the thrips remain hidden under the cover of leaf sheaths and bracts. The synthetic pyrethroids gave adequate control. Of the different synthetic pyrethroids, permethrin was the most effective.

Since the pest population is high during dry months from December to May, pesticide application during this period is important. Four sprays with either monocrotophos 0.025% or dimethoate 0.05% or quinalphos 0.025% during March- September gives effective control of thrips. Application of fish oil rosin soap also is beneficial to manage thrips damage.

Cardamom cv. Mudigere-1 (Malabar) is tolerant to thrips.

9.2.1.4. Shoot/Capsule Borer, *Conogethes punctiferalis*

i) Damage: It is a serious problem to cardamom growers of Kerala, Tamil Nadu and Karnataka. At the early stage of the crop, the caterpillars of this yellow coloured moth bore into the core of the aerial stem (Fig. 9.5) resulting in the death of central spindle, which appears as characteristic dead hearts. The caterpillar bores into the stem and feeds on the internal contents. This results in the decay of the central spindle and production of dead heart. Fecal matter of the caterpillar can be seen coming out through the holes.

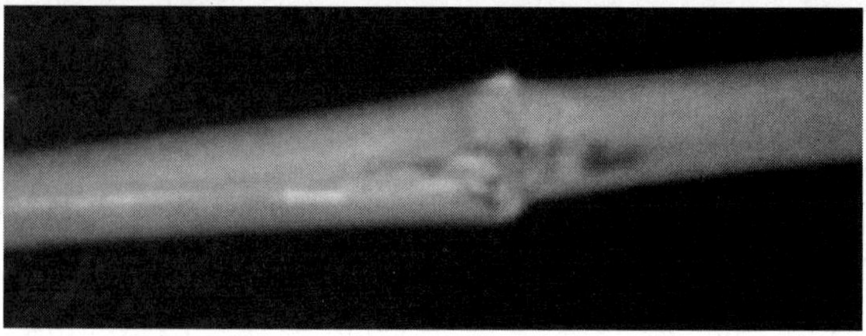

Fig. 9.5. Shoot borer infestation on cardamom.

At the time of flowering, when the caterpillars attack the panicles and spikes it may lead to flower shedding and drying up of the attacked portions. At a later stage of the crop, the caterpillars bore into the capsules, feed on the seeds and make them hollow. The presence of excreta at the region of attack indicates presence of the caterpillars in the pseudostem, inflorescence and pods.

Pest infestation is pronounced in three seasons viz. January-February, June and September-October.

ii) Management: Late stages of larvae bore into the pseudostem and remain there. For an effective management of the pest, the insecticides have to be targeted on early stages of the larvae, which are usually present within 15-20 days after adult emergence in the field. Spraying fenthion 0.05% or monocrotophos 0.05% is recommended during the months of February-March and September-October. Spraying with quinalphos 0.025%, carbaryl 0.1%, monocrotophos, fenthion or dimethoate at 0.05% or endosulfan or phenthoate at 0.1% is recommended against the shoot borer.

As soon as bore holes of stem borer are noticed, injection of *Bacillus thuringiensis* preparation into the bore hole (0.5 ml in 10 ml water) will kill the larvae so that subsequent resurgence can be reduced.

Cardamom cv. Mudigere-1(Malabar) is tolerant to shoot borer.

9.2.1.5. Cardamom White Fly, *Kanakarajiella (Dialeurodes) cardamomi*

i) Damage: It is a serious pest in cardamom growing tracts of Kerala. The adult is a small soft-bodied insect, about 2 mm long and having two pairs of white wings. The nymphs are elliptical and pale green. The nymphs secrete sticky honeydew, which drops on to lower leaves. On these, black sooty mould develops, which interrupts photosynthesis of the leaves.

ii) Management: The flies are attracted towards yellow colour. So metal sheets painted yellow and coated with sticky materials, such as castor oil or poly-vinyl butanol would serve as traps. By placing such yellow sticky traps between rows of cardamom plants, population of adults can be monitored and adults trapped to some extent. Nymphs are effectively controlled by spraying the lower surface of leaves with a mixture of neem oil (500 ml) and triton (500 ml) in 100 litres of water. Acephate 0.075% and triazophos 0.04% are equally effective. The spray may be repeated two or three times at 15 days interval.

Nymphs can be controlled by spraying neem oil with soft soap (500 ml neem oil and 500 g soft soap in 100 litres of water).

9.2.1.6. Root Grub, *Basilepta fulvicorne*

i) Damage: The grubs of a small, greenish blue beetle cause damage. The grubs are short, stout, pale white in colour and often assume a shape resembling 'C', which feeds on cardamom roots. The symptoms start as yellowing of leaves, which later result in the drying up and death of the plant.

ii) Management: Collect the beetle with hand nets or sticky traps at the time of mass emergence (March-April and August-September) and destroy. Early stages of the grub which are usually present in soil during May-June and September-October can be controlled either by drenching chlorpyriphos 0.04% at 3-4 litres per clump or by applying phorate at 2-4 g a.i./clump 10-15 cm around the plant.

The population of root rot grubs in cardamom field was reduced significantly by soil application of carbofuran 3G at 100 g/plant followed by spraying imidachloprid at 0.5 ml/litre of water.

9.2.1.7. Scale, *Aulacaspis* sp.

i) Damage: This scale insect is found on the lower surface of leaves, leaf sheath, panicles and fruit stalk. As a result of damage, capsules get shriveled, panicles become dry and the leaves become yellow. The pest is mostly seen during summer months.

ii) Management: Spray monocrotophos or fenthion at 0.05 % during the peak season.

9.2.1.8. Aphids, *Pentalonia nigronervosa*

i) Damage: Aphids (Fig. 9.6) are vectors of Katte virus disease. Colonies of aphids are seen concealed inside leaf sheath of old pseudostem. Aphids suck the sap and devitalize the plant.

Fig. 9.6. Aphid on cardamom leaf.

ii) Management: Spraying of 0.05% dimethoate gives effective management of thrips.

9.2.2 Vertebrate Pests

i) Damage: Wild boar, monkey, rodents and birds are important vertebrate pests of cardamom in South India. Rodents and squirrel constitute rodent pests incurring on an average 10% yield losses in cardamom. Cultivation of cardamom is being severely affected in Koppa, Sringeri and Agumbe in Ckikmagalur district; Thirthahalli in Shimoga district and at Madai, Monnagiri and Galibeedu in Kodagu district of Karnataka due to monkey menace. Damage due to monkeys varied from 2.80 to 17.50% to cardamom suckers.

ii) Management: Rodent pests can be controlled by clean cultivation, overlapping of panicles from August, timely harvests and mulching which saved 2.56, 0.21, 5.54 and 4.20% capsules, respectively. Locally made wooden snap traps baited with germinating paddy seeds reduced capsule damage from 8.2 to 4.90%. These practices are cheap, practicable and eco-friendly.

Cardamom nurseries can be protected from wild boar damage for 45 days using a mixture of 5 g each of phorate 10G and sand and suspending the same in punched polybags (10 cm x 10 cm size) along the borders of cardamom nursery (at 3 m distance and 2 m height). Selective shooting would ameliorate the wild boar problem.

Selective debranching of shade trees and watch and ward during early morning (before 7 a.m.) and late evening (after 5 p.m.) hours significantly reduced clump damage from 49 to 3.2% due to monkey menace in cardamom at Mudigere, Ckikmagalur, Karnataka.

Integrated Management: Three cultural practices, viz., dispersion of panicles in an overlapped fashion, clean cultivation (without weeds) and timely harvest together with snap trapping and mulching saved 14.40% cardamom capsules from rodent pests (Srihari and Chakravarthy, 1998). Suspending a mixture of 5 g of phorate 10G granules + 5 g sand in punched polybags along borders of cardamom nursery + good drainage + watch and ward reduced damage due to wild boar from 10-15% to 3-5% seedling damage during summer (Chakravarthy, 1993).

Integrated Pest Management

- Trashing the clumps of old peels, leaves and suckers during May-June, spraying with monocrotophos (June-July) 35% EC at 1.25 ml/litre up to 1.5 m plant height followed by another spray with Zolone 25% EC at 2ml/litre 30 to 40 days later, removal of border affected suckers leaving 5 cm from the ground, were found effective in suppressing thrips and other major pests of cardamom.
- Néemark 15G and Nimbecidine were at par with monocrotophos in reducing the thrips damage.
- Root grubs (*Basilepta* sp.) infestation in nursery could be effectively checked by incorporating carbofuran 3G at 15 to 20 kg/ha.
- Damage by rodents and birds could be considerably minimized by adopting trapping, clean cultivation, timely harvest and by baiting with Bromodiolone (0.005%) or Rodafarin cake.
- Use pest tolerant varieties for planting [Cardamom cvs. Mudigere (Malabar) against shoot borer and thrips].
- Use yellow sticky traps to control whiteflies at 15 to 20 traps/ha.
- Spray neem oil at 3-5 ml/litre against whiteflies.

9.3. GINGER

Ginger (*Zingiber officinale*), an herbaceous perennial, belonging to the family Zingiberaceae, is valued for its rhizome. The country of its origin is not definitely known but is presumed to be either from India or China. India is the largest producer (3,05,900 metric tonnes) and exporter of dry ginger in the world (export earning of Rs. 191 million). The other ginger producing countries are Jamaica, Sierra Leone, Nigeria, Southern China, Japan, Taiwan and Australia. In India, It is being cultivated in 85,100 hectares producing 0.306 million tonnes of ginger with an average yield of 3.6 tonnes per hectare.

9.3.1. Insect Pests

9.3.1.1. Shoot Borer, *Dichcrosic punctferalis*

i) Damage: The shoot-borer attacks the ginger crop in India. The larvae bore into the shoots. They can cause them to wilt and die. Caterpillar bores the rhizomes and pseudostem causing dead heart. Extrusion of frass through holes can be observed (Fig. 9.7).

Fig. 9.7. Shoot borer infestation on ginger.

ii) Management: Spraying 625 litres/ha of 0.07% endosulfan is effective for the management of shoot borer. The shoot borer can also be managed by spraying 0.1% malathion or 0.075% monocrotophos or 0.05% dimethoate or 0.3% Dipel (*Bacillus thuringiensis* product) at 21 days interval during July-October. The spraying has to be initiated when the first symptom of pest attack is seen on the inner most leaves on the pseudostem.

Pruning and destroying freshly infested pseudo stems (at fortnightly intervals) during July-August and spraying 0.1% malathion (at monthly intervals) during September-October is also effective against the pest.

9.3.1.2. Leaf Folder, *Udaspes folus*

i) Damage: Larvae of leaf roller cut and fold leaves and feed from within. The adults are medium size butterflies with brownish black wings with white spots. The larvae are dark green in colour.

ii) Management: Spraying with 0.1% carbaryl or 0.05% dimethoate once in 10 days is effective in management of the pest.

9.3.1.3. Scale, *Aspidiella hartii*

i) Damage: The scale insect has been found on rhizomes and do little damage, but may spoil the appearance of the sample. Scale insects have been reported attacking ginger plants in several countries. The scale insects suck the sap from rhizomes which shrivel and dry up (Fig. 9.8).

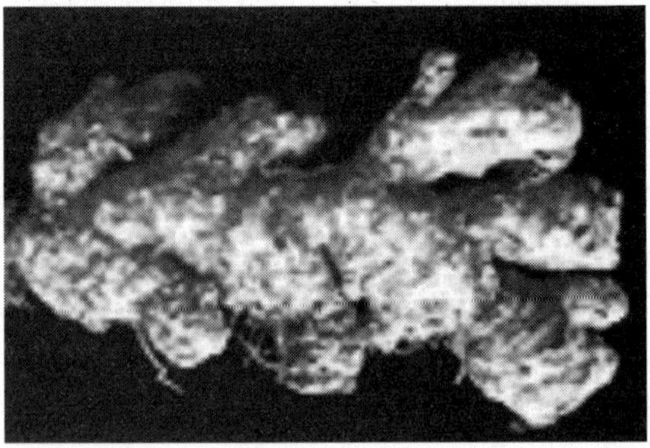

Fig. 9.8. Rhizome scale infestation on ginger.

ii) Management: Dipping of rhizomes in 0.1% quinalphos (5 ml/l of water) prior to sowing gives effective control.

Fumigating the rhizomes with carbon bisulphide, after which the scales can be removed in the process of cleaning, may kill them.

9.3.1.4. Cut Worm

i) Damage: The primary shoot is eaten just below or just above ground level. The first leaf may also be eaten. Feeding is done at night and larvae can be found in the ground at the base of the shoot during the day. Larvae grow up to 40 mm long, are grey, brown or pink and curl up when disturbed.

ii) Management: Spray when first damage is evident. Spraying in the late afternoon will give better control.

9.3.1.5. *Helicoverpa armigera*

i) Damage: Young shoots are eaten off at ground level. The caterpillar grows up to 40 mm long and varies in colour from pinkish-brown to light green, with a variable number of dark markings and longitudinal stripes along the body.

ii) Management : Control with a registered insecticide.

9.3.1.6. African Black Beetle

i) Damage: Ginger stems are gnawed approximately 5 cm below ground level, causing plant collapse. Adult beetles are dark brown to black and approximately 12 mm long.

ii) Management: Spray when first damage is evident. Spraying in the late afternoon will give better control. Insecticide should be applied with sufficient water to wet the soil down to 5 cm.

9.3.1.7. Stored-Products Pests, *Sitodrepa panacea, Lasioderma serricorne*

i) Damage: The common stored-product pests attack stored ginger. Paulose (1973) refers to them as the common drug store beetle, *Sitodrepa panacea* and the tobacco borer beetle, *Lasioderma serricorne.*

ii) Management: Fumigation of the store houses should be carried out regularly to control these pests.

9.3.2. Mite Pests

9.3.2.1. Mite, *Rhizoglyphus* sp.

i) Damage: A minute translucent mite of the genus *Rhizoglyphus* sometimes develops in large numbers in stored rhizomes, and while doing little damage may spoil the appearance of the produce.

ii) Management: A light dusting of flowers with sulphur will kill the mites and prevent reinfestation.

9.4. TURMERIC

Turmeric (*Curcuma longa*) is best known as an important spice crop. It is believed to have its origin in South-East Asia. Turmeric is valued for its rhizomes, often processed and used commercially. It is mainly cultivated in India and to a small extent in China, Indonesia, Peru and Jamaica. It is an important source of yellow dye and is also used as a colouring matter in food industries, confectionery and drugs. In India, it is cultivated in 1,49,400 ha with annual production of 5,29,000 tonnes during 2004-05. India's export earning from turmeric was Rs. 92 million annually.

9.4.1. Insect Pests

9.4.1.1. Rhizome Fly, *Mimegralla cocruleifrons*

i) Management: Disease-free, healthy rhizomes should be selected for planting. Proper drainage should be provided. Soil application of phorate or carbofuran at 1 kg a.i./ha or spraying twice with 0.05% monocrotophos gives effective control of the pest.

9.4.1.2. Rhizome Scale, *Aspidiella hartii*

i) Damage: In Africa and the South Pacific, the rhizome scale infests rhizomes in the field (at later stages) and in storage. Adult female scales are circular (about 1 mm in diameter) and light brown to grey and appear as encrustations on the rhizomes (Fig. 9.9). They feed on sap and when the rhizomes are severely infested, they become shriveled and desiccated affecting their germination.

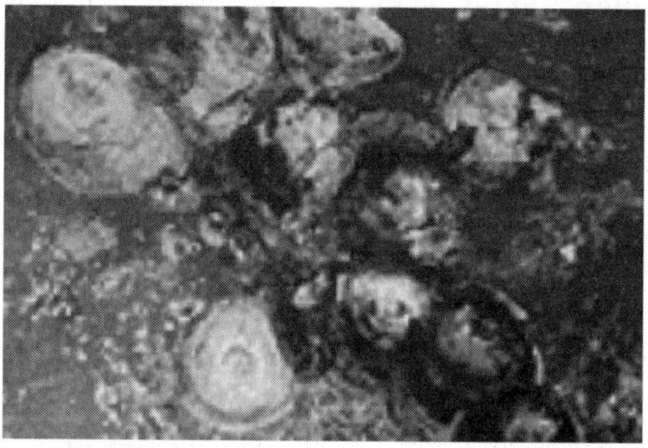

Fig. 9.9. Infestation of scale on rhizome of turmeric.

ii) Management: Dipping the seed rhizomes in 0.5% malathion or quinalphos for 30 min prior to storage and at the time of planting is effective for the management of scale. Rhizome treatment with 0.075% quinalphos or phosalone, monocrotophos for 20-30 min was more effective for obtaining a higher recovery of rhizomes, higher number of sprouts and lesser incidence of rhizome scale.

Two Hymenopteran parasites, *Physcus* sp. and *Adelencyrtus moderatus*, attack this pest.

9.4.1.3. Thrips, *Panchaetothrips indicus*

i) Damage: The thrips infest turmeric leaves causing them to roll, turn pale and gradually dry up. The pest infestation is more common during post-monsoon period especially in drier regions of the country.

ii) Management: Spraying 0.05% dimethoate or 0.1% phosphamidon is effective for the management of thrips.

9.4.1.4. Leaf Eating Caterpillar, *Udaspes folus*

i) Damage: In India, the hesperiid caterpillar ('grass demon') is also a serious foliage feeding insect. Larva folds the leaves, lives inside and feeds on the leaves.

ii) Management: Spraying 0.2% carbaryl or dimethoate or phosphamidon gives protection against leaf eating caterpillar.

9.4.1.5. Leaf Feeding Beetles, *Lema* spp.

i) Damage: Adults and larvae of leaf feeding beetles feed on leaves with parallel feeding marks on them.

ii) Management: The pest can be controlled by dusting with 4% endosulfan dust or spraying 0.1% malathion.

9.4.1.6. Shoot Borer, *Conogethes punctiferalis*

i) Damage: In Asia, caterpillars of the shoot borer attack the central shoot and killing it resulting in dead heart. Extrusion of frass through holes can be seen (Fig. 9.10).

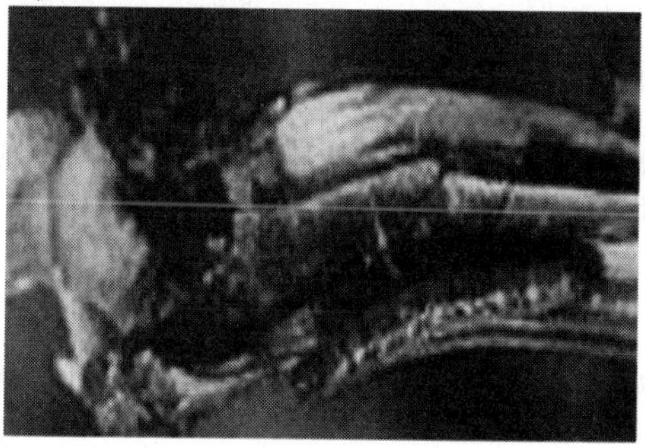

Fig. 9.10. Shoot borer infestation on turmeric.

ii) Management: Pruning and destroying freshly infested pseudo stems (at fortnightly intervals) during July-August and spraying 0.1% malathion or 0.05% endosulfan (at monthly intervals) during September-October resulted in lower incidence of the pest.

Spraying 0.075% monocrotophos or 0.3% Dipel (*Bacillus thuringiensis* product) at 21 days interval during July to October is effective in controlling the pest infestation. The spraying has to be initiated when the first symptom of pest attack is seen on the inner most leaves on the pseudostem.

9.4.1.7. Lace Wing Bug, *Stephanitis typicus*

i) Damage: In Asia, the tinged bug causes leaf discolouration by sucking the sap resulting in drying up of leaves. The pest infestation is more common during the post-monsoon period especially in drier regions of the country. Nymphs and adults suck sap in colonies, from under the leaf surface. Leaves turn yellow and black resinous excretions in the form of tiny spots may be seen between the damaged areas (Fig. 9.11).

Fig. 9.11. Lace wing bug infestation on turmeric leaves.

ii) Management: Spraying dimethoate or phosphamidon each at 0.05% is effective in managing the pest.

9.4.1.8. Root Grubs

i) Damage: Root grubs occasionally feed on tender rhizomes, roots and base of pseudo stems causing yellowing and wilting of shoots.

ii) Management: The pest can be controlled by soil drenching with 0.075% chlorpyriphos.

9.4.2. Mite Pests

9.4.2.1. Mite

i) Management: The mites can be controlled by foliar spraying with 0.05% methyl parathion.

9.5. CINNAMON

Cinnamon (*Cinnamomum cassia*) is a thin inner bark of a cinnamon tree. The bark contains about 1% essential oil. The active principles are eugenol, cineole and cinnamaldehyde. It is used in stick form in fruit preserves. It is used ground for cakes, cookies and puddings. It is also used in spicing sauces, pickles, pulav and biryani.

9.5.1. Insect Pests

9.5.1.1. Psyllid, *Trioza cinnamomi* (Homoptera: Triozidae)

i) Damage: A jumping plant louse, which is principally associated with the foliage of cinnamon leaves causing leaf galls. Rajapakse and Kulasekera (1982) reported that *T. cinnamomi* is the most important insect pest of cinnamon in Sri Lanka. Mani (1973) also reported that an unknown psyllid also causes formation of galls on cinnamon in India. The feeding by the psyllid causes abnormal cell development and the formation of galls. The developing galls enclose the feeding psyllid.

ii) Management: Regular pruning could control the *Trioza* infested gall-bearing shoots. Natural enemies often limit the psyllids. Quinalphos (0.05%) or dimethoate (0.06%) are also recommended to control psyllid infested leaves and shoots in India. The above insecticides should be sprayed thrice at 15 days interval during the vegetative bud stage to control the homopteran psyllid. If required, repeat the spray at 10-12 days interval.

9.5.1.2. Cinnamon Butterfly, *Chilasa clytia* (Lepidoptera: Papillionidae)

i) Damage: The cinnamon butterfly is the most serious pest of cinnamon especially in younger plantations and in the nursery and is generally seen during the post monsoon period. The adults are large sized butterflies and occur in two forms. One of the forms has blackish brown wings with white spots on outer margins; the other form has black wings with bluish white markings. Fully-grown larvae are pale yellow with dark stripes on the sides and measure about 2.5 cm in length (Fig. 9.12). The larvae feed on tender and slightly mature leaves; in severe cases of infestation, the entire plant is defoliated and only midribs of leaves with portions of veins are justify behind.

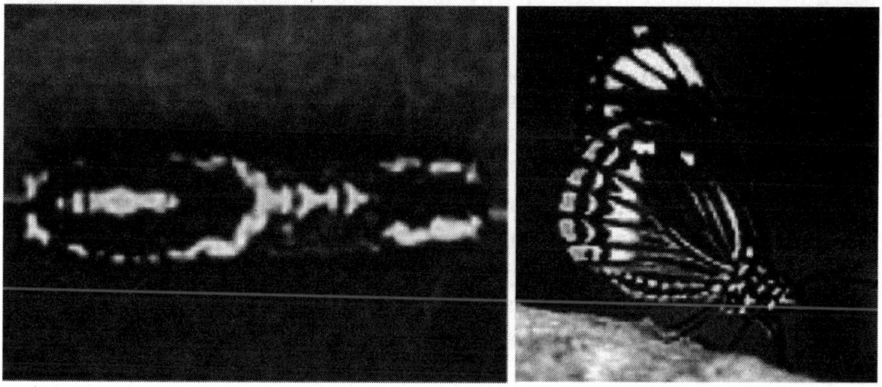

Fig. 9.12. Larva and adult of Cinnamon butterfly.

ii) Management: Eggs are heavily parasitized by the egg parasitoid *Telenomus remus*. Butani (1983) reported that hand removal of larvae and pupae in light infestations and spraying quinalphos 0.05% on tender and partly mature leaves in severe infestations can control the pest.

9.5.1.3. Leaf Miner, *Conopomorpha civica*

Infestation by the leaf miner is more common during the monsoon period and generally nursery seedlings are seriously affected. The adult is a minute silvery grey moth. The larvae are pale gray initially and later become pink measuring about 10 mm in length.

i) Symptoms: Larvae feed on the tissues between the upper and lower epidermis of tender leaves resulting in linear mines that end in 'blister' like patches. The infested leaves become crinkled and the mined areas dry up leading to the formation of large holes on the leaves.

ii) Management: Spraying 0.05% quinalphos during emergence of new flushes is effective in preventing the pest infestation.

9.5.2. Mite Pests

9.5.2.1. Mite, *Eriophyes boisi* (Acarina: Eriophyiidae)

i) Damage: A mite has been found to cause morphologically different galls in cinnamon leaves (Mani, 1973). Perera *et al.* (1985) reported that the mite is an important pest causing leaf galls in Sri Lanka, and the number of new galls found on young and medium sized leaves were greater than on old leaves. They also reported that number of new galls was found to increase rapidly after rainfall with the flushing of new leaves. These mite-infested galls can be found on the upper leaf surface and they are pinkish in color at the beginning and become green on maturity. The females of mites lay eggs singly or in clusters on the upper surface or lower surface on newly emerged tender leaves. The feeding of the nymphs on the sap stimulates the development of pale green to yellowish galls on the leaf surface. The feeding by the mite causes abnormal cell development and the formation of galls. The developing galls enclose the feeding mite.

ii) Management: The numbers of effective miticides for controlling eriophyid mites are limited. Two miticides that may be effective include abamectin and bifenthrin, but these pesticides need to be applied before the mite enter the leaves and cause injury.

Regular pruning could control the *Eriophyes* infested gall-bearing shoots. Application of 30 ml monocrotophos/25 litres of water at 4 days interval resulted in lowest number of galls (Rajapakse and Ratnasekera, 1997).

9.6. VANILLA, *Vanilla planifolia, V. andamanica*

9.6.1. Insect Pests

9.6.1.1. Leaf Feeding Beetles and Caterpillars

i) Damage: A few species of leaf feeding caterpillars and beetles feed on leaves and tender stems.

ii) Management: The pests can be controlled by spraying 0.05% quinalphos.

9.6.1.2. Sucking Bug, *Halymorpha* sp.

i) Damage: The adults and nymphs of the sucking bug infest tender shoot tips and emerging inflorescence resulting in their drying and rotting.

ii) Management: Spraying endosulfan or quinalphos each at 0.05% on tender shoot tips and emerging inflorescence is effective for the management of the pest.

9.6.1.3. Snails and Slugs

i) Damage: Snails and slugs feed and damage tender shoot tips and leaves especially in moist and shaded areas in the plantations.

ii) Management: Hand picking and poison baiting with snailkill helps in preventing the pest infestation.

9.7. NUTMEG, *Myristica fragrans*

Nutmeg is the dry hard wrinkled seed or pit of the nutmeg fruit. Mace is the orange red fleshy covering of the nutmeg kernel. They contain 7-14% essential oil. This essential oil contains highly toxic compound known as myristicin, used in small amounts to flavour puddings and fruit pie.

Myristicin present in nutmeg seeds could lead to delirium and deep stupor. Smaller amounts of nutmeg may cause vomiting and colic. Nutmeg has antimicrobial property. The volatile components of these spices are believed to be responsible for this property.

9.7.1. Insect Pests

9.7.1.1. Black Scale, *Saissetia nigra*

i) Damage: The black scale infests tender stems (pencil thick) and leaves especially in the nursery and sometimes young plants in the field. The scales are clustered together and are black, oval and dome-shaped. They feed on plant sap and severe infestation causes the shoots to wilt and dry up. The infested shoots invariably develop sooty mould cover.

ii) Management: Spraying quinalphos at 2 ml/litre of water gives effective management of the black scale insects.

9.7.1.2. White Scale, *Pseudaulcapsis cockerelli*

i) Damage: The white scale is grayish white, flat and shaped like a fish scale and occurs clustered together on the lower surface of the leaves especially in nursery seedlings. The pest infestation results in yellow streak and spots on the affected leaves. In severe infestations, the leaves wilt and dry.

ii) Management: Spraying quinalphos at 2 ml/litre of water gives effective control of the white scale insects.

9.7.1.3. Shield Scale, *Protupulvinaria mangiferae*

i) Damage: The shield scale is creamy brown and oval and occurs on tender leaves (Fig. 9.13) and stems especially in nursery seedlings. The pest infestation result in wilting of leaves and shoots.

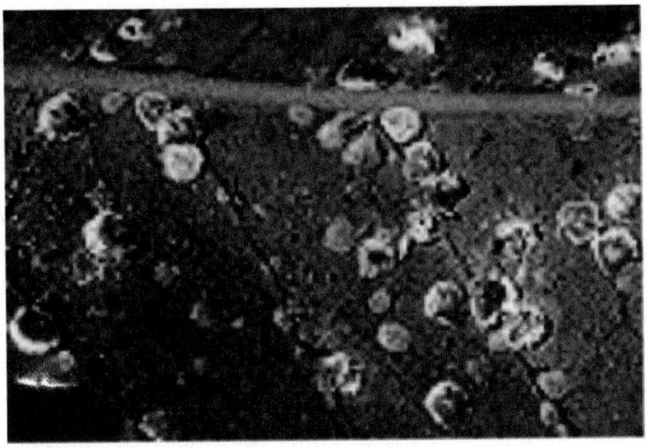

Fig. 9.13. Shield scale infestation on nutmeg leaf.

ii) Management: Spraying monocrotophos at 0.05% (1.25 ml/litre of water) is effective against shield scale insects.

9.8. CORIANDER

Coriander (*Coriandrum sativum*) is used as a common flavouring substance. The stems, leaves and fruits have a pleasant aroma. The whole young plant is used in preparing chutney. Its leaves are used for flavouring curries, sauces and soups. Dry seeds are extensively used in preparation of curry powder, pickling spices, sauces and seasonings. The seeds contain 0.5-1% essential oil which has an active principle, an isomer of geraniol. This plant contains Vitamin A. In medicines, its seeds are used as a carminative, refrigerant, diuretic, flatulence, vomiting and intestinal disorders. The decoction is used in indigestion, cold and other diseases. The oil from seeds is used in sweet meats, cocoa eatables, liquor and for agreeable smell and taste. Coriander seeds contain thalides which increase the levels of anticancer protective enzymes. The oil is also useful as an insecticide.

It is being cultivated in 0.282 million hectares producing 0.172 million tonnes of coriander seeds with an average yield of 0.6 tonnes per hectare. In India, coriander is cultivated in Andhra Pradesh, Madhya Pradesh, Karnataka, Tamil Nadu and Uttar Pradesh.

9.8.1. Insect Pests

9.8.1.1. Aphid

i) Damage: Aphid infestation affects the yield of the crop both in quality and quantity.

ii) Management: Spraying dimethoate or monocrotophos or chlorpyriphos each at 0.1% or 0.07% endosulfan at 10 days interval is effective for the management of the pest. Pollination in coriander is brought about by honeybees. Hence, care should be taken in choosing the insecticide that does not damage honeybees or at least causes minimum damage. Endosulfan has been considered relatively safe which can be sprayed in evening hours as the activity of the honeybees remain low.

9.8.2. Mite Pests

i) Damage: Frequently mites attack coriander at the seed formation stage. The infestation is more severe on the young inflorescence as well as leaves. The leaves of whole plant become whitish yellow and appear sickly. If the infestation is not checked timely, the yield is reduced and grains become shriveled.

ii) Management: Mites can be effectively controlled by spraying the crop with 0.02% ethion. The control will be more effective if it is used as prophylactic measure or at least applied at the initial stage of the pest attack. It can also be controlled by spraying 0.07% dicofol or 0.03% phosphamidon.

9.9. FENUGREEK

Fenugreek seed is a hard lentil seed. Its colour is dark fawn and has astringent aroma. Fenugreek seeds contain 5% bitter fixed oil. The oil has strong celery-like odour. Fenugreek seeds aid in maintaining the blood glucose levels in non-insulin dependent diabetes. The fibre present in the seed may be responsible for this. It is also used with the butter milk in the treatment of dysentery.

9.9.1. Insect Pests

9.9.1.1. Aphid

i) Damage: Aphid sucks the sap of tender parts of plants and affects the growth, adversely.

ii) Management: Spraying of 0.03% dimethoate or 0.025% methyl demeton or 0.04% monocrotophos is recommended to control the aphid. If the crop is grown for green (vegetable) purpose then spray malathion.

9.10. CUMIN, *Cuminum cyminum*

Cumin seed is pungent, sharp and astringent. It contains 2-4% essential oil. The active principle is an aldehyde cumino. It is used as a stimulant and carminative agent. Cumin seeds contain bio-active substances called pthalides which increase the levels of anticancer protective enzymes in the body.

9.10.1. Insect Pests

9.10.1.1. Aphid, *Myzus persicae*

i) Damage: Aphid is a major pest of cumin crop. It sucks the sap of tender parts and reduces the yield.

ii) Management: Spraying of 0.03% dimethoate or 0.025% methyl demeton or 0.04% monocrotophos is recommended to control the aphid. Two sprays with monocrotophos and acephate were effective for the management of aphids in cumin. Honey bees help in pollination of cumin, hence care must be taken in choosing the insecticide that does not affect honey bee activity or at least causes minimum damage. Endosulfan at 0.07% is relatively safe to honey bees and can be sprayed in evening hours when honey bee activity is low.

9.10.1.2. Leaf Eating Caterpillar

i) Damage: This pest causes damage to the foliage of plants reducing the yield of crop.

ii) Management: It can be controlled by spraying of 0.02% phosphamidon in the early stage of crop.

9.10.2. Mite Pests

i) Management: Mites can be effectively controlled by spraying the crop with 0.02% ethion. The systemic insecticides normally used against sucking pests might also be used for the control of mites.

9.11. CLOVE, *Eugenia caryophyllata*

It is the dried flower bud of the clove tree. It contains about 15% essential oil. The chief active principle is eugenol. Cloves are used whole in meats, pickling and fish. It is used ground in cakes, cookies and puddings. Used as flavouring agent in pulav, rice puttu and fruit cake. Cloves contain eugenol which has anti mutagenic effect. It is used for tooth pain.

9.11.1. Insect Pests

9.11.1.1. Stem Borer, *Sahyadrassus malabaricus*

i) Damage: The stem borer infests the main stem of young trees at the basal region. The larvae of the pest girdle the stem and bores downward into it. The girdled portion and bore hole are covered with a mat-like frass material. The infested trees wilt and succumb to the pest attack subsequently.

ii) Management: Inspect the base of clove trees regularly for symptoms of pest attack. Spray 0.1% quinalphos around the bore-hole and inject the same into bore-hole after removing the frass. Swab the basal region of the main stem with 0.2% carbaryl and keep the basin free of weeds.

9.11.1.2. Scale Insects

i) Damage: Many species of scales such as wax scale, *Ceroplastes floridensis*; shield scale, *Pulvinaria psidii*; masked scale, *Mycetaspis personata* and soft scale, *Kilifia accuminata* infest clove seedlings in the nursery and sometimes young plants in the field. The scales are generally seen clustered together on tender stems and lower surface of leaves causing wilting of shoots. The plants present a sickly appearance.

ii) Management: Spraying 0.05% monocrotophos is effective for the management of scale insects.

9.12. FENNEL, *Foeniculum vulgare*

9.12.1. Insect Pests

9.12.1.1. Yellow Willow Aphid, *Cavariella aegopodii*

i) Damage: This aphid causes serious problems primarily in the coastal regions. High populations of this aphid can stunt young plants, and can also deposit honeydew, which causes quality problems in fennel.

ii) Management : For the effective control of aphids on fennel, three sprays with methyl parathion or methyl oxydemeton each at 0.025% or DDVP at 0.05% or endosulfan at 0.07% or phosphamidon at 0.03% at 30 days interval commencing from 30 DAP should be given. Methomyl at 2.25 kg a.i./ha and permethrin at 0.315 kg a.i./ha were also effective in suppressing aphid populations.

9.12.1.2. Beet Army Worm, *Spodoptera exigua*

i) Damage: This insect causes serious damage to fennel, and is most prevalent during fall production in the desert regions. Reduced stands can result from high insect populations.

ii) Management : a) Biological Methods: The most common parasites of the beet armyworm are wasps (e.g., *Hyposoter exiguae* and *Chelonus insularis*), and the tachinid fly (*Lespesia archippivora*).

Predators also play a role in reducing populations of lepidopterous larvae. Two predators that feed on lepidopterous eggs are minute pirate bugs (*Orius* spp.), and bigeyed bugs (*Geocoris* spp.).

Viral diseases (e.g., nuclear polyhedrosis virus) also play a role in reducing populations of lepidopterous larvae.

b) Chemical Methods: Foliar spraying with methomyl at 2.25 kg a.i./ha, permethrin at 0.315 kg a.i./ha, spinosad at 0.225 kg a.i./ha, tebufenozide at 0.315 kg a.i./ha or thiodicarb at 1.85 kg a.i./ha gives effective management of the pest.

9.12.2. Mite Pests

i) Damage: Mites sometimes attack fennel. The infestation is more severe on the young inflorescence as well as leaves. The leaves of whole plant become whitish yellow and appear sickly. If the infestation is not checked timely, the yield is reduced and grains become shriveled.

ii) Management : Mites can be effectively controlled by spraying with 0.02% ethion. The systemic insecticides normally used against the sucking pests can also be used for the control of mites.

REFERENCES

Agarwal, G.P. and Hasija, S.K. 1967. *Alternaria* rot of citrus fruits. *Indian Phytopath.* **20**: 259-260.

Agnihotri, N.P. 1999. Monitoring of pesticide residues. In *Pesticide Safety Evaluation and Monitoring* (N.P. Agnihotri, ed.), pp. 94-118. All India Co-ordinated Res. Project on Pesticide Residues, New Delhi.

Agnihotrudu, V. and Mithyantha, M.S. 1978. *Pesticide Residues - A Review of Indian Work*. Rallis India Ltd., Bangalore.

Ahmed, M. 1990. Deviation in feeding behaviour of *Achaea janata* Linn. (Lepidoptera: Noctuidae). *Ann. Ent.* **8**(2): 15-17.

Ahuja, A.K., Mohapatra, S., Sharma, D. and Awasthi, M.D. 1998. Monitoring of vegetables for pesticide contamination at harvest. *Proc. Nat. Symp. Pest Mangmt. Hort. Crops*, Bangalore, pp. 243-246.

Anandaraj, M. and Devasahayam, S. 2004. Pests and diseases of cinnamon and cassia, *In. Cinnamon and Cassia* (P. N. Ravindran, K. N. Babu and M. Shylajah, eds.), pp.139-258. CRC Press, New York.

Anandaraj M., Devasahayam, S., Krishnamoorthy, B., Mathew, P.A. and Rema, J. 2001. *Cinnamon*. Extn. Pamphlet, Indian Inst. of Spices Res., Calicut, Kerala, India.

Anon. 1979. *Tropical Legumes: Resources for the Future*. National Acad. Sci., Washington DC, pp. 117-121.

Anon. 1987a. Pesticide residues in fruits - A study conducted jointly by S.V. College, New Delhi and FRSL, Ghaziabad (Un-published).

Anon. 1987b. Annual Report of citrus blackfly scheme. Nagpur Univ., Nagpur, India, 27pp.

Anon, 1987c. Improvement of betel vine cultivation in Mahoba. *Progress Report 1986-87*. National Botanical Res. Inst., Lucknow, India.

Anon. 1990. Biological suppression of pests and diseases in fruit crops. In *Proceedings of Indo-USSR Joint Workshop on Problems and Potentials of Biocontrol of Pests and Diseases*, Project Directorate of Biological Control, Bangalore, pp. 26-28.

Anon. 1996. Annual Report. All India Network Project on Agricultural Ornithology, Andhra Pradesh Agri. Univ., Hyderabad, 32 pp.

Anon. 1996. *Cinnamon; Cultivation and Processing*. Tech. Bull. 5, Dept. of Export Agriculture, Ministry of Agriculture, Lands and Forestry, Sri Lanka, pp.7-8.

Anon. 2000. *Bean Fly Information System. Database on Bean Flies*. Extn. Folder No. 73, Indian Inst. of Hort. Res., Bangalore, 6 pp.

AVRDC. 1996. *AVRDC 1995 Progress Report*. Asian Vegetable Res. and Devp. Centre, Shanhua, Taiwan, 187 pp.

Awasthi, M.D. 1986. Effect of chemical treatments on the decontamination of synthetic pyrethroids residues from brinjal fruits. *Pestic. Sci.* **17**: 89-90.

Awasthi, M.D. 1993a. Pesticide residues in fruits. In *Advances in Horticulture – Fruit Crops, Vol. 3* (K.L. Chadha and O.P. Pareek, eds.), pp. 1639-1647. Malhotra Publishing House, New Delhi.

Awasthi, M.D. 1993b. Pesticide residues in vegetable crops. In *Advances in Horticulture –Vegetable Crops, Vol. 6* (K.L. Chadha and G. Kalloo, eds.), pp. 921-976. Malhotra Publishing House, New Delhi.

Awasthi, M.D. 1998. Pesticide residues in food chain in India – Fruits and vegetables. In *Pesticides, Man and Biosphere* (O.P. Shukla, Omkar and A.K. Kulshrestha, eds.), pp. 121-160. APH Publishing Corp., New Delhi.

Awasthi, M.D. 2001. Pesticide residues in fruits and vegetables. In *Integrated Pest Management in Horticultural Ecosystems* (P. Parvatha Reddy, A. Verghese and N.K. Krishna Kumar, eds.), pp.263-278. Capital Publishing Co., New Delhi.

Awasthi, M.D. and Mohapatra, S. 1998. Insecticide residues in ripe mango fruits following post-harvest treatments. *Proc. Nat. Symp. Pest Mangmt. Hort. Crops*, Bangalore, pp. 221-224.

Awasthi, M.D. and Sharma, D. 1997. Uptake of fungicide residues and their persistence on ripening mango fruits from post-harvest treatments. *Pesticide Res. J.* **9**: 41-45.

Babu, T.K. and Azam, K.M. 1987. Studies on the biology, host spectrum and seasonal population fluctuations of the mealy bug, *Maconellicoccus hirsutus* (Green) on grapevine. *Indian J. Hort.* **44**: 284-288.

Baketia, D.R.C., Brar, K.S and Gill, R.S. 1986. Mortality factors in the natural population of *Holotrichia consanguinea* (Coleoptera: Scarabidae). *J. Soil. Biol. Ecol.* **4**: 62-64.

Bath, D.S. and Singh, D. 1989. Studies on the economic threshold level of mustard aphid, *Lipaphis erysimi* (Kalt.) on the radish seed crop in India. *Trop. Pest Mangmt.* **35**: 154-156.

Batra, R.C., Brar, S.S., Singh, R. and Kapoor, S. 1980. Carbaryl for the control of *Rhipophorothrips cruentatus* Hood and thinning of grape berries. *Sci. Cult.* **46**: 75-76.

Bhardwaj, L.N. and Agarwala, R.K. (1986a). In *Advances in Research on Temperate Fruits* (T.R. Chadha *et al.*, eds.), pp. 397-401. Y.S. Parmar Univ. of Hort. & Forestry, Himachal Pradesh, Solan, India.

Bhardwaj, L.N. and Agarwala, R.K. (1986b). In *Advances in Research on Temperate Fruits* (T.R. Chadha *et al.*, eds.), pp. 403-405. Y.S. Parmar Univ. of Hort. & Forestry, Solan, Himachal Pradesh, India.

Bhardwaj, S.P. 1987. Investigations on the response of lepidopteran sex pheromones of temperate fruit pests in Himachal Pradesh. *Indian Agr. Ecosyst. Environ.* **19**: 87-91.

Bhardwaj, S.P. 1988. *Indian J. Agric. Sic.* **58**: 655-656.

Bhatnagar, R.K., Srivastava, K.P., Agnihotri, N.P., Jotwani, M.G. and Gajbhiye, V.T. 1982. Efficacy of bird repellent in sorghum at maturity stage of crop. In *Management of Problem Birds in Aviation and Agriculture* (R.A. Agarwal and R.K. Bhatnagar, eds.), PP. 224-226. Indian Agri. Res. Inst., New Delhi, 242 pp.

Bhatnagar, V.S., Lateef, S.S., Sithanathan, S., Pauler, C.S. and Reed, W. 1982. Research on *Heliothis* management at ICRISAT. *Proc. Int. Workshop on Heliothis Management*, Patancheru, Hyderabad, pp. 385-397.

Bijay Kumar. 2007. *Indian Horticulture Database 2006.* National Hort. Board, Gurgaon, Haryana, India, 324 pp.

Bindra, O.S. and Kalra, R.L. 1973. A review of work done in India on pesticide residues. In *Progress and Problems in Pesticide Residue Analysis* (O.S. Bindra and R.L. Kalra, eds.), pp. 1-44. Punjab Agri. Univ., Ludhiana.

Bindra, O.S. and Singh, J. 1971. Biological observations on the *Holotrichia consanguinea* Blanchard (Coleoptera: Scarabidae) and the relative efficacy of insecticides against the adults. *Indian J. Ent.* **33**: 225-227.

Bose, P.C. and Singh, C. 1980. *Pesticides* **14**(8): 30-31.

Butani, D.K. 1975c. Insect pests of fruits and their control-16: Fig. *Pesticides* **9**(4): 32-36.

Butani, D.K. 1976. Insect pests of fruits and their control-21: Pomegranate. *Pesticides* **10**(6): 23-26.

Butani, D.K. 1978a. Pests and diseases of jackfruit in India and their control. *Fruits* **33**: 351-356.

Butani, D.K. 1978b. Insect pests of tamarind and their control. *Pesticides* **12**(11): 34-41.

Butani, D.K. 1979. *Insects of Fruits.* Periodical Export Agency, Vivek Vihar, Delhi, 415 pp.

Butani D. K. 1983. Spices and pest problems 2. Cinnamon. *Pesticides* **17**(9): 32-33.

Butani, D.K. and Jotwani, M.G. 1984. *Insects in Vegetables.* Periodical Expert Book Agency, New Delhi, India, 356 pp.

Cardet, C., Kanji, T., Delobel, A. and Dantu, P. 1998. Efficacy of neem and groundnut oil in protecting leguminous seeds against seed beetles in the Sahel. *Agroforestry Systems* **40**: 39-40.

Chakravarthy, A.K. 1993. Vertebrate Pest Management. Final Report of ICAR ad-hoc Project, Univ. of Agri. Sci., Bangalore, 98 pp.

Chakravarthy, A.K., Srihari, K. and Chandra Prakash, J. 1998. Management of bird pests in oil palm plantations in Karnataka. In *Advances in IPM for Horticultural Crops* (P. Parvatha Reddy, N.K. Krishna Kumar and A. Verghese, eds.), pp. 183-187. Assoc. for Advancement of Pest Mangmt. in Hort. Ecosystems, Bangalore.

Chander, R. 1984. Biology and control of plum fruit moth (Abstr.). *Nat. Symp. Temperate Fruits*, Solan, India, p. 81.

Chander, R. and Dogra, G.S. 1977. A technique of granular application for wooly apple aphid control. *Entomol. Newsl.* **7**(9/10): 42.

Chander, R. and Dogra, G.S. 1983. Lepidopterous insects infesting apple in Himachal Pradesh. In *Proc. Workshop on High Altitude Ent. and Wildlife Ecol.*, Zool. Survey of India, Calcutta.

Chandramohan, N. 1994. Seasonal incidence of diamondback moth, *Plutella xylostella* L. and its parasitoids in Nilgiris. *J. Biocontrol* **8**: 77-80.

Chang, T. T. 1992. Decline associated with brown *Phellinus noxius. Plant* **2**: 90-95.

Chari, M.S. 1970. *The Bionomics and Control of Cluster Bean Pod Gall Midge,* Asphodylia cyamopsii *N. Sp. (Cecidomyiidae: Diptera).* Ph. D. thesis, Sardar Patel Univ., Anand, 76 pp.

Chaturvedi, A.N. 1985. *Firewood Farming on the Degraded Lands of the Gangetic Plain.* U.P. Forest Bull. No. 50, Lucknow, Govt. of India Press (I), 286 pp.

Coronel, R.E. 1991. *Tamarindus indica* L. In *Plant Resources of South East Asia, Wageningen; Pudoc. No. 2. Edible Fruits and Nuts* (E.W.M. Verheij and R.E. Coronel, eds.), pp. 298-301. PROSEA Foundation, Bogor, Indonesia.

Dalvi, C.S., Dumbre, R.B. and Khanvilkar, V.G. 1992. Record of mango hopper species and their composition in the konkan region. *J. Maharashtra Agri. Univ.* **17**: 404-406.

Garg, S.C. and Kasera, H.L. 1984. *Indian Perfumer* **28**(2): 95-97.

George, E.A.O. and Kandasamy, C. 1986. Effect of various host plants on the biology of the psyllid, *Diaphorina citri* Kuwayama (Homoptera: Psyllidae). *Pest Mangmt.* **1**: 36-40.

Ghai, S. 1976. Phytophagous mites and their control in India. *Nat. Acad. Sci.*, India **46-B**: 49-59.

Ghosh, L.R. and Verma, K.D. 1988. A new host record of *Brachycaudus helichrysi* (Kaltenbach) in India (Homoptera: Aphididae). *J. Aphidol.* **2**: 66-68.

Gopalakrishnan, C. 1998. Studies on Entomopathogenic Fungus Paecilomyces farinosus (Holmskiold) Brown and Smith for the Control of Cabbage Diamond-back Moth, Plutella xylostella L. Ph. D. thesis, Bangalore Univ., Bangalore.

Gopalakrishnan, C. and Mohan, K.S. 1993. Epizootics of *Erynia neoaphidis* (Zygomycetes: Entomophthorales) in field population of *Brevicoryne brassicae* (Homoptera: Aphididae) on cabbage. *Entomon* **18**: 21-24.

Gopalakrishnan, C. and Mohan, K.S. 1996. Field efficacy of entomopathogenic fungus *Nomuraea rileyi* against Noctuid pests of tomato (Abstr.). *Natl. Symp. Hortil. Biotechnol.*, Bangaore, p. 65.

Griffith, R. 1977. A species of bacterium pathogenic to the palm weevil, *Rhynchophorus palmarum.* 1976 Ann. Rept., Red Ring Res. Divn., Minst. of Agri. Land and Fisheries, Trinidad and Tobago, pp. 31-35.

Gunasena, H.P.M. and Hughes, A. 2000. *Tamarind.* International Centre for Underutilized Crops, Southampton, UK, 171 pp.

Gupta, B.P., Joshi, N.K. and Joshi, R. 1984. *Prog. Hort.* **15**: 295.

Gupta, B.P., Joshi, R. and Tripathi, G.M. 1986. Control of peach leaf curling aphid, *Brachycaudus helichrysi* (Kalt.) with granular formulations of some insecticides. *Prog. Hort.* **18**: 129-131.

Gupta, B.P. and Ray, K.M. 1971. Biology and control of *Eulecanium (Lecanium) coryli* (Linn.) (Coleoptera: Coccidae). *Prog. Hort.* **2**(4): 21-30.

Gupta, P.R., Thakur, J.R. and Dogra, G.S. 1988. Management of peach leaf curl aphid, *Brachycaudus helichrysi* (Kalt.) infesting stone fruits in Himachal Pradesh. *J. Tree Sci.* **7**: 26-30.

Hagley, E.A.C. 1962. The palm weevil *Rhynchophorus palmarum* L. a probable vector of red ring disease of coconuts. *Nature* **193**: 499.

Handa. 1992. Monitoring of pesticide residues in Indian environment. In *Pest Management and Pesticides – Indian Scenario* (B.V. David, ed.), pp. 221-224. Namratha Publn., Madras, India.

Hocking, D. 1993. *Trees for Drylands,* Tamarindus indica *L. Family Leguminosae: Caesalpinodeae.* Oxford & IBH Publishing Co., New Delhi, pp. 305-307.

Hoyle, J.C. 1968. A simple method for poisoning coconut trees. *J. Agri. Sci. Trinidad* **68**: 459-465.

Jaffee, B.A., Gaspard, J.T., Ferris, H. and Muldoon, A. 1988. *Soil Biology and Biochemistry* **20**: 631-636.

Jayanthi, P.D.K. and Verghese, A. 1998. Hourly trap catch of the mango fruit fly (*Bactrocera dorsalis* Hendel) using methyl eugenol bottle trap. *Insect Envt.* **4**(2): 60.

Jayanthi, P.D.K. and Verghese, A. 1999. Report of the occurrence of banana weevils in Bangalore. *Insect Envt.* **5**: 148.

Kakar, D.K. and Sharma, J.P. 1988. Feasibility of *Trichogramma* spp. for the biological suppression of fruit borers of pomegranate. *National Symp. on Tree Prot.,* Dr. Y.S. Parmar Univ. Hort. & Forestry, Solan, Himachal Pradesh, India.

Kakar, K.L., Dogra, G.S and Nath, A. 1987. Incidence and control of pomegranate fruit borer, *Virachola Isocrates* (Fabr.) and *Deudorix epijarbas* (Moore). *Indian J. Agric. Sci.* **57**: 749-752.

Kakar, K.L., Nath, A. and Dogra, G.S. 1986. Studies on the feeding behaviour of birds and their control under mid-hill conditions of Himachal Pradesh. In *Advances in Research on Temperate Fruits* (T.R. Chadha, V.P. Butani and J.L. Kaul, eds.), pp. 389-395.

Kaphalia, B.S., Takroo, R., Malhotra, S., Nigam, V. and Seth, T.D. 1990. Organochlorine pesticide residues in different Indian cereals, pulses, spices, vegetables, fruits, milk, butter, desi ghee and edible oils. *J. Assoc. Official Analytical Chemists* **53**: 501-512.

Kapoor, V.C. (ed.). 1993. *Indian Fruit Flies (Insecta: Diptera: Tephritidae).* Oxford & IBH Publishing Co. Pvt. Ltd., New Delhi, 228 pp.

Kennedy, J.S., Day, M.F. and Estop, V.F. 1982. *A Conspectus of Aphids as Vectors of Plant Viruses.* Commonwealth Inst. Entomol., 114 pp.

Krishnaiah, K., Jagan Mohan, N. and Prasad, V.G. 1981. Efficacy of *Bacillus thuringiensis* Ber. for the control of Lepidopterous pests of vegetable crops. *Entomon* **6**: 87-93.

Krishna Kumar, N.K. 1998. Final report of ICAR Adhoc Scheme on Bioecology and Management of the Serpentine Leaf miner, Liriomyza trifolii (Burgess) on Vegetable Crops. Indian Inst. of Hort. Res., Bangalore, 62 pp.

Krishna Kumar, N.K., Krishna Murthy, P.N., Easther, A.P. and Rama, N. 1997. Seasonality and management of the gall midge, *Aspondylia* sp. on brinjal and sweet peppers. *Proc. First Natl. Symp. Hortil. Crops*, Bangalore, pp. 98-101.

Krishna Kumar, N.K., Krishna Murthy, P.N., Rama, N. and Easther, A. P. 1997. Bitter gourd, an unsuitable host plant of serpentine leaf miner. *Insect Envt.* **2**: 125-126.

Krishna Kumar, N.K., Krishna Murthy, P.N., Rama, N., Easther, A. P. and Srinivasan, K. 1998. Assessment of crop loss and seasonality of the accidentally introduced pest, *Liriomyza trifolii* (Burgess), in India. *Proc. Int. Conf. Trop. Spices, Communities Ecosystems,* Tiruvananthapuram.

Krishna Kumar, N.K. and Srinivasan, K. 1988. Seasonality of *Eurytoma* sp. a parasitoid on okra petiole maggot, *Melanagromyza hibisci* Spencer. *Trop. Pest Mangmt.* **34**: 368-369.

Krishnamoorthy, A. 1993. *Bionomics and Development of Integrated Management Strategy for California Red Scale,* Aonidiella aurantii *(Maskell).* Ph. D. thesis, Univ. of Agri. Sci., Bangalore, India.

Krishnamoorthy, A. and Mani, M. 1985. Investigations on *Telenomus remus* Nixon and *Apanteles marginiventris* Cress. against *Spodoptera litura* (F.) on cabbage. *Entomon* **10**: 277-280.

Krishnamoorthy, A. and Mani, M. 1996a. Suppression of brinjal mealy bug *Coccidohystrix insolita* with *Cryptolaemus montrouzieri*. *Insect Envt.* **2**: 50.

Krishnamoorthy, A. and Mani, M. 1996b. Biosuppression of *Helicoverpa armigera* (Hubn.) on tomato using two egg parasitoids, *Trichogramma brasiliensis* (Ashm.) and *T. pretiosum* (Riley). *J. Ent. Res.* **20**: 37-41.

Krishnamoorthy, A. and Mani, M. 1997. The efficacy of green lacewing Chrysoperla carnea in controlling Myzus persicae on capsicum. Proc. First Natl. Symp. Pest Mangmt. Hortil. Crops: Envtl. Implications and Thrusts, Bangalore.

Krishnamoorthy, A. and Singh, S.P. 1987. Biological control of citrus mealy bug with an introduced parasite, *Leptomastix dactylopii* in India. *Entomophaga* **32**: 143-148.

Krishna Moorthy, A. and Singh, S.P. 1988. Observational studies on the occurrence of parasitoids of *Papilio* spp. in citrus. *Indian J. Pl. Prot.* **16**: 79-81.

Krishna Moorthy, P.N. and Krishna Kumar, N.K. 2002. Advances in the use of botanicals in the IPM of major vegetable crops. *Proc. Intn. Conf. Vegetables,* Bangalore, pp.

Krishna Moorthy, P.N., Krishna Kumar, N.K., Selvaraj, C. and Daniel, J.S. 1998. Neem seed kernel extract applications for diamondback moth management: Transfer of technology for mechanized farming. *Pest Mangmt. Hort. Ecosystems* **4**: 128-130.

Krishna Moorthy, P.N. and Srinivasan, K. 1989a. Petiole mining thresholds and potassium chloride sprays for management of bean fly (Diptera: Agromyzidae) in snap beans and cowpea. *J. Econ. Ent.* **82**: 246-250.

Krishna Moorthy, P.N. and Srinivasan, K. 1989b. Distribution and larval behaviour of *Lampides boeticus* in vegetable lablab. *Insect Sci. Applic.* **10**; 491-496.

Krishna Moorthy, P.N. and Srinivasan, K. 1990. Diagnosis and management of bean flies in beans. In *Horticulture – New Technologies and Applications* (J. Prakash and R.L.M. Pierik, eds.).

Krishna Moorthy, P.N. and Srinivasan, K. 1997. Evaluation of neem products and standard chemicals for the management of bean fly, *Ophiomiya phaseoli* (Tryon) in French beans. In *Neem and Environment, Vol. 1* (R.P. Singh, M.S. Chari, A.K. Raheja and W. Kraus, eds.), pp. 433-438. Oxford and IBH Publishing Co. (Pvt.) Ltd., New Delhi.

Krishna Moorthy, P.N., Tewari, G.C. and Srinivasan, K. 1988. Action thresholds for the management of pod borer, *Adisura atkinsoni* Moore on field bean, *Lablab purpureus. Insect Sci. Applic.* **9**; 489-492.

Kumar, N.S. and Ananthakrishnan, T.N. 1984. Predators-thrips interaction with reference to *Orius maxidentex* Ghauri and *Caryonocoris indicus* Muraleedharan (Anthocoridae:Heteroptera). *Proc. Indian Nat. Sci. Acad.* **50B**: 139-145.

Kulkarni, K.A. 1971. Bionomics of grape flea beetle, *Sclodonta strigicolis. Mysore J. Agri. Sci.* **5**: 308-318.

Lakra, R.K., Gupta,D.S. and Bhanot, J.P. 1980. Nature and extent of damage and varietal susceptibility to peach borer, *Sphenopter lafertei* Thomson. *Indian J. Ent.* **42**: 233-239.

Lakra, R.K. and Singh, Z. 1984. Calendar of losses due to fruitfly *Carpomyia vesuviana* Costa in different *Zizipus* spp. *Indian J. Ent.* **46**: 261-269.

Lakra, R.K. and Singh, Z. 1985. Seasonal fluctuations in the incidence of ber fruitfly, *Carpomyia vesuviana* Costa under agro-climatic conditions of Hissar. *Haryana Agri. Univ. J. Res.* **15**: 42-50.

Lakshminarayana, V. and Menon, P.K. 1972. Exposure of general population in Hyderabad to DDT. *J. Food Sci. Tech.* **9**: 82-85.

Lall, B.S. and Rahman, M.F. 1975. Studies on the bionomics and control of the Erinose mite, *Eriophyes litchi* Keifer. *Pesticides* **9**(2): 49-54.

Lim, T.K., Sijam, K. and Ong, E.T. 1991. *Acta Hort.* **291**: 317-323.

Loubser, J.T. 1985. *Deciduous Fruit Grower* **35**: 285-290.

Loubser, J.T. and Klerk, C.A.D.E. 1985. *South African J. Enology and Viticulture* **6**: 31-33.

Malaguti, G. 1953. Pudricon del cogello de la palmera de accite Africana (*Elaeis guineensis* Jacq.). *Agr. Trop. Venezuela* **3**: 13-21.

Mani, M. 1993a. Biological suppression of oriental mealybug *Planococcus lilacinus* (Homoptera: Pseudococcidae). *National Seminar on Optimization of Productivity on Acid Lime*, Periyakulam, Tamil Nadu, India.

Mani, M. 1993b. Studies on mealybugs and their natural enemies in ber orchards. *J. Biol. Control* **7**: 75-80.

Mani, M. 1994a. Effectiveness of the exotic parasitoid *Leptomastix dactylopii* in the control of *Planococcus citri* in guava orchards. *J. Ent. Res.* **18**: 351-355.

Mani, M. 1995. Studies on the natural enemies of the wax scale *Drepanococcus chiton* (Green) (Homoptera: Coccidae) on ber and guava in India. *Entomon* **20**: 55-58.

Mani, M. and Krishnamoorthy, A. 1989a. Impact of insect predators in the control of *Aphis gossypii* Glover on guava. *J. Biol. Control* **3**: 128-129.

Mani, M. and Krishnamoorthy, A. 1989b. Occurrence of mealy bugs and their natural enemies on custard apple around Bangalore, South India. *J. Biol. Control* **3***: 77.*

Mani, M. and Krishnamoorthy, A. 1990a. Evaluation of the exotic predator *Cryptolaemus montrouzieri* Muls. (Coccinellidae, Coleoptera) in the suppression of green shield scale, *Chloropulvinaria psidii* (Maskell) (Coccidae, Hemiptera). *Entomon* **15**: 45-48.

Mani, M. and Krishnamoorthy, A. 1990b. Outbreak of mealy bugs and record of their natural enemies on pomegranate. *J. Biol. Control* **4**: 61-62.

Mani, M. and Krishnamoorthy, A. 1994. Natural enemies of *Siphoninus phillyreae* (Homop., Aleyrodidae) and *Aphis punicae* (Homoptera, Aphididae) on pomegranate. *Entomon* **20**: 31-34.

Mani, M. and Krishnamoorthy, A. 1994. Selectivity of pesticides to the parasitoid *Aphelinus* sp. of the green peach aphid, *Myzus persicae* infesting sweet pepper. *J. Aphidol.* **8**: 79-87.

Mani, M. and Krishnamoorthy, A. 1995. Outbreak of *Aphis punicae* and record of its natural enemies of pomegranate in India. *Indian J. Pl. Prot.* 23: 89-90.

Mani, M. and Krishnamoorthy, A. 1996a. First record of *Ooencyrtis papilionis* on pomegranate butterfly, *Deudorix isocrates* Fabr. (Lycaenidae, Lepidoptera). *Entomon* **21**: 275-276.

Mani, M. and Krishnamoorthy, A. 1996c. Record of two insect pests and their natural enemies on phalsa. *J. Insect Sci.* **9**: 182.

Mani, M. and Krishnamoorthy, A. 1997b. Suppression of *Planococcus citri* (Risso) on sapota. *Pest Mangmt. in Hort. Ecosystems* **3**: 45-47.

Mani, M. and Krishnamoorthy, A. 1998a. Biological control studies on the mango green shield scale, *Chloropulvinaria polygonata* (Ckll.) (Homoptera: Coccidae) in India. *Entomon* **23**: 105-110.

Mani, M. and Krishnamoorthy, A. 1998c. Outbreak of ash whitefly, *Siphoninus phillyreae* and record of natural enemies on pomegranate. *Entomon* **23**: 197-198.

Mani, M. and Krishnamoorthy, A. 1999c. First record of *Trichogramma chilotraeae* Nagaraja and Nagarakatti on pomegranate butterfly, *Deudorix isocrates* (Fabr.). *Entomon* **24**: (In Press).

Mani, M. and Krishnamoorthy, A. 1999e. The nigra scale *Parasaissetia nigra* and its natural enemies on custard apple in Karnataka. *Insect Envt.* **5**: 20.

Mani, M. and Krishnamoorthy, A. 2001. Biological suppression of the mealybugs *Planococcus citri* (Risso) and *P. lilacinus* (Ckll.) on pomegranate in India. *Entomon* **26**: (In Press).

Mani, M., Krishnamoorthy, A. and Dinesh, M.S. 2000. Biological control studies on the spiraling whitefly *Aleurodicus dispersus* Russel. *Entomological Congress: Perspectives for New Millenium*, Trivandrum, India.

Mani, M., Krishnamoorthy, A. and Pattar, G.L. 1995. Biological control of the mango mealybug, *Rastrococcus iceryoides* (Green) (Homoptera: Pseudococcidae). *Pest Mangmt. Hort. Ecosystems* **1**: 15-20.

Mani, M., Krishnamoorthy, A. and Singh, S.P. 1989. The impact of the predator, *Cryptolaemus montrouzieri* Muls. on the pesticide resistant populations of the striped mealybug, *Ferrisia virgata* (Ckll.) on guava in India. *Insect Sci. Appl.* **11**: 167-170.

Mani, M. and Thontadarya, T.S. 1987. Record of mealybug species on grapevine in Karnataka. *Curr. Sci.* **59**: 102.

Mani, M.S. 1973. *Plant Galls of India.* Macmillan Co. India Ltd, New Delhi, India, 354 pp.

Manjunath, T.M. 1985. *Maconellicoccus hirsutus* on grapevine. *FAO Pl. Prot. Bull.* **33**: 74.

Mann, G.S. 1981. Chemical control of *Synclera univolalis* Walker on jujube at Ludhiana. *Indian J. Agric. Sci.* **51**: 517-520.

Mann, G.S. 1986. House sparrows (*Passer domesticus* Linn.) causing severe damage to peach buds in residential areas at Ludhiana, Punjab. *Trop. Pest Mangmt.* **32**: 43.

Mann, G.S. and Singh, D. 1981. Population studies on jujube leaf webber on *Zizyphus mauritiana* Lamark at Ludhiana (Punjab). *Entomon* **6**: 351-355.

Matthews, G.A. 1984. *Pest Management.* Longman, 231 pp.

Mathur, A.C. and Tandon, P.L. 1973. Pineapple mealy bug and its control. Extn. Leaflet No. 5, Indian Inst. of Hort. Res., Bangalore.

Mishra, R.C., Verma and Gupta, P.R. 1987. Insecticides for the control of apple blossom thrips. *Indian J. Hort.* **44**: 115-120.

Mital, V.P. and Khanna, H.C.L. 1978. Efficacy of some fumigants against adults of tamarind bruchid, *Caryedon serratus* F. *Bull. Grain Technol.***16**: 147-148.

Mortensten, J.A. 1985. *Proc. of Florida State Hort. Soc.* **98**: 166-169.

Nagarajan, R., Neelanarayanan, P. and Kanakasabai, R. 1994. Rodent damage in pineapple gardens in Coleroon river Bank in Tamil Nadu. *Rodent Newsl.* **18**(4): 4.

Nagarkatti, S. and Jayanth, K.P. 1982. Population dynamics of major pests of cabbage and their natural enemies in Bangalore district (India). *Proc. Int. Conf. Pl. Prot.*, Kaula Lumpur, Malaysia, pp. 325-347.

Nair, M.R.G.K. 1975. *Insects and Mites of Crops in India.* Indian Council of Agri. Res., New Delhi, India, 404 pp.

Nanjan, K. and Kumar, S. 1983. Notes on the control of fruit borer in pomegranate. *South Indian Hort.* **31**: 104-105.

Narang, M.L. and Chandel, J.S. 1995. Preliminary studies on bird damage to apple orchards in Himachal Pradesh (Abstr.). In *Environment and Birds*, Secretariat Box 45, Srinagar, Garhwal, Uttar Pradesh, pp. 54-55.

Narayanan, K. 1985. Control of *Heliothis armigera* through nuclear polyhedrosis virus. In *Microbial Control and Pest Management* (S. Jayaraj, ed.). Tamil Nadu Agri. Univ. and ICAR Publication.

Narayanan, K. 1987. Field efficacy of nuclear polyhedrosis virus of *Adisura atkinsoni* Moore on field beans. *J. Biol. Control* **1**: 44-47.

Narayanan, K. and Gopalakrishnan, G.1988. Integrated control of *Heliothis armigera* (Hbn.) with their pathogens and pesticides in certain horticultural crops. *Proc. Natl. Workshop* Heliothis *Mangmt.* Tamil Nadu Agri. Univ., Coimbatore, pp. 165-176.

Narayanan, K., Jayaraj, S. and Subramanian, T.R. 1970. Control of three species of Lepidopteran insects with the pathogen, *Bacillus thuringiensis* Berliner. *Madras Agric. J.* **57**: 665-673.

Patel, B.K. 1967. *The Bionomics of the Mustard Sawfly,* Athelia lugens proxima *Klug. (Trenthredinae: Hymenoptera).* M. Sc. Thesis, Sardar Patel Univ., Anand, 76 pp.

Patel, G.T. 1981. *Biology and Chemical Control of Chiku Bud Borer,* Anarsia *sp. (Lepidoptera: Gelichiidae), A Pest of Chiku.* M. Sc. (Agri.) thesis, Gujarat Agri. Univ., Sardarkrushinagar, Gujarat, India.

Patel, G.T., Patel, C.B. and Shah, A.H. 1984. Biology of Sapodilla bud borer, *Anarsia achrasella* (Lepidoptera: Gelichiidae). *Indian J. Agric. Sci.* **54**: 695-697.

Patel, J.R.1980. *The Bionomics and Control Measures of the Potato Tuber Moth,* Phthorimaea (Gnorimoschema) operculella *Zeller (Gelechidae: Lepidoptera).* M. Sc. Thesis, Gujarat Agri. Univ., Anand Campus, Anand, 76 pp.

Patel, J.R., Patel, R.C. and Amin, P.R. 1983. Effect of systemic insecticides on thrips and mites and its relation to leaf curl disease of chilli. *Gujarat Agri. Univ. Res. J.* **8**: 129-133.

Patel, M.K., Moniz, L. and Kulkarni, Y.S. 1948b. *Curr. Sci.* **17**: 180-190.

Patel, N. 1993. Vertebrate pest damage to pomegranate in arid horticulture. *Rodent Newsl.* **17**(2): 6.

Patel, N.C., Patel, J.J., Jayani, D.B. and Patel, J.R. 1997. Evaluation of insecticide schedules for controlling chilli thrip, *Scirtothrips dorsalis* Hood infesting chilli, *Capsicum annuum* L. *Indian J. Agric. Sci.*

Perera, H. A. S., Sritharan, R., and Perera, K. P. 1985. Some studies of cinnamon galls in Sri Lanka. *Sri Lankan J Agric. Sci.* **22**: 23-27.

Pruvost, O., Couteau, A. and Luisetti, J. 1989. *Fruits* **44**: 343-350.

Pruvost, O., Couteau, A. and Luisetti, J. 1990. *Fruits* **45**: 125-140.

Pruvost, O., and Luisetti, J. 1991. *Acta Hort.* **291**: 314-327.

Rai, K.M. and Tripathi, G.M. 1984. Resistance of various rootstocks to the attack of apple woolly aphid, *Eriosoma lanigerum* (Hausmann). *Prog. Hort.* **16**: 181-185.

Rai, R., Mukherjee, I.N. and Singh, J. 1993. Preliminary studies on mango leaf coating mite, *Cisaberoptus kenyae* Keifer (Acari: Eriophydae). *Entomon* **18**: 193-197.

Rajamohan, N. and Jayaraj, S. 1978. Field efficacy of *Bacillus thuringiensis* and some other insecticides against pests of cabbage. *Indian J. Agric. Sci.* **48**: 672-675.

Rajapakse, R.H.S. and Kulasekera, V. L. 1982, Some observations on insect pests of cinnamon in Sri Lanka. *Entomon* **7**: 221- 223

Rajapakse, R. H. S. and Ratnasekera, D. 1997. Studies on the distribution and control of leaf galls in cinnamon caused by *Trioza cinnamoni* Boselli. in Sri Lanka. *Int. J. Trop. Agric.* 15: 53-56.

Ramakrishnan, T.S. and Damodaran, S. 1956. *Proc. Indian Acad. Sci.* **43**: 213-222.

Rao, R.P.V., Lakshiminarayanan, K. and Subharatnum, G.V. 1983. Residual toxicity of certain insecticides for grapevine flea beetle. *Entomon* **8**: 395-396.

Rao, R.P.V., Lakshiminaryanan, K. and Subharatnum, G.V. 1984. Seasonal occurrence of grapevine flea beetle, *Scelodonta stringicollis*, around Hyderabad. *Entomon* **9**: 395-396.

Rawat, U.S., Pawar, A.D. and Singh, R. 1989. Establishment of *Aphelinus mali* (Hald.) in cold and dry zones of Kinnaur district in Himachal Pradesh. *J. Inesct Sci.* **2**: 109-113.

Rawat, U.S., Thakur, J.N. and Pawar, A.D. 1988. Introduction and establishment of *Chilocorus bijugas* Mulsant and *Pharsoscymnus flexibilis* Mulsant, predatory beetles of San Jose scale at Thanedar areas in Himachal Pradesh. *Curr. Sci.* **57**: 1250-1251.

Rom, R.C., Dozier, W.A, Knowles, J.W., Carlton, C.C., Arrington, E.H., Wehunt, E.J., Yadava, U.L., Doud, S.L., Ritchie, D.F., Clayton, C.N., Zehr, E.L., Gamrbell, C.E., Britton, J.A. and Lockwood, D.W. 1985. *Compact Fruit Tree* **18**: 85-91.

Sachan, J.N. 1984. Varietal susceptibility of ber fruits to the damage of fruitfly, *Carpomyia vesuviana* . *Bull. Ent.* **25**: 198-199.

Sahai, B. and Joshi, L.D. 1965. Bionomics and biological control of San Jose scale (*Quadraspidiotus perniciosus* Comstock) in UP. *Punjab Hort. J.* **5**: 37-43.

Sandhu, P.S. and Dhindsa, M.S. 1995. Orchard birds – A review. In *Environment and Birds*, Abstracts, Secretariat Box 45, Srinagar, Garhwal, Uttar Pradesh, p. 57.

Shah, A.H., Patel, C.B. and Patel, G.M. 1986. Chiku bud borer, *Anarsia achrasella* sp. Nov., a new pest of chiku orchards in South Gujarat. *Indian J. Ent.* **48**: 122-127.

Sharma, D.D. 1985. Major pests of litchi in Bihar. *Indian Fmg.* **35**(2): 25-26.

Sharma, D.D. and Lall, B.S. 1983. Comparative efficacy of insecticides to *Actrocercops transitans* Smallen (Lepidoptera: Gracilleridae). *Indian J. Agric. Sci.* **53**: 181-188.

Sharma, D.D. and Rahman, M.F. 1982. Control of litchi mite, *Aceria litchi* (Keifer) with particular reference to evaluation of problem and post-bloom application with different insecticides. *Entomon* **7**: 55-56.

Sharma, D.D., Singh, S.P. and Akhori, K.K. 1986. Relationships between the population of *Aceria litchi* Keifer on litchi and weather factors. *Indian J. Agric. Sci.* **56**: 59-63.

Sharma, H.K. 1986. *Influence and Control of* Sphenoptera laferti *Thomson in Stone Fruits.* M. Sc. thesis, Y.S. Parmar Univ. of Hort. and Forestry, Solan, Himachal Pradesh, India.

Sharma, J.P. and Dogra, G.S. 1982. Biology and seasonal history of plum scale, *Eulecanium* sp. ? *tiliae* (L.) in Himachal Pradesh. *Bull. Ent.* **23**: 74-84.

Sharma, J.P. and Dogra, G.S. 1986. Studies on the control of the plum scale, *Eulecanium* sp. ? *tiliae* (L.) (Homoptera: Coccidae) through chemical and natural elements. *Indian J. Ent.* **48**: 258-263.

Sharma, P.L., David, A.C. and Bhalla, O.P. 1963. Flat headed borer, *Sphenoptera lafertei* Obb. and its control in Himachal Pradesh. *Himachal Hort.* **4**: 1-5.

Shivankar, V.J., Rao, C.J. and Singh, S. 2008. *Citrus Entomology A Deacade at NRCC.* Tech. Bull. No. 11, National Res. Centre for Citrus, Nagpur, Maharashtra, India, 78 pp.

Shivankar, V.J.and Singh, S. 1998a. Citrus leaf miner – *Phyllocnistis citrella* – A pest of citrus to recon with (Abstr.). *Intn . Conf. on Pest & Pesticide Mangmt. for Sustainable Agri.,* CS Azad Univ. Agri. & Technol., Kanpur, India, pp. 11-13.

Shivankar, V.J.and Singh, S. 1998b. *Management of Insect Pests in Citrus.* Tech. Bull. No. 3, National Res. Centre for Citrus, Nagpur, India, 99 pp.

Shukla. R.P. and Prasad, V.G. 1985. Population fluctuations of the oriental fruit fly, *Dacus dorsalis* Hendel, in relation to hosts and abiotic factors. *Trop. Pest Mangmt.* **31**: 273-275.

Simwat, G.S. and Siddu, A.S. 1974. Biology of fig leaf roller, *Phycodes minor* Moore (Glyphipterygidae: Lepidoptera). *Indian J. Ent.* **36**: 149-150.

Singh, A. and Chawla, R.P. 1978. Contamination of market grape with insecticide residues. *Indian J. Pl. Prot.* **6**: 95-96.

Singh, H. and Singh, G. 1975. Biological studies on *Heliothis armigera* (Hubner) in Punjab. *Indian J. Ent.* **37**: 154-164.

Singh, M.P. and Vashista, B.B. 1985. Field screening of some ber cultivars for resistance to ber fruit fly, *Carpomyia vesuviana. Indian Pl. Prot.* **12**: 55-56.

Singh, S.P. 1993. Biological control of pests. In *Advances in Horticulture: Fruit Cops* (K.L. Chadha and O.P. Pareek, eds.), pp. 1591-1616. Malhotra Publishing House, New Delhi, India.

Singh, S.P. 1995. *Technology for Production of Natural Enemies.* Tech. Bull. No. 4, Project Directotate of Biological Control, Bangalore, India, 221 pp.

Singh, S.S., Srivastava, P.M. and Phogat, K.P.S. 2001. Collar-zone weevil – An emerging problem of apple growers in U.P. hiils. *Indian Hort.* **46**(2): 11-12.

Singh, V., Dubey, 0. P., Nair, C. P. R. and Pillai, G. B. 1978. Biology and bionomics of insect pests of cinnamon. *J. Plantn. Crops* **6**: 24- 27.

Smith, F.F., Webb, R.E. and Dusly, R.F. 1979. Greenhouse whitefly control by sticky yellow board traps. *9th Int. Cong. Pl. Prot.*, Washington DC, Abstr. 654.

Srihari, K. and Chakravarthy, A.K. 1998. Ecological approaches for the management of vertebrate pests in Western Ghats region of Karnataka. *Indin J. Ecol.* **25**: 55-61.

Srinivasan, K and Krishna Moorthy, P.N. 1991. Indian mustard as a trap crop for management of major lepidopterous pests on cabbage. *Trop. Pest Mangmt.* **37**: 26-32.

Srinivasan, K and Krishna Moorthy, P.N. 1993. Evaluation of neem products and other standard chemicals for the management of major pest complex on cabbage: Comparison between standard spray regime and IPM involving mustard as a trap crop. In *Neem and Environment, Vol. I* (R.P. Singh, M.S. Chari, A.K. Raheja and W. Kraus, eds.), pp. 447-458. Oxford and IBH Publishing Co. (Pvt.) Ltd., New Delhi.

Srinivasan, P.M. 1958. Control of fruit borer, *Heliothis armigera* (Hub) on tomato. *Indian J. Hort.* **16**: 187-188.

Srivastav, U.K. and Patel, N.T. 1990. Pesticide residues in foods – Present situation. In *Pesticide Industry in India.* CMA Monograph No. 142, Oxford & IBH Publishing Co. (Pvt.) Ltd., New Delhi, pp. 89-105.

Srivastava, R.P. and Verghese, A. 1985. Fogging as a new method of hopper control in mango. *Indian J. Ent.* **47:** 113-115.

Subramanyam, T.M., Sadasivam, R., Raman, N.Y. and Gowder, R.B. 1973. *Madras Agric. J.* **60**: 90-92.

Sundarababu, P.C. 1968. Studies on the efficacy of Thuricide in the control of the leaf caterpillar, *Hymenia recurvalis* on amaranthus. *S. Indian Hort.* **16**: 17-19.

Sundarababu, P.C. 1969. Note on the possible use of *Bacillus thuringiensis* Berliner for the control of certain lepidopterous pests. *Madras Agric. J.* **56**: 824-826.

Sundarababu, P.C. 1970. Comparative efficacy of Thuricide against the spotted beetle, *Epilachna vignitioctopunctata* F. on brinjal (*Solanum melongena*). *S. Indian Hort.* **18**: 85-88.

Surulivelu, T., Chellaiah, S. and Balasubramanian, G. 1977. Efficacy of certain insecticides and *Bacillus thuringiensis* in the control of lab-lab pod borers. *Madras Agric. J.* **65**: 120-121.

Tandon, P.L 1985. Spatial Distribution Sampling Technique and Population Dynamics of Citrus Green Scale Coccus viridis (Green) on mandarin. Ph. D. thesis, Univ. of Agri. Sci., Bangalore, India.

Tandon, P.L. and Lal, B. 1979. Effectiveness of different banding materials for preventing migration of *Drosicha mangiferae* (Green) (Margarodidae, Homoptera). *Indian J. Hort.* **36**: 323-326.

Tandon, P.L. and Lal, B. 1980. Comparative efficacy of synthetic garlic oil with some modern insecticides against *Drosicha mangiferae* (Green). *Prog. Hort.* **12**: 62-65.

Tewari, G.C. and Krishna Moorthy, P.N. 1984. Yield loss in tomato caused by fruit borer. *Indian J. Agric. Sci.* **54**: 341-343.

Tewari, G.C. and Krishna Murthy, P.N. 1986. *Eurytoma* sp. A new parasite on *Asphondylia* sp. infesting egg plant. *Entomon* **12**: 227-228.

Tewari, G.C., Krishna Murthy, P.N. and Sardana, H.R. 1987. Nature of damage and chemical control of gall midge, *Asphondylia* sp. infesting eggplant. *Indian J. Agric. Sci.* **57**: 745-748.

Thakur, J.R. and Dogra, G.S. 1980. Woolly apple aphid, *Eriosoma lanigerum*, research in India. *Trop. Pest Mangmt.* **26**: 8-12.

Thyagaraj, N.E., Chakravarthy, A.K. and Srihari, K. 1996. Vertebrate pests of cocoa and their management practices in hill region of Karnataka. *IPM and Sustainable Agriculture* **6**: 214-218.

Timyan, J. 1996. *BWA Yo: Important Trees in Haiti.* Southeast Consortium for International development, Washington DC, USA.

Toor, H.S. 1982. Problem birds and their management in Punjab. In *Management of Problem Birds in Aviation and Agriculture* (R.A. Agarwal and R.K. Bhatnagar, eds.), pp. 132-142. Indian Agri. Res. Inst., New Delhi.

Tripathi, R.S., Jain, A.P., Kashyap, N., Rana, B.D. and Prakash, I. 1992. Rodents in North-western desert. In *Rodents in Indian Agriculture* (I. Prakash and P.K. Ghosh, eds.), pp. 357-395. Scientific Publishers, Jodhpur.

Umanarasimhan, A. and Chacko, M.J. 1988. *Rastrococcus* spp. (Hemiptera, Pseudococcidae) and their natural enemies in India as potential biocontrol agents for *R. invadens* Williams. *Bull. Ent. Res.* **78**: 703-708.

Verghese, A. 1986. Supervised control and scouting help better insect pest management in mango. *Indian Hort.* **30**(4): 17-18, 24.

Verghese, A. 1993. Report submitted to Indian Institute of Horticultural Research, Bangalore.

Verghese, A. 1997. Crop loss due to flea beetle in grapes. *Drakshavritta:* 147-148.

Verghese, A. 1998a. Management of mango leaf webber: A vital package for panicle emergence. *Insect Envt.* **4**(1): 3.

Verghese, A. 1998b. Non-destructive control of the bat, *Cynopterus sphinx* Vahl (Chiroptera: Pteropodidae) in grapes (*Vitis vinefera* Linnaeus) in India. *Intn. J. Pest Mangmt.* **44**:81-85.

Verghese, A. 1998c. Status of stone weevil and fruit fly with reference to management for export of mango (Abstr.). *National Symp. on Mango Prodn. and Export*, Central Inst. of Subtropical Hort., Lucknow.

Verghese, A. 1999. Efficacy of neem cake extract and selected insecticides against the grape berry webber, *Adoxophyes privatana* (Walker) (Lepidoptera: Noctuidae). *Pest Mangmt. Hort. Ecosystems* **5**: 99-101.

Verghese, A. 2000a. Effect of imidachloprid, lamda cyhalothrin and azadirachtin on the mango hopper, *Idioscopus niveosparsus* (Leth.) (Homoptera: Cicadellidae). *Acta Hort.* **509**: 733-736.

Verghese, A. 2000b. Potential of botanicals in the management of the pomegranate aphid, *Aphis punicae* Passerini. *Insect Envt.* **5**: 147-148.

Verghese, A. and Jayanthi, P.D.K. 1999. Lepidopteran pest complex on mango inflorescence. *Insect Envt.* **5**: 51-52.

Verghese, A., Jayanthi, P.D.K. and Sudha Devi, K. 2000. Developing an IPM based export protocol and forewarning model for mango fruit fly, *Bactrocera dorsalis* Hendel. *National Seminar on Hitech Hort.,* Bangalore.

Verghese, A. and Ramachander, P.R. 1998. Inter-specific association among prey, *Chloropulvinaria psidii* (Maskell) predator, *Cryptolaemus montrouzieri* Mulsant and ant, *Camponotus compressus* Fab. in a guava ecosystem. *Pest Managmt. Hort, Ecosystems* **4**: 21-24.

Verghese, A. and Rao, G.S.P. 1985. Spatial distribution of the mango leaf gall, *Procontarinia matteiana* Keif. and Cocc. on mango cv. Dashaheri. *Indian J. Hort.* **24**: 139-143.

Verghese, A. and Sudha Devi, K. 1998. Do fallen leaves indicate gall population on mango tree? In *Advances in Horticultural Crops* (P. Parvatha Reddy, N.K. Krishna Kumar and A. Verghese, eds.), pp. 13-14. Assoc. for Advancement of Pest Mangmt. in Hort. Ecosytems, Bangalore, India.

Verghese, A. and Tandon, P.L. 1989. Management of mealy bug in grapevines with special reference to Thompson Seedless. *Drakshavritta*: 79-82.

Verghese, A. and Tandon, P.L. 1994. Management of defoliating beetles in grapevine. *Drakshavritta:* 145-148.

Verghese, A. and Tandon, P.L. 1995. Shot hole borer – A potential pest of grape. *Drakshavritta*, pp. 168-169.

Varma, G.C. and Gill, G.S. 1976. Laboratory studies on the comparative efficacy of biotic insecticides for the control of *Plutella xylostella* (Linnaeus) (Plutellidae: Lepidoptera). *J. Res. PAU* **14**; 3104-308.

Verma, A.N., Khurana, A.D. and Singh, R. 1974. Chemical control of bark eating caterpillar, *Indrabela quadrinotata* (Walker) infesting pomegranate. *Indian J. Ent.* **36**: 297-301.

Verma, A.N. and Singh, R. 1975. Chemical control of bark eating caterpillar, *Indrabela quadrinotata* (Walker) by painting the bark with insecticides. *Haryana J. Hort. Sci.* **4**: 145-149.

Verma, K.L. and Chander, R. 1988. Root borer, *Dorysthenes hugelii* Robtenbacher (Cerambicidae: Coleoptera) and its management in apple orchard in Himachal Pradesh (Abstr.). *Nat. Symp. Tree Prot.*, Y.S. Parmar Univ. of Hort. and Forestry, Solan, Himachal Pradesh, India, p. 34.

Verma, K.L. and Singh, M. 1989. Evaluation of different predators and parasites as biological control agents on peach leaf curl aphid. *Pesticides* **23**(1): 29-32.

Verma, T.D. 1985. *Studies on Tortricids Infesting Apple with Special Reference to Biology and Control of* Archips pomivora *Meyrick.* Ph. D. thesis, Y.S Parmar Univ. oHort. and Forestry, Solan, Himachal Pradesh, India.

Verma, T.D. 1985a. Incidence and chemical control of bark eating caterpillar, *Indrabela quadrinotata* Walker on plum trees. *Indian J. Agric. Sci.* **55**: 131-132.

Vevai, E.J. 1969. Know your crop and its pest problems and control-9: Grapevine. *Pesticides* **3**(6): 31-37.

Virakthamath, C.A., Tewari, G.C., Srinivasan, K. and Gupta, M. 1993. American serpentine leaf miner is a new threat to crops. *Indian Fmg.* **43**(2): 10-12.

Weiss, E. A. 2002. *Essential Oil Crops.* CAB international, 191 pp.

Zaka-ur-Rab, M. 1982. Predation of the plum scale, *Eulecanium coryli* (L.) (Homoptera: Coccinellidae) in Kashmir. *J. Bombay Nat. Hist. Soc.* **79**: 436

SUBJECT INDEX